Advance Praise for
EarthEd: Rethinking Education on a Changing Planet

"*EarthEd* is a welcome and urgently needed addition to the education of children. With their entire futures at risk from our current path, youth will benefit greatly from the wisdom and perspectives in these essays. Adults are often reluctant to change their lifestyles, but when their children see the world through the lens of environmentalism and ask parents to protect their futures, that can galvanize change."

  —**David Suzuki**, award-winning scientist, environmentalist, and broadcaster

"The strength of *EarthEd: Rethinking Education on a Changing Planet* is the timeliness of its topic and the appeal of up-to-date scholarship and thinking coupled with the pedagogical implications and best practice applications. The book provides numerous, diverse, and exciting models of learning rooted in Indigenous ways of knowing, in learning *from* and *in* nature, in innovative collaborations between schools and governments, and in programs designed to make goals of sustainable development tangible and meaningful."

  —**Thomas Walker**, Director of the Environmental Studies Master's Program, Goucher College

"This book examines how we can rethink education to foster students' natural development of their best selves—to allow them to play, to let them connect with nature, to encourage creativity through the arts, and to engage with issues through meaningful 'real world' experiences. The good news is that these steps are achievable in schools. And the more individual educators do them, the more chance we collectively have of helping shift education to prepare students for the realities of the future they face. Though it's easy to be overwhelmed by the challenges facing our planet, this volume outlines actionable, viable, and satisfying shifts that we can make in the education of the next generation of leaders."

  —**Buffy Cushman-Patz**, Founder and School Leader of the School for Examining Essential Questions
    of Sustainability (SEEQS)

"The timing and theme of *EarthEd: Rethinking Education on a Changing Planet* is critically important to all of humanity. The planet's survival rests on works such as this to inspire a new generation of educators to help students navigate the pathways toward a just and sustainable future."

  —***Lee F. Ball Jr.***, University Sustainability Director, Appalachian State University

"Never has Worldwatch's classic *State of the World* series been as critical as it is today. The 2016 U.S. elections clearly demonstrated how broken educational systems are with regard to the existential threats faced by civilization. *EarthEd: Rethinking Education on a Changing Planet* provides guidance on how that disaster might be avoided in the future—a textbook for all educators and those wishing to fix the system."

  —**Paul R. Ehrlich**, President, Center for Conservation Biology, Stanford University

About Island Press

Since 1984, the nonprofit organization Island Press has been stimulating, shaping, and communicating ideas that are essential for solving environmental problems worldwide. With more than 1,000 titles in print and some 30 new releases each year, we are the nation's leading publisher on environmental issues. We identify innovative thinkers and emerging trends in the environmental field. We work with world-renowned experts and authors to develop cross-disciplinary solutions to environmental challenges.

Island Press designs and executes educational campaigns in conjunction with our authors to communicate their critical messages in print, in person, and online using the latest technologies, innovative programs, and the media. Our goal is to reach targeted audiences—scientists, policymakers, environmental advocates, urban planners, the media, and concerned citizens—with information that can be used to create the framework for long-term ecological health and human well-being.

Island Press gratefully acknowledges major support of our work by The Agua Fund, The Andrew W. Mellon Foundation, The Bobolink Foundation, The Curtis and Edith Munson Foundation, Forrest C. and Frances H. Lattner Foundation, The JPB Foundation, The Kresge Foundation, The Oram Foundation, Inc., The Overbrook Foundation, The S.D. Bechtel, Jr. Foundation, The Summit Charitable Foundation, Inc., and many other generous supporters.

The opinions expressed in this book are those of the author(s) and do not necessarily reflect the views of our supporters.

STATE OF THE WORLD

EarthEd:
Rethinking Education
on a Changing Planet

Other Worldwatch Books

State of the World 1984 through *2016*
(an annual report on progress toward a sustainable society)

Vital Signs 1992 through *2003* and *2005* through *2015*
(a report on the trends that are shaping our future)

Saving the Planet
Lester R. Brown
Christopher Flavin
Sandra Postel

How Much Is Enough?
Alan Thein Durning

Last Oasis
Sandra Postel

Full House
Lester R. Brown
Hal Kane

Power Surge
Christopher Flavin
Nicholas Lenssen

Who Will Feed China?
Lester R. Brown

Tough Choices
Lester R. Brown

Fighting for Survival
Michael Renner

The Natural Wealth of Nations
David Malin Roodman

Life Out of Bounds
Chris Bright

Beyond Malthus
Lester R. Brown
Gary Gardner
Brian Halweil

Pillar of Sand
Sandra Postel

Vanishing Borders
Hilary French

Eat Here
Brian Halweil

Inspiring Progress
Gary Gardner

EarthEd:
Rethinking Education
on a Changing Planet

Erik Assadourian, *Project Director*

Cem İskender Aydın
Pamela Barker
Nicole Bell
Monica Berger-González
Marvin W. Berkowitz
Bill Bigelow
Eve Bratman
Kate Brunette
Kelli Bush
Melissa Campbell
Liliana Caughman
David Christian
Dwight E. Collins
Jonathan Dawson
Rafael Díaz-Salazar
Robert Engelman
Joshua Farley
Kei Franklin
Russell M. Genet
Josh Golin
Dag O. Hessen

Andrew J. Hoffman
Craig Holdrege
Daniel Hoornweg
Nadine Ibrahim
Esbjörn Jorsäter
Mona Kaidbey
Lesley Le Grange
Laura Lengnick
Carri J. LeRoy
Chibulu Luo
Helen Maguire
Michael Maniates
Monica M. Martinez
Amanda McCloat
Amy McConnell Franklin
Dennis McGrath
Alexander Mehlmann
Marilyn Mehlmann
Asher Miller
Melissa K. Nelson
Simon Nicholson

Sabine O'Hara
Hugo Oliveira
Ali Değer Özbakır
Jessica Pierce
Raquel Pinderhughes
Olena Pometun
Luis González Reyes
Mark Ritchie
Jacob Rodenburg
Bunker Roy
Michael Sanio
Deirdre Shelly
David Sobel
Michael K. Stone
Linda Booth Sweeney
Takako Takano
Vanessa Timmer
Joslyn Rose Trivett
David Whitebread
Nancy Lee Wood
Tetsuro Yasuda

Erik Assadourian and Lisa Mastny, *Editors*

ISLANDPRESS

Washington | Covelo | London

For additional related and updated content, videos, blog posts, and links, and to learn more about Worldwatch's EarthEd Catalyst Network, visit www.EarthEd.info.

This book is dedicated to the teachers of the world: those connecting students to the living, breathing cycles of the Earth; those teaching character, life skills, and the ability to think critically and creatively about the future; and, most importantly, those teaching students to be bold leaders who will defend and heal the Earth in the tumultuous centuries ahead.

Contents

PART TWO: HIGHER EDUCATION REIMAGINED

List of Boxes, Tables, and Figures

BOXES

TABLES

FIGURES

Units of measure throughout this book are metric unless common usage dictates otherwise.

Foreword

The word "system" is the most radical word spoken in any language. It is radical in the true sense because it points to our inescapable rootedness in the fabric of life, from microbes that inhabit our bodies to the air we breathe. The word symbolizes our implicatedness in the world and our dependence on things beyond ourselves. The modern celebration of individualism stands at the other extreme as an assertion of autonomy and independence from the friends, families, communities, societies, and ecologies on which we depend. Systems thinkers, in contrast, see the world as networks of interdependence, not merely as a stage for individual performance.[1]

One result of a systems perspective ought to be gratitude for the things that have been given to us that owe nothing to our individual efforts. In large measure, we are the result of our genes, upbringing, local conditions, teachers, cultures, and the particular places that nurture every moment of our lives, inside and out. We live, in other words, within a web of obligations and relationships that transcend the conventional boundaries by which we organize academic disciplines and bureaucracies.

Thinking of the world as a network of systems begins in natural history, ecology, and the study of biophysical conditions, both within and without. It likely begins early in life, in a child's curiosity about what is connected to what. It is grounded in the physical sciences, but it extends through every discipline in the curriculum. The tools of systems thinking range from complicated computer modeling to intuition and the vague hunch that something is missing.

Systems thinking leads to the recognition of the counterintuitive results of human action, to an awareness of the unpredictability of events, and, in turn, to the necessary precaution that leaves wide margins for error, malfeasance, and acts of God. But the scope, scale, and technological velocity of change now threaten the future of civilization. This gives us every reason to avoid making irrevocable and irreversible system changes without due diligence and a great deal of careful thought. Applied to policy and law, systems thinking would cause us to act with greater precaution and foresight.

The idea of systems is fundamentally political, because it underscores our interrelatedness and mutual dependence. The political community and the ecological community are one and indivisible, but they are not equal. The human community, in all of its manifestations, is a subset of the larger web of life. But the essential questions of politics—who gets what, when, and how—pertain throughout the entire system. The millions of human decisions that have appropriated the majority of the planet's net primary productivity for human use are political choices that cross species lines. The preservation of half of the Earth as a sanctuary for biodiversity, as proposed by biologist Edward O. Wilson, would be a political choice as well.[2]

This is familiar ground to most of the readers of Worldwatch's annual *State of the World* reports. But it is not well known or comprehended by the great majority of people in the United States, Europe, or elsewhere—a failure of education that has large consequences. The elections of 2016 in Western democracies, for example, showed the fault lines emerging in our civic culture. They are not, first and foremost, the standard disagreements between liberals and conservatives about the size and role of governments and markets. Rather, they are a dispute between advocates of competing paradigms about the possible and desirable scale of human domination of the ecosphere and who benefits and who loses.

The upshot is that recent political events in the United States and Europe reveal large disparities in scientific knowledge and in the command of factual evidence about Earth systems, ecology, oceans, and so forth. We might expect that, under growing ecological stress, there also would be a rise in demonization of "others," hatred, fear, demagoguery, and violence. In such circumstances, public ecological literacy will become increasingly important to inform and moderate political discourse and to improve governance under conditions of what political theorist William Ophuls once described broadly as "ecological scarcity."[3]

This is neither an academic issue nor an argument for one political party or another. It is to say that a great deal depends on whether our leadership at the highest levels—and all of their advisers, cabinet members, and voters, of whatever party—have a worldview that corresponds to the actual physical realities of our planet, from ecology, climate, Earth systems, and biogeochemical cycles down to the small things that govern life processes in soils and our guts alike.

Do these decision makers understand how rapid climate change might adversely affect agriculture, hydrology, biological diversity, and coastal cities;

human contrivances of economies, governments, institutions, and technology; and thus issues of war and peace, poverty and prosperity, health and disease, social stability and chaos? Do they understand that ecological mismanagement can contribute to drought, famine, illness, war, and death? Do they understand the intimate connection between the environment and justice? Do they know that both are central to the proper and wise conduct of the public business?

The 2016 U.S. election, and recent elections in other democracies, suggest that many people are ignorant of such things. In place of an ecologically informed world, we witness the eruption of a calcified, bulletproof, crackpot ecological ignorance, hardened into a contradictory hodge-podge of promises, fantasies, and ill logic that could lead to irrevocable planetary disaster. The "true believers" in this group say that climate change is a hoax that has no scientific basis, and so, at this late hour, they propose mining and burning more coal, extracting more oil, and enlarging the fossil fuel empire. They plan to terminate clean energy plans, eviscerate environmental protections, and withdraw from international climate agreements. Whether or not this is the final phase in the denouement of civilization, I do not know. But I do know that "they know not what they do."

Could better education have made a difference? If, as children or young adults, the ideologues and extremists assuming power in the United States and elsewhere had been exposed to educators and authors such as those writing in this book, might it have made a difference? If they had spent more time outdoors as children, might they have bonded to nature and acquired a deeper love of and respect for life? Had some of them read Aldo Leopold instead of Ayn Rand, would they know better how to separate truth from nonsense?

Might they even have become ecologically literate? Would a lasting acquaintance with soils, animals, water, forests, and plants, tutored by ecologically grounded teachers, have broadened and deepened their attachment to the Earth and to the other species with whom we share this planet? Might better science courses have sparked their curiosity about what is connected to what and why those connections matter?

If they had read more widely, say the writing of Loren Eiseley or Thomas Berry, might they have comprehended the larger narrative in which humankind is a bit player? Had they read Edward O. Wilson or Elizabeth Kolbert, might they have understood the peril of the "sixth extinction" and the importance of preserving large parts of the Earth as a safe haven for biological diversity? Had they read economist Herman Daly, would they have understood why thermodynamics is more fundamental than economic theories?

The fact is that an ecological education does matter, and it matters a great deal whether or not the leaders and followers of whatever, wherever, and whenever understand the fine print of human life on a planet with a biosphere as reflexively as they understand the laws of gravity. If humans are to persist on Earth and thrive, ecology must become the default setting for ethics, farming, forestry, land management, living, building, governance, politics, investment, and business. We need a revolution that must begin with a sea change in our thinking.

We are a half century into what has been called the age of ecology. We—educators, scholars, researchers, legislators, and activists alike—have accomplished a great deal. But it is not nearly enough relative to the scope and scale of the changes that are under way, and the time available to make up the difference is very short. The authors in this volume are among the leaders at the forefront of environmental education. The stories, research, and insight that follow will inspire by the examples, models, and sheer dedication and perseverance. But all of us need to do more, particularly in connecting the science of ecology and Earth systems to a new, ecologically informed civic culture, politics, and system of governance.[4]

David W. Orr
Paul Sears Distinguished Professor Emeritus and Counselor to the President, Oberlin College

Acknowledgments

Every year, *State of the World* requires the work of many hands to ensure its growth from conception to a fully actualized entity—not too different than educating a child (although, admittedly, a lot easier).

This year, I have many to thank for their help in giving life to the report— our thirty-fourth! First, a sincere thank you goes to Worldwatch's staff and fellows, who make production of this report possible year after year. And a very special thanks to the members of our board whose commitment to the Worldwatch community has been invaluable.

I'd also like to thank Worldwatch's many supporters, both the kind individuals who have invested in our ideas and research over our many years and the foundations and institutions that help us continue to actualize our work. Thanks especially to: the 1772 Foundation; Ray C. Anderson Foundation; Aspen Business Center Foundation; Brighter Green; Del Mar Global Trust; Folk Works Fund, Fidelity Charitable; White Pine Fund, Fidelity Charitable; Garfield Foundation; German Federal Ministry for the Environment, Nature Conservation and Nuclear Safety (BMU) and the International Climate Initiative; GIZ (German Agency for International Cooperation) with Meister Consultants Group, Inc.; J. W. Harper Charitable Fund, Schwab Charitable; Hitz Foundation; La Caixa Banking Foundation and the Club of Rome; Steven Leuthold Family Foundation; Michl Fund of the Community Foundation of Boulder County; National Renewable Energy Laboratory (NREL), U.S. Department of Energy; New Horizon Foundation; Paul and Antje Newhagen Foundation of the Silicon Valley Community Foundation; Overseas Development Institute (ODI) with the U.S. State Department; Robert Rauschenberg Foundation; Shenandoah Foundation; The Laney Thornton Foundation; True Liberty Bags; U.S. Agency for International Development (USAID) with Deloitte Consulting LLP; and the Weeden Foundation Davies Fund.

I am also indebted to the sixty-three authors who contributed to this report. With this project, we had the ambition to greatly expand the conversation about what education for sustainability and resilience entails. Our authors—as educators and practitioners with centuries of experience among them—certainly rose to the occasion. Special thanks also goes to David Orr, one of the founding fathers of environmental education, for writing the Foreword.

Many thanks to editor Lisa Mastny for helping to unify the voice of the twenty-five chapters and thirty text boxes; and to graphic designer Lyle Rosbotham for bringing beauty to the words with his book design and photo selections.

I also want to acknowledge the hard work of our *State of the World 2017* interns, Lydia Nagelhout, Kei Franklin, and Drew Walsh, who all played an important role in early research and organizational efforts for the project—thank you!

Much gratitude goes to the book's publisher, Island Press. To David Miller, Julie Marshall, Emily Turner Davis, Jaime Jennings, Maureen Gately, Sharis Simonian, and the entire team, thanks for helping to bring *State of the World* from concept to final form over these past six years.

A special thanks also goes to our international publishing partners: Turkiye Erozyonla Mucadele, Agaclandima ve Dogal Varliklari Koruma Vakfi (TEMA), and Kultur Yayinlari IsTurk Limited Sirketi (both in Turkey); FUHEM Ecosocial and Icaria Editorial (Spain); Paper Tiger Publishing House (Bulgaria); WWF-Italia and Edizioni Ambiente (Italy); Earth Day Foundation and the Vojvodina Foundation (Hungary); Worldwatch Institute Europe (Denmark); Worldwatch Brasil; China Social Sciences Press; Korea Green Foundation Doyosae (South Korea); Taiwan Watch Institute; Worldwatch Japan; and the Centre for Environment Education (India). You make it possible for us to share this report with people around the world, for which we are especially grateful.

Thanks to Gaelle Gourmelon, Worldwatch marketing and communications director, who ensures each year that the findings of *State of the World* spread far and wide in the media and global blogosphere.

And finally, above all, I want to thank the Earth. Looking back on earlier acknowledgments in these volumes, including the several I've written over the years, I realize that we never have thanked our living planet for providing the breath of life that allows us to do this work and to revel in being alive, and, more concretely, for providing the trees, water, minerals, and energy to print this report—all in hopes of inspiring more people to join the essential work of bringing about a more sustainable future. We hope that your sacrifice is not in vain.

Erik Assadourian
EarthEd Project Director

Introduction

EarthEd: Rethinking Education on a Changing Planet

Erik Assadourian

What is education for? Education—the process of facilitating learning—has been an integral part of human societies since before we were even human. After all, humans are not the only species that transmits knowledge from one individual to another. Chimpanzees and dolphins, for example, both teach their young specialized foraging and hunting techniques that are known only to their communities and pods. Learning has been documented in numerous species, even in plants and bacteria. Because learning is a natural part of being alive—and increases the odds of staying alive—at its very root, the role of education may be to facilitate survival, both for the individual that is learning and for the social group (and species) of which it is a part.[1]

As humans evolved—going beyond day-to-day survival and developing systems of writing, arts, tools, and the like—complex cultural systems formed and helped to shape educational priorities. As anthropologists David Lancy, John Bock, and Suzanne Gaskins explain, "the end points of learning . . . are culturally defined." In other words, education prepares children for life in the cultures into which they are born, giving them the tools and knowledge that they need to survive in the physical and social realities in which they most likely will spend their entire lives.[2]

This might have been fine throughout most of human history, where cultural knowledge correlated strongly with the knowledge that was needed to survive and thrive in the immediate environment (for example, how to identify which plants and animals are dangerous and which are edible; how to make fire, tools, clothing, and shelter; and how to coexist with neighboring

Erik Assadourian is a senior fellow at the Worldwatch Institute and director of *State of the World 2017* and Worldwatch's EarthEd Project.

populations). But the cultures that most humans are now born into are variations of consumer cultures—cultures that, through their profligate use of resources and promotion of unsustainable levels of consumption, are rapidly undermining the Earth's systems to the point that they now threaten the very survival of countless species and human communities around the world.[3]

For humans to thrive in the future, we will need to systematically rethink education, helping students learn the knowledge that is most useful for their survival on a planet that is undergoing rapid ecological changes. We must provide them with the tools and strategies that they need to question the current sociocultural reality and to become bold leaders who will help pull us back from the brink of ecocide and usher in a sustainable future. But even that is not enough. Considering how much damage human civilization has already done to the Earth, students also must learn how to prepare for and adapt to the ecological shifts that are already locked in to their future—and ideally do this in ways that help both to restore Earth's systems and to preserve their own humanity.

State of the World 2017 explores how education—particularly formal education—will need to evolve to prepare students for life on a changing planet. Some priorities will not change much in this new "Earth Education" or "EarthEd" context: basic literacy, numeracy, multilingualism—these skills will continue to be as important in the future as they are today. But many new educational priorities must emerge: ecoliteracy, moral education, systems thinking, and critical thinking, to name a few. Without these and other key skills, today's youth will be ill-equipped for the dual challenges that they face of building a sustainable society and adapting to a changing planet.

Our Changing Planet

Over the past few hundred years, as humans have harnessed coal, oil, and natural gas to generate heat, steam power, electricity, liquid fuels, and new materials, we have unleashed the start of a climate shift that has never before been experienced in human history, with temperatures today already higher than during our last eleven thousand years of civilization. Moreover, we have enabled a massive spike in the human population, thanks to discoveries ranging from germ theory to the scientific developments behind the Green Revolution. As early innovations solidified into a complex industrial economic system based primarily on fossil fuels, humanity's impact on the planet has grown exponentially—to the point where most of the Earth's ecosystem services are now degraded or are being used unsustainably.[4]

Worse yet, we have created a series of positive feedback loops that are further accelerating the damage. This includes the $579 billion a year spent around the world to promote the ever-increasing consumption of consumer goods—from fast food, soft drinks, and coffee to cars, computers, and smartphones. Amazingly, many of these goods are no longer seen as luxuries but as necessities, even entitlements—indicators of a basic level of prosperity—despite the planetary resource constraints that make it impossible for all Indians or Chinese, let alone the entire human population, to live like Americans or even Europeans. In the process of normalizing the consumer economy—and actively spreading it to people around the world (including to 220 million Chinese over the past fifteen years)—we have locked in a frightening series of ecological changes, whose tragic impacts are only starting to manifest today.[5]

Let's look at climate change. In the past, as the Earth emerged from episodic ice ages, temperatures tended to rise 5 degrees Celsius over periods spanning some five thousand years. Now, models project that temperatures will increase 2 to 6 degrees Celsius in the next century and will continue rising beyond that. This translates to many meters of sea-level rise, rapid acidification of the world's oceans, and dramatic changes in rainfall patterns, causing, in turn, droughts, disasters, and famines—all within a very short time frame (from a human history perspective, let alone a geological perspective). In all probability, this will be catastrophic to human civilization as we know it today.[6]

And climate change is not the only worrisome change looming. We are crossing several other planetary boundaries as well: disrupting the phosphorus and nitrogen cycles, depleting biodiversity, and spewing enormous amounts of chemicals into the air, soil, and water, to the point that we have brought about a new, human-dominated, geological epoch: the Anthropocene. Meanwhile, the human family is adding 83 million members each year. At current projections—assuming that ecological catastrophes do not slow this growth—the global population is projected to reach 9.7 billion by 2050. Of course, businesses and marketers will continue to work hard to sell this growing population ever more stuff, putting ever-greater pressure on Earth's overtaxed systems.[7]

We have hit a point where climate scientists now question whether civilization—whether their own children and grandchildren—will actually survive. "It's clear the economic system is driving us toward an unsustainable future, and people of my daughter's generation will find it increasingly hard to survive," says Will Steffen, director of the Climate Change Institute at The Australian National University. "History has shown that civilizations have risen, stuck to their core values, and then collapsed because they didn't change. That's where we are today."[8]

The defining quest for humanity today is how we will be able to provide fulfilling lives for 8–10 billion people even as Earth's systems are declining rapidly. These cannot be consumer lives, ecologically speaking, but decent lives that offer access to vital services, such as basic health care and education, to livelihood opportunities, and to essential freedoms. Unfortunately, few people today understand the urgency or magnitude of this quest—some even deny it—and few fully grasp the changes that are necessary to succeed. Far fewer have the skills that are required to help with this transition or, at least, to survive the ecological shifts if the quest for a sustainable future fails. Education will be essential in changing this.

Educational Reform on a Planetary Scale

Unfortunately, schooling today tends to ignore the massive changes that are looming and offers little in the form of preparation for slowing those changes or coping with them. Worse yet, many would argue that schools are often designed to "train children to be employees and consumers," only exacerbating our current problems. This comes as little surprise, given that consumerism is the dominant cultural context in which most students now grow up. Socializing them for that reality may be the "natural" role for education, even if, in the long term, it is maladaptive.[9]

This maladapted role of education is made far worse when governments change the law to make it easier to mislead students about climate change, as lawmakers in the U.S. states of Tennessee and Louisiana have done, or when school boards allow corporations to shape the curriculum. In Chapter 13 of this book, Josh Golin and Melissa Campbell of the Campaign for a Commercial-Free Childhood discuss the expanding foothold that corporations have in schools around the world, from the oil giant Chevron sponsoring science education to fast-food purveyor McDonald's recruiting teachers to host school fundraisers in its restaurants. There are long lists of how students are indoctrinated into becoming unquestioning consumers in schools (let alone through the six or more hours on average that American youth spend watching television and interacting with computers, tablets, and smartphones each day). But even when schools guard themselves from these types of infiltrations, they are still doing very little to prepare students for the social and ecological realities that they will soon inherit.[10]

Considering the present moment in history, it is clear that most schools are forgoing their responsibility to question the status quo—whether this is the dark history of colonization and genocide on which industrial civilization is founded, or the horrific ecological and societal abuses on which the consumer economy continues to be built. The current role of schools will have to change if we are to prepare students to slow down—and survive—the ecological transition ahead.

Specifically, we will need to redesign education to teach students to become sustainability champions: those who are willing to boldly step out of current realities and commit themselves to drive social, political, economic, and cultural change so that human societies can live sustainably on the planet. Almost as importantly, education must make students more resilient to the changes that are locked in to their future—offering them a variety of life skills (particularly skills that will increase in value as the consumer era comes to an end) and coping skills, such as social and emotional learning, which will enable them to more sanely navigate the tumultuous, conflict-ridden future. Ideally, given the limited hours in the school day, curricula will need to be designed around lessons and projects that maximize both education for sustainability and education for resilience, whenever possible. (See Figure 1–1.)

This is the necessary path forward, given that the precise future that the next generations will inherit remains uncertain. Will governments, corporations, and civil society find the will to significantly scale back economic and population growth, consumption, and the use of fossil fuels in order to stabilize the climate? Will agreements be "too little, too late" to stop climate change, but at least keep the transition to a hot state manageable (whatever that means)? Or will negotiations break down entirely, with business-as-usual and climate denial driving us to a rapid and out-of-control shift to a 4 degree or even 6 degree Celsius apocalyptic future, marked by devastating famines, inundated

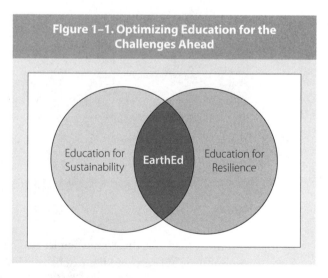

Figure 1–1. Optimizing Education for the Challenges Ahead

Education for Sustainability

EarthEd

Education for Resilience

cities, mass migrations, and climate wars? Even in the best scenario of intentional economic degrowth, the skills and knowledge that students will need will be very different than what they are being taught today.[11]

Principles of Earth Education

For humanity to get through the coming century, our schools must emphasize a new set of proficiencies—a Common Core-equivalent that will enable us to survive life on a changing planet. These Earth Education Core Principles, or EarthCore, include six broad tenets, each building on the former (although with considerable interlocking, as all sturdy construction has). (See Figure 1–2.) Redesigning education so that these principles are present in essentially every aspect of the school experience—from class lessons and field trips to lunch menus and school infrastructure—can ensure that students are better prepared both to become leaders in the sustainability transition and to navigate the disrupted future ahead.

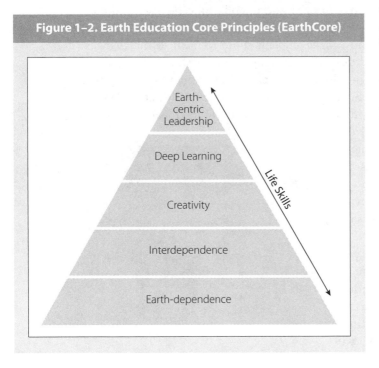

Figure 1–2. Earth Education Core Principles (EarthCore)

Earth-centric Leadership

Deep Learning

Creativity

Interdependence

Earth-dependence

Life Skills

Many of these EarthCore principles are being taught already to some degree or another, but rarely to the extent needed, nor typically in combination with one another. The challenge will be finding ways to integrate these principles in education as quickly, and to as great an extent, as possible.

Principle 1: Earth-dependence

At the base of the EarthCore pyramid is a deep understanding that humanity, as a species and as a civilization (in all of its cultural variations), is completely and utterly dependent on the Earth, a lesson that most people seem to have forgotten in the modern era. This understanding—and the corresponding humility and awe (in both the joyous and fearful sense of the word)—is essential, for without this foundation, the pyramid, and the institution of education and human civilization, will collapse.

But how does one teach "Earth-dependence"? Ecoliteracy is a key piece of the puzzle. Without a strong understanding of both the environmental sciences (which includes the underlying basic sciences such as biology, ecology, chemistry, and physics) and the limits to growth, children will grow up with unrealistic expectations for what life in our closed planetary system can provide. The good news, as Michael K. Stone describes in Chapter 3, is that ecoliteracy can manifest in all aspects of education—even the school cafeteria, which Luis González Reyes explores in more depth in Chapter 6.[12]

Ecoliteracy is not just a curricular add-on. It can, and must, be taught in in-depth ways and be embedded fully in the core curriculum. At the School for Examining Essential Questions of Sustainability (SEEQS), a middle school in Honolulu, Hawai'i, students devote two hours a day, four days a week, to exploring an "Essential Question of Sustainability," and they focus a full semester on a topic such as, "What are ways to restore and preserve native habitats in Hawai'i?" Rather than passively exploring this question by memorizing facts and figures, students learn this material actively, working on collaborative projects, receiving mentoring from local experts, and presenting their findings to the larger community. Through this, students not only learn about, but deeply internalize, the challenges—and solutions—to the sustainability crisis they face.[13]

But learning about our dependence on the Earth academically is not enough. As David Sobel discusses in Chapter 2, before we can get children inspired to "save the rainforests," we need to nurture their own relationships with local forests, streams, and meadows. Nature-based and place-based learning opportunities—such as the forest schools now present in many countries around the world—are leading the way in creating educational experiences that cultivate deeper relationships with the broader ecological community. At the Wald Kindergarten (Forest Kindergarten) in Langnau am Albis, Switzerland, a score of four- to seven-year-olds spends all day in the woods—rain, shine, or snow—playing, learning, and connecting directly with the local ecosystem. This extended time in nature deeply affects children's development, from reducing attention-deficit disorders to improving confidence, cognitive functioning, and self-control. Most importantly, it helps reveal nature's role as "ultimate teacher," an insight that Indigenous education encapsulates and continues to provide today, as Melissa K. Nelson discusses in Chapter 4.[14]

Finally, it is one thing to be ecologically literate and bonded to the Earth or a local environment, but another to remain dutiful in sustaining it—even in the face of social and cultural pressures to do the opposite. Cultivating stewardship, as Jacob Rodenburg and Nicole Bell discuss in Chapter 5, is essential. This cultivation occurs in many ways, from teaching young children to know their animal and plant neighbors in their "neighborwood," to getting teens to reach out to local conservation groups and volunteer with them.[15]

Principle 2: Interdependence

Given that many of the challenges of the future will center on the equitable distribution of increasingly constrained resources and ecosystem services, education must cultivate a deeper understanding of our interdependence with our fellow humans, irrespective of differences in culture, creed, color, gender, or sexual orientation. Interdependence can be taught in myriad ways, but three elements are fundamental.

First, moral or "character" education must be front and center. There is a real value in actively teaching children to be good people: humble, helpful, kind, just, and, above all, aware that they have moral obligations to others, including future generations, other species, and the Earth itself. In Chapter 7, Marvin W. Berkowitz describes how, in places around the world where character education has been integrated into schools—from the United States to Singapore—it can have tremendous impact, improving student and teacher behavior, student satisfaction, and academic results.[16]

Second, schools need to teach social and emotional learning skills, helping students develop empathy; mindfulness; the ability to recognize, understand, and deal constructively with others' emotions; and resilience when confronted with difficult obstacles. In Chapter 8, Pamela Barker and Amy McConnell Franklin discuss how more schools are integrating training in social and emotional learning—both directly into curricula, such as through mandatory empathy training in Danish schools, and through innovative approaches such as replacing detention with meditation, as has occurred at the Robert W. Coleman Elementary School in Baltimore, Maryland.[17]

Finally, an important part of interdependence is being able to live together peacefully and respectfully—what educators of the Escuela Nueva movement, a program founded in Colombia in the 1970s and now serving 5 million children in nineteen countries, call *convivencia*, or the "art of living together." In the Escuela Nueva program, children typically work in small groups (with the teacher playing a background role as facilitator rather than instructor) and learn to cooperate and to negotiate differences. This is reinforced with an intentional "emphasis on the formation of democratic and participatory values," such as through sharing responsibilities for maintaining and managing the school, and through electing each student to a committee. Training children in the art of living together, including democratic decision making and how to gain consensus, will prove a very valuable skill. With strong cultivation of character education, social and emotional learning, and *convivencia*, students should be far better prepared for the increased conflict and difficult ethical choices that will be part of their future.[18]

Principle 3: Creativity

Because the challenges ahead are going to be complex and will require fresh ideas on how to solve them, education also should prioritize creativity. Play is a key element in developing this creativity—one that, unfortunately, has increasingly been excised from education, whether through reduced recess time or through the heightened focus of early education on academics. Play expert David Whitebread warns that, "We put our children's future at risk, and their ability to deal with the many difficulties that the human species will confront through the twenty-first century and beyond, if we do not recognize the importance of play and begin to develop policies, both in relation to our domestic arrangements and our schooling systems, that support and nurture their natural and adaptive playfulness."[19]

A "junk playground" in Berlin, Germany.

Mads Bødker

In Chapter 9 of this book, Whitebread explores the health and social benefits of play and points to ways both within and outside schools to allow for greater time and opportunity to play, from city streets that provide safe and traffic-free environments to playgrounds full of "junk" that stimulate creativity and independence. While play is disappearing in many countries, including the United States and the United Kingdom, in others it is well protected. In Finland, a country often touted for its high educational attainment, "playful learning" is so important that formal schooling does not start until the age of seven, and elementary schools provide an average of seventy-five minutes of recess each day, three times that of U.S. schools.[20]

As children get older and opportunities for free play inevitably shrink, play in the form of the arts, scientific experimentation, and project design (such as presentations, stories, and films) offers new opportunities to be creative. In Chapter 10, Marilyn Mehlmann and her colleagues describe how comic art is a particularly adept way to provide an opportunity to play, through a medium that encourages drawing, storytelling, writing, and group collaboration and that can be readily focused on themes of sustainability and social change. In countries from Belarus to India, sustainability-focused comics programs have helped students playfully explore ways to get to a more livable future.[21]

Principle 4: Deep Learning

As important as creativity is cultivating the ability to "learn how to learn"— what is known in the field of artificial intelligence as deep learning. Just as computers of the future should be able to apply one set of knowledge to another realm, so too should we as people. This makes us more flexible, more adaptable, and thus better able to manage the surprises that the future throws at us.[22]

One of the foundations of deep learning is systems thinking. Understanding that the world is made up of interconnected, nested systems, many of which follow similar rules, is key to fully grasping the challenges ahead. In Chapter 12, Linda Booth Sweeney discusses how teachers in a variety of disciplines are adding systems thinking to their lessons and how national governments are starting to integrate systems thinking into broader learning goals.[23]

In Chapter 11, Dennis McGrath and Monica M. Martinez explore how deep learning can be made even "deeper" by prioritizing certain competencies, such as critical thinking, working collaboratively, and having students direct their own learning and apply it across disciplines and to the real world. From middle schoolers in Maine studying invasive species and presenting solutions to the local city council, to high school students in Philadelphia learning about law and history through the design of their own citizen lobbying project, deeper learning can play an important role in preparing students for civic life and leadership.[24]

As biologist Edward O. Wilson has observed, today "we are drowning in information, while starving for wisdom. The world henceforth will be run by synthesizers, people able to put together the right information at the right time, think critically about it, and make important choices wisely." Synthesizers are those who can take learning from one realm and apply it to another. Better integrating systems thinking, critical thinking, and deeper learning into school curricula will help develop a new generation of synthesizers who are better able to understand the Gordian sustainability knots that they are inheriting and to discover novel ways of untying them.[25]

Principle 5: Life Skills

Perhaps one of the most important elements of the EarthCore principles is the learning of life skills—so integral that, in the diagram, it interweaves through all other aspects, like rebar does in concrete. Many basic life skills are acquired in nature. Life skills require a mix of critical thinking, social and emotional intelligence, and creativity. But unlike the other categories, life skills arguably can be seen as both ends as well as means. Without providing life skills, education is failing to prepare children for life, especially on a planet in such a rapid state of change. Life skills include a wide variety of proficiencies, including basic survival skills, such as cooking and gardening; language learning; comprehensive sexuality education; and vocational training.

In Chapter 14, Helen Maguire and Amanda McCloat demonstrate how home economics plays an important role in teaching students essential skills, such as nutrition, cooking, sewing, and balancing a household budget—all of which have benefits for both sustainability and resilience. For example, choosing healthier, less-processed foods requires cooking skills, which fewer people now possess. Gaining this knowledge will help address the obesity epidemic today as well as prepare students for a future where processed foods may be harder to procure.[26]

Multilingualism is also a key life skill that research reveals offers cognitive, health, social, and economic benefits. It even "rewires" the brain, with studies of bilingual individuals demonstrating improved multitasking ability, cognitive flexibility, and resistance to dementia. With population movements all but guaranteed to accelerate in the future—as droughts, floods, disasters, and conflicts increase—knowing multiple languages will boost both employability and adaptability.[27]

Comprehensive sexuality education, which Mona Kaidbey and Robert Engelman discuss in Chapter 15, is also an essential life skill with which to graduate. This is not just for the obvious reason of preventing unwanted pregnancies and thus reducing population growth, but also to improve gender relations, prevent unwanted sexual attention, and make life more enjoyable. "Sexuality is a dimension of who we are as human beings," Kaidbey and Engelman explain, and not being aware of or comfortable with one's own sexuality can greatly reduce one's well-being.[28]

Finally, vocational training—gaining skills that can lead to a future livelihood—is an essential element that education cannot afford to overlook. In Chapter 18, Nancy Lee Wood discusses how vocational training can provide affordable learning opportunities that both lead to employment and help in our transition to a more sustainable society— whether through direct training in renewable energy engineering, building repair and retrofitting, and regenerative agriculture, or through fields such as peace and conflict management, which will help in navigating the turbulent times ahead. The Mechai Pattana School in Thailand may offer the best example of a school integrating life skills into its core curriculum: at this high school, students are trained in social entrepreneurship, community forestry, and gardening and manage a variety of school businesses, from selling eggs and off-season limes to assembling small solar panel kits to market to Thai villagers.[29]

Kei Franklin

A student at the Mechai Pattana School in Thailand assembles portable solar panels to sell to surrounding communities.

Principle 6: Earth-centric Leadership

At the pinnacle of the EarthCore pyramid is Earth-centric leadership. Earth-centric leadership is the full actualization of education, of empowering and emboldening students to be reverent Earth citizens who work energetically to build a sustainable future and to help their fellow beings survive the coming changes. How does one teach Earth-centric leadership? To some extent, it will stem organically from teaching the other EarthCore principles— but it is too important to not teach actively.

Teaching Earth-centric leadership first requires schools to teach students what educational thinker Paulo Freire calls "critical consciousness." This will enable students to perceive the hypocrisies embedded in the social, political, economic, and cultural systems of which they are a part and that will need to be corrected if they are to help create a just and sustainable society. These corrections will not necessarily be easy to make, but being critically conscious of the need for them will be an essential prerequisite of Earth-centric leaders.[30]

Second, schools will need to teach and mobilize students to take an active role as advocates, organizers, social entrepreneurs, and leaders of all types. Students will need to assess the need for change and then take action bravely and strategically, embracing their role as change agents, even in the face of significant resistance. Although this type of engagement is rare in schools today, some institutions are leading the way. In Toronto, Canada, the Grove Community School has framed its elementary education around community activism, social justice, and sustainability education. Embedding these values directly in the school's curriculum, the teachers are empowering their students to be activists by discussing current events in class, writing letters to government officials, and even joining local environmental protests.[31]

In Bali, Indonesia, the sustainability-oriented Green School also is taking an active role in encouraging students to solve real-world problems. This has led to real-world results: in 2013, sisters Melati and Isabel Wijsen organized a plastic-bag-ban campaign in the province as a school project. Two years later, after a petition that included thousands of signatures and a brief hunger strike, they and their fellow students got a commitment from the governor of Bali to ban the bags province-wide by 2018. Their efforts have since gone global, and their organization, Bye Bye Plastic Bags, now has chapters in nine countries around the world.[32]

While few schools have integrated activism and Earth-centric leadership training into their teaching practices, many individual educators have done so, working within their school systems to confront injustices in their local communities. Rebecca Jim, a guidance counselor at an Oklahoma high school near Tar Creek, one of the worst hazardous waste sites in the United States, worked with teachers to integrate study of Tar Creek into ten classrooms across a variety of subject areas. Through this, students not only learned about the Superfund site near them, but played an active role in advocating for its cleanup. Jim—who supported students' efforts to analyze the health and environmental impacts of the site, present findings to the community and the media, and even publish two volumes of poetry and essays on Tar Creek—was central in both mobilizing her students to become sustainability-minded civic leaders and getting the U.S. Environmental Protection Agency to better address the site's toxic legacy.[33]

Fortunately, not all students will have an environmental disaster near them to serve as a focal point for their leadership training, but projects abound everywhere. Whether the conservation of a local stream or wetland, the creation of a new park, lobbying the local government to support sustainable living, or the establishment of a sustainable community social enterprise, there is infinite opportunity for teachers and students to develop projects that cultivate Earth-centric leadership.

Earth Education will prove essential in adequately preparing students for the challenges ahead. Ideally, EarthEd and the EarthCore principles will be advanced by school systems at an administrative level. But the teacher can, and does, play an essential role in bringing EarthCore to the fore. For example, a teacher in Vermont, inspired by a documentary on forest schools, can set up "Forest Fridays," bringing her students out into the woods one day each week. An art teacher can have her students design comics for sustainability, and a history teacher can add systems thinking to his lesson to better explain the dynamic interactions underlying historic events. A home economics teacher can teach sustainable gardening, cooking, and nutrition classes, and on and on. As the world faces an accelerating rate of change, providing students with these EarthEd fundamentals will be critical in building the more sustainable and resilient future they deserve.[34]

Higher Education Reimagined

To this point, the discussion of Earth Education has focused mainly on primary and secondary schooling. But education is a lifelong endeavor, and the focus must extend throughout all stages of learning. Moreover, if students attain an Earth-centric education when they are young, only to "graduate" to a neoliberal economics program or a traditional business school where they learn that maximization of profit is the primary fiduciary responsibility of a business leader, then the effort to rethink education for life on a changing planet will have failed. Hence, the second part of *State of the World 2017* explores how higher education also must be centered on EarthEd.

Here, higher education should be understood in its broadest context. University is just one path, even if the currently preferred one. Vocational training, apprenticeships, folk high schools, and shorter, more-targeted trainings play an important role in higher education—especially in a future where resource constraints may limit access to cost-intensive university education.

Higher educational opportunities also should be available to underserved populations, from rural villages to prison inmates. Barefoot College, for example, provides schooling and nondegree vocational training to thousands of Indian villagers each year and is also training female village elders to be solar engineers, which has helped provide both employment and development opportunities in more than sixteen hundred villages worldwide. (See Chapter 18.) And, as Joslyn Rose Trivett and her colleagues discuss in Chapter 19, providing science and sustainability education in prisons can play an important role in empowering inmates, improving their psychological well-being and offering livelihood opportunities post-release.[35]

For those who do attend university, the role of the institution, writ broadly, must be to help usher in a sustainable future. All disciplines will need to account for the ecological realities in which they, and human societies, are embedded. Some colleges, like College of the Atlantic (COA) in Bar Harbor, Maine, have done this to their very core. Regardless of what students choose to study, COA offers only one major: human ecology, or "the exploration of relationships between humans and their natural, cultural, and built environments." Within that context, students take an active role in designing their own course of study, whether choosing to study the arts, environmental sciences, sustainable business, languages, or many other concentrations.[36]

Although few schools have gone as far as COA, many are taking great strides in integrating sustainability more directly into their operations, infrastructure, research, and curricula. Although much of the emphasis today is on eco-efficiency, renewable energy, and building standards, more universities are starting to focus their attention on greening their curricula. In the Association for the Advancement of Sustainability in Higher Education's (AASHE) Sustainable Campus Index, for example, Appalachian State University received a top score of 96 percent for its curriculum, forty points higher than the U.S. average. The university not only offers undergraduate and graduate programs on sustainability, but all of its academic departments offer at least one course that includes exploration of sustainability, and all students graduate from programs that have adopted at least one of the university's three sustainability learning goals.[37]

Unfortunately, just as primary and secondary education has been captured to some degree by corporate interests in the United States, so too has higher education—from research funding to campus life. This limits the degree to which universities are free to reinvent themselves in the ways that our changing planet requires. In Chapter 16, Michael Maniates describes how universities have grown up in an era of economic growth and play an integral role in its promotion. Moving forward, higher education will need to prioritize redirecting the university away from this role and toward preparing human society for a post-growth future.[38]

In Chapter 17, Jonathan Dawson and Hugo Oliveira discuss redesigning the classroom, where instruction today focuses overwhelmingly on delivering an unquestioned set of information. To improve education, schools and universities need to better ground education in community, in dynamic knowledge, in students' experiences, and even in their bodies, by making the intellectual tangible—whether by going on nature walks, acting in classroom plays, or doing local service education projects.[39]

Another challenge is to infuse specific disciplines that are critical for developing a sustainable future with an EarthEd orientation. Whether in agriculture, economics, or engineering, there is no shortage of work ahead. In Chapter 22, Daniel Hoornweg, Nadine Ibrahim, and Chibulu Luo describe how, with the world's economies, populations, and cities on a trajectory to grow well into the future, more engineers will be needed. However, our future engineers will need to have a better grasp on sustainability if they hope to provide the required infrastructure in ways that do not accelerate climate change and ecological decline.[40]

In Chapter 21, Laura Lengnick explores how future farmers will need to have a strong grasp of sustainable, restorative, and resilient agricultural practices to feed the 9.7 billion people projected to be on the planet by 2050, all while climate change reduces yields and as the spread of consumerism expands the demand for more ecologically taxing meats and processed foods.[41]

Joshua Farley makes the point in Chapter 20 that until economics is grounded in the real world, our myopic view of economic activity will continue to taint effective economic planning and policy making. Fortunately, there is an upswelling of effort by students and professors of economics to integrate the real world into their discipline.[42]

Finally, professional schools will need to upgrade to be Earth-centric as well. As Andrew J. Hoffman describes in Chapter 23, the business school curriculum has to evolve from being focused just on reducing *un*sustainability in business operations to reexamining the role of business in society in order to create greater sustainability. In Chapter 24, Jessica Pierce describes how medical schools, too, must adapt to the times as pollution and environmental disruptions lead to more diseases, and as our current medical system invests disproportionate time and energy in treating symptoms rather than preventing illness. Although few medical schools have even broached these challenges, it is essential that they do.[43]

Moving Forward

The gap between Earth Education and where schools are today is about as wide as the gap between human civilization's current climate policies and what science requires of us to get to a sustainable future. In societies where sustainability is typically an afterthought at best, is there any chance that we can get to a truly Earth-centric education system?

The models certainly exist: the Mechai Pattana School, Barefoot College, SEEQS, College of the Atlantic, and forest schools all point the way. More pioneers are needed to take these experiments even further. And more reformers are needed to bring good ideas into existing institutions.

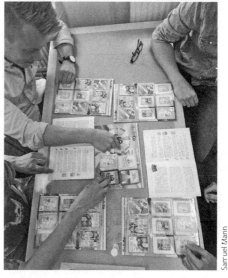

Students pursuing a masters degree in Sustainable Media Technology at the Royal Institute of Technology in Stockholm, Sweden, play a sustainability board game, GaSuCo.

Of course, both pioneers and reformers will need support. Global programs such as Eco-Schools—which now boasts more than forty-nine thousand registered schools (see Chapter 3)—and AASHE's Sustainable Campus Index can help provide incentives, support, and a clear path to ratchet up efforts.[44]

Philanthropy also can play a role in nudging schools toward an EarthEd model, although today foundations appear to be focusing more on short-term educational reforms that improve conventional academic performance. Even if their efforts succeed, however, without embedding these reforms in an Earth-centric "theory of change," students will remain underprepared for the challenging times ahead.[45]

Most important is shifting government policy to better orient schools on sustainability, as this is where the bulk of funding and education policy direction comes from. Although many governments have made some effort to integrate education for sustainable development (ESD) into their national curricula, few have deeply integrated sustainability to the extent needed to tackle the challenges ahead. Sweden, for example, requires every level of education to promote sustainable development, and, since the mid-1990s, the country's

National Agency for Education has been integrating ESD into the curriculum, supporting this with teacher training, sharing of best practices, and regular assessments of ESD implementation. Yet while leading the way, even Sweden's efforts are far from comprehensive: for example, no assessment of ESD learning is included in Swedish national testing, signaling ESD's secondary role in national learning objectives.[46]

Perhaps with implementation of the new United Nations Sustainable Development Goals—Goal 4 of which aims to "ensure inclusive and quality education for all and promote lifelong learning"—governments will increase their attention to sustainability and resilience education. After all, if education does not teach about sustainability or how to live on a changing planet, it is hard to argue that it is quality education that promotes lifelong learning. Perhaps reassuringly, a more-specific target of Goal 4 is: "By 2030, ensure that all learners acquire the knowledge and skills needed to promote sustainable development, including, among others, through education for sustainable development and sustainable lifestyles, human rights, gender equality, promotion of a culture of peace and non-violence, global citizenship, and appreciation of cultural diversity and of culture's contribution to sustainable development."[47]

If governments strive to achieve this goal, education could look quite different in the future. Chapter 25 of this book envisions what education could look like in 2030 if educators, administrators, and others both comprehend *and act on* this need for a fundamental shift in educational priorities—focusing on Earth Education and establishing new types of schools that specialize in teaching local ecological knowledge, activism, and Earth-centric leadership. On a changing planet, it will be essential for education to evolve, not simply to expand to reach more people.[48]

Education alone will not save humanity, but it may play an essential role in enabling people to get through the turbulent times ahead with their humanity intact. It also may help train a new generation of leaders who can slow the ecological crisis to a speed to which humanity may be able to adapt. For education to play this role, however, it will take bold leadership on the part of educators and the administrators and policy makers that support them. If they can summon up this leadership, then perhaps tomorrow's students will be not only better equipped for surviving the challenges ahead, but also well on the way to building a sustainable future.

PART ONE

Earth Education Fundamentals

CHAPTER 2

Outdoor School for All: Reconnecting Children to Nature

David Sobel

One of the salient problems facing us today is children's alienation from the natural world. They are too creeped out to touch earthworms, they don't know where their food comes from, and they are afraid to walk in the forest alone. Or, if they are walking in the forest, they can't see the forest for their iPhones. We, and our children, are easily seduced by the panoply of digital treats. It is so much easier to be a couch potato than to plant potatoes. The result is that twenty-first-century children spend eight hours a day interacting with digital media, and only thirty minutes a day outside.[1]

When interviewed about their computer use about fifteen years ago, children in Putney, Vermont, described how this happened. One girl sheepishly admitted: "Before we had a computer, I used to read a lot and go outside more to be in the neighborhood. Now, it's so easy to go exploring on the computer, it's like it's too much work to go outside." Another boy agreed: "I'll be playing a really cool computer game, and I'll think, 'Wow, it's beautiful outside, I should really go outside.' But I can't stop myself from playing—it's kind of like I'm addicted." A third student summarized: "For me, I learned to love nature before I did computers, and so it doesn't really affect me. But if I started to use computers when I was really young, it might have kept me from getting into nature." Today, the computerization of childhood is so complete that not even this level of awareness exists for most children.[2]

What's to be done? Numerous overlapping educational initiatives hold promise for reconnecting children and nature. At one extreme, a U.S. public television program on "The Future of Education in the 21st Century" was

David Sobel is the author of eight books on childhood and nature and is senior faculty in the Education Department at Antioch University New England in Keene, New Hampshire.

giddily enthusiastic about students in a constant relationship with computers and other screens—using them in classrooms, for homework, and in their leisure time. At the other extreme, the kindergarten-through-seventh grade Environmental School in Maple Ridge, British Columbia, has no formal school building. Although they use some indoor facilities around the community, they are outdoors, through cold, wet, and snowy Canadian winters, most of the time. There is a right balance between these two approaches.

Essential for a liveable future is education for sustainability. It is a paradigm shift that aspires to educate students to make decisions that balance the preservation of healthy ecosystems, vibrant economies, and equitable social systems in this generation and in all generations to come. Around the world, numerous overlapping movements have this goal in mind: "green schools," education for sustainable development, environmental education, community-based education, nature and place-based education, the farm-to-school movement, and more.

Unfortunately, some of these approaches tend to get too heady too fast. Before you know it, we are trying to get children to save the rainforest, understand energy flow in ecosystems, reduce the carbon footprint of their schools, and address problems of income inequity in their communities. Too quickly they leave behind the primacy of children caring for animals, digging to China, or playing capture-the-flag at dusk. In other words, the cognitive, problem-solving, technology-based aspirations to save the world have to be balanced with the physical, socioemotional, immersive values of making mud pies. In fact, one may lead to the other, if we implement education practices that honor the insights of developmental and conservation psychology.

In a landmark 2012 chapter on the development of conservation behaviors, environmental psychologists Louise Chawla and Victoria Derr synthesize the research on the relationship between childhood experience and adult environmental behavior. They conclude that, "if societies seek to achieve a sustainable world where people will not only act to protect the biosphere today, but future generations will also value this goal and work for its achievement, then children need to be provided with regular access to nature." They add, "Research has linked a background of childhood play in nature with every form of care for the environment: informed citizen action, volunteerism, public support for pro-environmental policies, environmental career choices, and private-sphere behaviors like buying green products, conserving energy, and recycling."[3]

In other words, opportunities for nature play and learning need to be an integral part of cultivating adult environmental behavior. Children cannot just learn about the environment through virtual simulations; they need to get wet and dirty in order to fall in love with the Earth. Other factors become important later in the developmental trajectory. As Chawla and Derr explain, "In middle childhood and adolescence, young people need opportunities to extend their environmental knowledge and skills in more formal ways. In addition to a tapestry of nature in different spaces of their lives, children need people who can help them appreciate and understand what they find there."[4]

With these indications in mind, one can consider a developmental continuum of promising practices to connect children to nature in the twenty-first century: from early childhood, to elementary school, to middle and high school. If these practices are nurtured, they could shape a generation of young adults that are grounded in nature, selectively mature in their use of technology, and committed to environmental preservation.

Early Childhood: Immersion in Nature

As public schools in the United States and elsewhere have put greater emphasis on standardized testing, a counterpoint grassroots movement has arisen to "naturalize" early childhood programs. Since the original Earth Day in 1970, nature preschools have been growing steadily in the United States, and forest kindergartens have spread rapidly across Europe from their origins in Scandinavia. Although Denmark has conventional indoor early childhood programs, about 10 percent of the country's children participate in outdoor schools, and the older participants attend five days a week from 8:30 a.m. to 4:00 p.m. Germany has about one thousand forest kindergartens, and the schools also are popular in Australia, Japan, New Zealand, and Taiwan.

Parents of forest kindergartners appreciate this opportunity for immersion in the natural world. It is the antidote to the urbanization and antiseptic lives of many post-modern children. Early childhood programs offer an at-oneness with nature that constitutes the roots of empathy for ferns, hedgehogs, and wild places. Paul Doolan, a father who sent his daughter to a Swiss forest kindergarten (wonderfully documented in the film *School's Out: Lessons from a Forest Kindergarten*), articulates the unique quality of the experience:

> For two years, my little girl went to kindergarten in the forest. Not a school in the forest, just the forest. No walls, no roof, no heating—only the forest, a few tools, and incredibly dedicated teachers. One day, she came home from a day of particularly vicious downpours, her feet inevitably

soaked, her eyelashes caked in mud, her cheeks ruddy with the cold, and her eyes sparkling with fire, and I said to her it must have been tough being outside all morning in such weather. She looked at me in genuine incomprehension, looked out the window: "What weather?" she asked.[5]

Nature-based early childhood programs aspire to that old progressive education chestnut of balancing head, heart, and hands. Although most nature-based programs share common aims—they honor the primacy of children immersed in nature, and they support self-directed play— their different styles reflect the poles of teaching practices that we need to balance. Nature pre-schools, for example, work from a cognitive readiness mindset, which is reflected in the beautiful facilities, desks, and somewhat greater emphasis on formal literacy and numeracy, while getting kids outside for one-third to one-half of the day. Forest kindergartens, meanwhile, embrace an initiative/resiliency mindset, with an emphasis on minimizing indoor facilities, being out in all weather, and giving children opportunities to solve problems on their own.

Betsy Beneke/USFWS

A preschool class snowshoes in below-freezing temperatures at Sherburne National Wildlife Refuge, Minnesota.

Advocates of nature-based early childhood are convinced that their approach provides a win-win solution. They contend that their children are cognitively well-prepared for formal schooling and are also healthier and bonded with nature in a way that will incline them toward conservation values and behaviors. They accomplish the same goals as conventional indoor programs, and more. A recent case study documents how the Chippewa Nature Preschool in Midland, Michigan, led parents to advocate for nature kindergarten in the local Bullock Creek public schools. The success of the nature kindergartens has led to the creation of nature first grades. Parents and administrators realize that children are happier, more engaged in school, and have increased vocabularies and science knowledge as a result of this change in teaching practice.[6]

A similar grassroots movement has emerged in the state of Vermont. After watching the *School's Out* video in 2012, two innovative teachers in the town of Quechee decided to do a Forest Day, one full day a week in the woods where students solve outdoors math problems, climb ledges and trees, and learn to identify local plants and animals. Today, dozens of Vermont schools have Forest Days programs. The North Branch Nature Center in the capital, Montpelier, supports these efforts and is in the process of providing year-round curriculum materials for similarly inclined teachers. All of this has happened under the radar, without state mandates or targeted funding. It is a backlash to the academification, digitalization, and indoorification of young children's lives.

Elementary Education: Exploration and Connection to Place

Whereas early childhood programs are about immersion in nature, elementary programs focus more on exploring the physical world, becoming adept in it, and connecting to the nearby natural and cultural worlds. The desire to venture beyond the bounds of one's yard and neighborhood has always been an inherent part of childhood and is well represented in children's literature. Christopher Robin explores the Hundred Acre Wood with Winnie the Pooh. Caddie Woodlawn runs wild with her brothers exploring the Wisconsin woods and rivers surrounding their farm. On all continents—from Sioux children in North America to iKung children in Africa—children historically have had freedom to roam. But this freedom has been curtailed as a result of urbanization, parental anxieties, and digitalization: it is now a rarity to hear a mother encouraging her children to explore outside all day and "be home by when the streetlights come on."

Insightful schools understand the physical and cognitive benefits of these natural childhood instincts and try to recreate these opportunities across the elementary grades. The Adventure Education curriculum at The College School of Webster Groves, near St. Louis, Missouri, is a perfect example of a sound developmental approach. Outdoor adventures begin in kindergarten with the annual Day in the Woods, when the children study pond water, climb hills, and complete a half-mile hike. The first graders' one-night camping trip in a county park includes a ten-foot rock climb, a creek exploration, and fossil hunting. The second graders begin cave explorations, and third graders embark on a three-day wilderness camping experience. In fourth and fifth grade, the students move back in time as well, participating in a historical reenactment of a pioneers camp at an old prairie homestead.[7]

An integral aspect of this "expanding horizons" approach is the development of traditional wilderness living skills. The camping movement in the United States at the beginning of the twentieth century exhorted boys and girls to get out into the open and heed nature's everlasting voice. Back then, the call to nature was a response to the rapid industrialization and deforestation in the later part of the nineteenth century. The creation of U.S. national parks and national forests marked the beginning of the conservation movement. Today, in the early twenty-first century, we are heeding the same call, yet we are now impelled to conserve not just nature, but also the connectedness between children and nature.

One promising sign is the quiet growth of nature mentoring programs. Whereas environmental education programs have a curricular and cognitive orientation toward teaching about food webs, nutrient recycling, and environmental problems, nature mentoring programs teach shelter construction, wild edibles, bow and arrow fabrication, and even basketweaving. After a Wilderness Youth Project program in Santa Barbara, California, one child enthused, "Three hours isn't enough for these trips. We should do five hours, we should do all day, we should do twenty-four hours. We should build forts and live out here." A certain gusto is developed in nature mentoring programs that can carry over into the classroom, reflecting a synthesis of both cognitive goals and the socioemotional/resilience goals of forest kindergartens.

In the United Kingdom, the forest kindergarten movement has morphed into the forest schools movement, extending to the elementary grades. Forest Schools Canada, with the same orientation, was founded in 2013, and interest in place-based education and forest schools is particularly strong in the province of British Columbia. The Davis Bay Elementary School north of Vancouver (motto: "Where the outdoors is in!") reflects a schoolwide commitment to a nature- and place-based curriculum. Students in the three multi-age classrooms—Hive K/1, Cocoon 1/2/3, and Rookery 3/4/5—spend as much as half of each day outside, engaged in activities such as writing poetry in the forest, doing beach cleanups, and digging clams.

Strawberry Vale Elementary in Victoria, British Columbia, is a good example of a school that has transformed its monoculture schoolyard over the past fifteen years with gardens, natural play structures, a pond, fruit trees and berry bushes, a shrub labyrinth, climbable ledges, rain gardens, and more. This emphasis on naturalized schoolyards, signified more broadly by the creation of the International School Grounds Alliance in 2011, suggests that schools are opening their doors in recognition that the outdoors constitutes a significant component of the learning environment.

In Australia, bush schools are cropping up from coast to coast. A distinctive component of this approach is the acknowledgement of Indigenous ways of knowing and being in the landscape, based on a recognition that this was, first and foremost, Aboriginal land, and that students can learn from these First Peoples. Aboriginal elders are sought out to serve as mentors, and sites of spiritual significance are honored. As in forest schools at the elementary level, the emphasis moves from play to skills development and then to the development of literacy, numeracy, and other core curriculum content. At one bush school, after a shelter-building experience, students engaged in a complex mathematical problem-solving discussion about ratios, percentages, and surface area as they discussed building a larger shelter that they could use in the future.[8]

This same honoring of First Peoples is manifest at many Australian and Canadian professional development events for educators. Each event begins with the lead speaker stating, "I wish to acknowledge the ancestral, traditional, and unceded Aboriginal territories of the Coast Salish Peoples, and in particular, the Squamish, Musqueam, and Tsleil-Waututh First Nations in Metro Vancouver on whose territory we stand." This commitment to place is a core curricular principle at The STAR (Service To All Relations) School north of Flagstaff, Arizona, on the edge of the Navajo Reservation. This first completely off-the-grid, solar-powered school serves a population of Navajo and Hopi children and integrates a place-based curriculum with immersion in the cultural traditions of Native peoples.[9]

The Green School in Bali, Indonesia, synthesizes many of the same elements. Each class has vegetable and flower gardens, and the experiential learning process includes composting, worm farms, water conservation, lifecycles, and permaculture. The school embraces cultural traditions such as rice harvest festivals and Nyepi (Balinese New Year), and the teachers create projects in affiliation with community organizations such as Yayasan Senyum (Smile Shop) and the Bali Animal Welfare Association.[10]

This devolution away from a mandated, homogenized curriculum toward one that focuses on distinctive local places and cultures is a promising international phenomenon. Preservation of the firsthand experience of nature in childhood parallels the commitment to preserving fading languages in Ireland and the Amazon basin. Preserving the opportunities for firsthand experience in childhood creates the commitment and resilience to deal with challenging issues in later life.

Middle School and High School: Rites of Passage and Social Action

Connectedness with the natural world changes when children pass through puberty and enter adolescence. The free play of early childhood and the exploration of middle childhood need to be deepened with rite-of-passage experiences and then transformed into productive social action and the development of leadership skills in environmental activism. Adolescent nature experience is enhanced in affinity groups of peers with adult mentors who exemplify values of environmental behavior and nature preservation (think Scouts, 4-H Clubs, church youth groups, school green teams, and Outward Bound wilderness expeditions).

In the adventure curriculum at The College School near St. Louis, Missouri, sixth graders embark on a five-day wilderness trip in a national forest, complete with an overnight solo—a true rite of passage for many young adolescents. Seventh graders take part in an urban exploration where they use local transportation, sleep in churches, do community service, and delve into ethnic neighborhoods. By eighth grade, students travel to Okefenokee Swamp and the southern mountains for bicycling, canoeing, and orienteering.[11]

This movement from close and familiar in the early grades to distant and strange in later years accurately mirrors the developmental transitions unfolding in a child's psyche. The solo wilderness or urban experience for children around the age of puberty mirrors the timing of the rite of passage in many land-based traditional cultures. Adolescence is a time to forge oneself through physical challenges and to apply oneself in solving knotty social and environmental problems. Appalachian Mountain Club's Youth Opportunities Program does the same thing for inner-city youth of color. They provide training in backpacking, campfire cooking, rock climbing, and group safety, then take the youth on wilderness outings throughout northern New England. Rites of passage in the wilderness are an environmental justice issue: everyone should have access.[12]

Kroka, a wilderness education program in the state of New Hampshire, conducts a semester program that manifests many of these components. It is like a semester abroad for high school students, but it is in their wild backyards. One version of the trip begins in January at a base camp where the students make their own backpacks and sleeping bags, dry their own food, and learn cross-country skiing. They then ski the length of the Catamount Trail, with a solo winter camping experience toward the end. The students hunker down in northern Vermont to craft their own paddles and pack baskets. As the spring snowmelt comes, they paddle south and arrive back where they started in May. Along the way, they participate in service projects in towns along the Connecticut River.[13]

In Canada, the Rediscovery program for First Nations youth in British Columbia is founded on the idea of reviving fading Indigenous cultural traditions. Challenged by substance abuse, juvenile delinquency, and other forms of family disruption, the local Native and non-Native communities set up a dynamic youth project in 1978 on the remote shores of the Queen Charlotte Islands (Haida Gwaii). Rediscovery staff, working with Haida Elders, open adolescents' eyes to a renewed contact with the land and heritage around them. As Canadian environmentalist David Suzuki comments: "Clearly, we need a renewed sense of Earth as home; belonging to the land, connected to all other living things. Youngsters today have few, if any, opportunities to experience the enormity and beauty of the wilderness—Rediscovery offers them that." Similar programs have multiplied throughout Canada and have even been adapted by Indigenous peoples in Thailand.[14]

Tenth-graders paddle a student-built canoe seventy kilometers on New England's Lake Champlain as part of an eight-day Kroka Odyssey Expedition during which they also study Homer.

Ming Wei Koh, senior specialist for ecoliteracy at Pacific Resources for Education and Learning (PREL), brings this argument full-circle. At PREL, Koh develops climate change education projects, place-based teacher trainings, and curricula to build community resilience in island nations across the Pacific basin. Many of these islands are being inundated with rising seas, and their cultures are disappearing with their coastlines. As Koh explains, she is creating "place-based education that utilizes Indigenous and Western knowledges and pedagogical practices to improve ecoliteracy. Improving ecoliteracy . . . will lead to a personal relationship with nature and thus, care and nurturing of the land/ocean continuum."[15]

Koh's work illustrates the synthesis that exemplary educational initiatives should be striving for: integrating developmentally appropriate natural world challenges for adolescents with the honing of the concepts and skills that will help them solve current ecological problems. Some of the communities that she works with are living and reviving the art of traditional navigation, with apprentices watching and studying master navigators to learn more about the movements of the waves, birds, and stars. As Koh explains: "Transmission of knowledge in this way has been going on for generations. But recently, communities like Waan Aelõñ in Majel (Marshall Islands) and the Polynesian Voyaging Society (Hawai'i) are using traditional navigation and canoe building to address contemporary community needs: providing life skills for youth, perpetuating cultural practices, and bringing attention to our global need for sustainability."[16]

The crux of the rite-of-passage experience has always been to take the individual into the wilderness to both craft personal identity and transform the child into the socially responsible adult. In the Nootka tradition in British Columbia, initiates were dropped off in the ocean at night as girls to take a long swim back to shore. When they arrived after this ordeal, they emerged on the beach as women. This is what we should be striving for in adolescence: natural world encounters that develop the grit and resilience to be adult problem solvers in an uncertain world.

Outdoor School

In November 2016, Oregon voters approved a ballot initiative, known as Initiative 67 ("Outdoor School for All"), that would enable every fifth or sixth grader in the state to attend a weeklong, overnight outdoor school. Outdoor school is a decades-old tradition in Oregon, and more than 1 million schoolchildren have attended since 1957. Cost prevents some schools from participating, however, and the ballot initiative would make the program accessible to all. Program directors emphasize the value of scientific investigation in the curriculum, but they note that the more important component may be the social learning aspect—learning acceptance for those who are different, or bravery in the face of a first night away from family.[17]

The program appears to have long-lasting effects. According to one analysis, "[c]hildren who attend outdoor school in Multnomah County—especially boys, Asian students, and students whose first language is Spanish—are more likely to show up in school afterward." High school students who get to be outdoor counselors "report being more confident at public speaking, more interested in other volunteer opportunities, and even more likely to use conflict mediation skills with their peers." One current middle school teacher who was a student and later a counselor at an outdoor school commented that during solo time, sitting and reflecting, he realized, "he wanted to stop acting out in class, stop picking fights with other kids, and choose a direction for his life." He was the first person in his family to go to college and is now a big advocate for Outdoors School for All. "It's not just a week outside," he says, "It's the chance of a lifetime."[18]

The Children & Nature Network, founded in the United States but now with membership on six continents, is a leading organization advocating for naturalizing children's lives. The network's Natural Families program supports the creation of Family Nature Clubs where parents and children get together regularly to visit local parks. And the new Cities Connecting to Nature initiative helps cities devise strategic plans for getting children outside. In Grand Rapids, Michigan, one of seven target cities, the Parks and Recreation Department, the Grand Rapids Public Schools, local nonprofit organizations, and the mayor are invigorating parks and playgrounds and creating outdoor learning labs.[19]

At the national level, the North American Association for Environmental Education and other advocacy groups have been successful in making environmental education and environmental literacy programs eligible for U.S. federal funds in the new Every Student Succeeds Act. This act provides funding for environmental education, professional development, and field-based service learning. The newly formed International Association of Nature Pedagogy aspires to the same kind of professional development and advocacy in Europe and Asia. The association's Erasmus project is examining nature education practices in Australia, the Czech Republic, England, and Scotland to articulate the different cultural heritages that shape practice in these diverse settings. The goal is to create video evidence and documentation of the benefits of learning in and with nature.[20]

What about going further and extending the one week that the Outdoor School for All initiative aspires to, and making it multiple weeks? Or having your child spend most of the day outside in a forest kindergarten? Maybe we can aspire to having elementary-age children spend one or even two days a week in an outdoor school, and then build in expeditions and rites of passage three times a year during middle school. What about regular opportunities to participate in ecological restoration, watershed monitoring, woodlot management, and daily farm chores during high school? Let's aspire to a world of twenty-first-century students that are both sophisticated in technology and rooted in the Earth.

CHAPTER 3

Ecoliteracy and Schooling for Sustainability

Michael K. Stone

It was 1992. Laurette Rogers, a fourth-grade teacher in San Anselmo, California, had shown her students a film about rainforest destruction. Distressed, they asked what they could do about it. "I just couldn't give a pat answer about writing letters and making donations," Rogers recalls. Instead, she took the advice of a trainer for a former Adopt-a-Species program: "Pick any species. Find out all about it, and you'll fall in love with it."[1]

Rogers wanted a local species, and she wanted it to be obscure, to counter bias toward beautiful and charismatic species. Her class chose an endangered shrimp that lived in only fifteen streams within a few kilometers of the school. They studied the ecology and lifecycle of the shrimp, which they learned are one strand of a web that encompasses insects, songbirds, streams, dairy ranches, watersheds, and, ultimately, the San Francisco Bay. They discovered that habitat restoration on behalf of the shrimp—planting willows and blackberries while ranchers built bridges and fencing to keep cattle out of the streams—required nurturing a network of people who sometimes see themselves as adversaries: ranchers and environmentalists; for-profit companies and public officials; teachers, students, and parents.[2]

They persevered, prospects for the shrimp improved, and the California Freshwater Shrimp Project evolved into STRAW (Students and Teachers Restoring a Watershed), cosponsored by The Bay Institute and the Berkeley-based Center for Ecoliteracy. STRAW has since expanded to address additional watershed issues, and celebrated its five-hundredth restoration in 2015. Some

Michael K. Stone is senior editor at the Center for Ecoliteracy, coeditor of *Ecological Literacy: Educating Our Children for a Sustainable Future* (Sierra Club Books), author of *Smart by Nature: Schooling for Sustainability* (Watershed Media), and winner of the Green Prize for Sustainable Literature.

forty thousand students—kindergarten through high school—have restored more than fifty-six kilometers of creek banks. And it's been good for more than shrimp. As one of the original fourth graders later reflected, "I think this project changed everything we thought we could do. . . . I feel it did show me that kids can make a difference in the world, and we are not just little dots."[3]

STRAW is a powerful example of education for ecoliteracy. The need is evident to prepare students as they inherit a host of environmental challenges: climate change, biodiversity loss, the end of cheap energy, resource depletion, gross wealth inequities, and more. This generation will require leaders who can understand the interconnectedness of human and natural systems and who have the knowledge, will, ability, and courage to act.

Responses to this imperative go by many names: ecological literacy, education for sustainability, eco-schools, green schools. They follow no blueprint or one-size-fits-all formula. Some embrace "sustainability" as a goal, while others find the concept problematic. Ecological literacy, as understood by the Center for Ecoliteracy, lies at the junction of schooling for sustainability and Earth-centric learning.[4]

Guiding Principles of Ecoliteracy

Physicist and systems theorist Fritjof Capra, a Center for Ecoliteracy cofounder, notes that common definitions of sustainability, such as "satisfying needs and aspirations without diminishing the chances of future generations," are important moral exhortations, but they do not give much practical guidance. However, he says, we can look to nature: "Since the outstanding characteristic of the biosphere is its inherent ability to sustain life, a sustainable human community must be designed in such a manner that its ways of life, technologies, and social institutions honor, support, and cooperate with nature's inherent ability to sustain life."[5]

Ecological literacy, then, is the ability to understand the basic principles of ecology—the processes by which the Earth's ecosystems sustain the web of life—and to live accordingly. Among the ecological principles articulated by Capra: diversity assures resilience; one species' waste is another species' food; matter cycles continually through the web of life; and life (as suggested by authors Lynn Margulis and Dorion Sagan) did not take over the planet by combat but by networking.[6]

The Center for Ecoliteracy has distilled its decades of experience into four guiding principles: "Nature Is Our Teacher," "Sustainability Is a Community

Practice," "The Real World Is the Optimal Learning Environment," and "Sustainable Living Is Rooted in a Deep Knowledge of Place." These principles begin with nature, as Capra suggests, and reflect work with thousands of educators.

Nature Is Our Teacher

Accepting nature as our teacher has several implications:

Focusing on nature as we encounter it integrates teaching across disciplines and between grade levels, providing an antidote to the fragmentation and narrowing of subject matter. "We do not organize education the way we sense the world," writes author and educator David W. Orr. "If we did, we would have Departments of Sky, Landscape, Water, Wind, Sounds, Time, Seashores, Swamps, Rivers, Dirt, Trees, Animals, and perhaps one of Ecstasy. Instead, we have organized education like mailbox pigeonholes, by disciplines that are abstractions organized for intellectual convenience." Orr argues that "at all levels of learning, K through Ph.D., some part of the curriculum [should]

One species' waste is another species' food: a flightless dung beetle with a ball of dung, Addo Elephant National Park, South Africa.

be given to the study of natural systems roughly in the manner in which we experience them. . . . doing so requires immersion in particular components of the natural world—a river, a mountain, a forest, a particular animal, a lake, an island—*before* students are introduced to more advanced levels of disciplinary knowledge."[7]

Nature also teaches us to think in terms of systems. Living beings, watersheds, or ecosystems cannot be fully understood apart from the systems they contain and the systems in which they are nested; neither can schools, communities, or economies. Ecological literacy entails thinking systemically, in terms of relationships, connectedness, and context—for example, understanding the complexities of the food web. (See Box 3–1.)

That, in turn, changes how schools operate. Author and farmer Wendell Berry contrasts bad solutions—which solve for single purposes and act

Box 3–1. Using Food-web Ecology to Help Teach Sustainability

Sustainable development demands healthy ecosystems that provide a range of ecosystem services on which we depend. Food provisioning is one of these key services and is embedded in food-web ecology. But how well is this perspective represented in contemporary education?

One of the most successful and intuitive terms in ecology is that of the *food chain*, which is frequently illustrated by the link from phytoplankton (tiny marine plants) to zooplankton (tiny marine animals) to plankton-eating fish, and so on up the trophic ladder. Food chains, and the food webs of which they are a part, are key concepts of any ecology textbook and are related to key issues such as species diversity, ecosystem productivity, and stability, as well as management-related issues such as harvesting. For example, how will ecosystems be affected by the removal of top consumers such as whales, tunas, or wolves? Food webs and food chains also are instrumental for understanding biomagnification, the concentration of toxic substances as they are ingested along successively higher trophic levels.

Due to its conceptual appeal, the idea of the food web is commonly introduced in primary school as a three-level ladder from plant to plant-eater to carnivore. Although highly simplified, this presents the important insight that affecting one species may have consequences for others. Secondary-school and university textbooks bring more complexity to the table and extend the simple three-level food chain to complex food webs. They also discuss fluxes of energy and matter within food webs, and some draw attention to the application of food-web insights, such as managing fisheries and wildlife.

But few science textbooks provide a thorough analysis of the widespread consequences of human influences on the food web, for example the (over)harvesting of top predators such as large marine mammals or terrestrial carnivores. In the context of sustainability, there is an underdeveloped opportunity among textbook authors, teachers, and scientists to use ecology curricula to highlight and discuss these issues and to convey key related ecological insights.

One issue that would be even more important to address directly is *where humans could best harvest within the food web to maximize calories in a sustainable manner*. Harvesting at the top of food webs is unsustainable, and the removal of top-level organisms can have strong cascade effects on the rest of the community. Harvesting lower down the trophic ladder, in contrast, could help maintain populations of critical species at healthier and more-sustainable levels.

Overall, greater awareness of the key role of food-web insights for sustainability should be a goal for the next generation of science and biology textbooks at all levels.

—Dag O. Hessen, Professor of Biosciences, University of Oslo

destructively on the patterns in which they are contained—with good solutions in harmony with their larger patterns that result in ramifying sets of
solutions. Farm-to-school programs are examples of good solutions that beget
other solutions: they increase access to healthy food, teach children where
their meals come from, support small-scale farmers, and keep money circu-
lating in the local economy. As of October 2016, the National Farm to School
Network reported programs in 42,587 schools in the United States.[8]

Learning from nature is often best accomplished by learning *in* nature. The
London Sustainable Development Commission reviewed sixty-one studies on
children's experiences with nature. They judged as "well-supported" findings
that time spent in natural environments as a child is associated with adult
pro-environment attitudes and feelings of being connected with the natural
world. They found a "well-supported" conclusion that experience of green
environments is associated with greater environmental knowledge.[9]

Forest schools and nature schools were introduced in Scandinavia in the
1950s. The Association of Nature Schools, which comprises ninety-one schools
throughout Sweden, calls a nature school "not a building or a place, but a
method of learning" that can be applied to teaching all subjects. "*Naturskolan*
offers teachers opportunities to explore and develop different ways of teaching
science and education for sustainable development (ESD) primarily using outdoor education in the community," according to one teacher. (See Chapter 2.)[10]

Sustainability Is a Community Practice

Sustainability is a practice, both in that it represents actions rather than a static
condition and in that people learn those actions by practice, just as they would
learn to play a musical instrument or develop skills in sport. And sustainability
is best practiced in community. Many of the principles of ecology are variations
on a single fundamental pattern: nature sustains life by creating and nurturing communities. Animals, plants, and microorganisms live in webs of mutual
dependence. Qualities that keep natural ecosystems vibrant and resilient, such
as diversity and interdependence, shape human communities as well.

A healthy network of relationships that includes all members of the community makes the practice more sustainable. When teachers, students, parents, and trustees decide and act collaboratively, students acquire skills of
leadership and community decision making needed by effective agents of
change. The Sustainability Academy at Lawrence Barnes, a magnet elementary
school in a high-poverty neighborhood in Burlington, Vermont, celebrates the
ties between community well-being and sustainability. A mural at one of the

school's entrances is topped by a quote from theologian Matthew Fox: "Sustainability is another word for justice, for what is just is sustainable, and what is unjust is not."[11]

Fifteen years ago, morale and academic performance were low at Lawrence Barnes, and the school was in danger of being closed. Then, in 2001, the school partnered with sustainability education pioneer Shelburne Farms, which was offering teacher workshops in response to Vermont's becoming the first U.S. state to adopt sustainability education standards. The resulting Sustainable Schools Project remade the curriculum around "big ideas of sustainability," which resemble principles of ecology articulated by Fritjof Capra, among them systems, diversity, interdependence, and community. The faculty identified "essential questions" that transcend grade level and subject matter (What does it mean to be a citizen in our community? What connections and cycles shape our Lake Champlain ecosystem?) and emphasized place-based education, community connections, and civic engagement.[12]

Andy Duback

A Day of Service event in the gardens of the Sustainability Academy at Lawrence Barnes in Burlington, Vermont.

In a few years, Barnes was transformed, test scores improved dramatically, and, in 2010, the school district designated Barnes as the first U.S. public elementary school with sustainability as its "framing lens," integrating academics, the campus, and day-to-day operations. Installing solar panels changed not only the buildings but also the school's teaching, by inspiring an energy literacy curriculum. Through a Healthy Neighborhoods/Healthy Kids program, students brainstorm about quality of life, then walk their neighborhoods and present report cards to the school, local agencies, and public officials, with tangible results. Once, Barnes students even found a park that the city had forgotten. They had contacted the Parks and Recreation Department to suggest installing lights in the park, because they did not feel safe. "We don't have a park on South Champlain Street," said the Department. "Yes, you do," the students responded, "There's a sign there that says, 'Parks & Rec Department.' We want to tell you about it."[13]

The Real World Is the Optimal Learning Environment

Instead of reading about natural processes or looking at simplified drawings, students can encounter nature in the rich, messy ways in which it actually exists. When they plant and harvest a garden or watch a creekside come back to life, they experience different time scales from those of a video game, a perspective needed for addressing persistent ecological challenges. (The students in the Shrimp Project stayed focused even after learning that it could take fifty to one-hundred years for their restoration to have a significant impact on the shrimp. They talked about taking their grandchildren to see their work and telling them, "We did that.")[14]

Whether restoring a habitat, monitoring local pond health for the city water commission, or lobbying for better-lighted city parks, students learn more when their actions have meaning and matter outside the classroom. Project-based learning is effective when learning is necessary to accomplish something students care about. Seeing themselves—and being seen—as effective instills confidence and motivation for responsible, active citizenship.

Schools are also part of the real world. Educators from the South Pacific atoll of Yap once visited the Center for Ecoliteracy and left a poster proclaiming "Curriculum Is Anywhere Learning Occurs." Schools teach—whether they are conscious of it or not—by how they relate to neighbors and invest in resources; by who participates in decisions; and by how they provision themselves with food, energy, materials, and water. This "hidden curriculum" reveals schools' understanding of their relationship with the world. Some of the most lasting lessons are learned, for better or worse, when students compare the school's words and its actions. The Center for Ecoliteracy became involved in school food reform, for instance, after discovering that teaching about good nutrition in the school gardens that it supported was being undermined as soon as students walked into the cafeteria.[15]

Sustainable Living Is Rooted in a Deep Knowledge of Place

Students discover what communities value by collaborating with people who were living there before they arrived and who will still be there long after they graduate. By working closely with community members, students learn to recognize and utilize community resources. The Healthy Neighborhoods/Healthy Kids program succeeds as students understand, and then work to improve, their city. The STRAW program, meanwhile, has added a new place-based dimension, incorporating into its restorations the planting of up to twenty drought-resistant species to test which are best suited to different locales as the climate changes.[16]

Relocalization is becoming a powerful strategy for sustainability. "What has served our species well in the past could serve us well in the future if we only relinquish the modern tendency to impose universal solutions upon the infinite variability of both people and the planet," write educators David Gruenewald and Gregory Smith. "Local diversity lies at the heart of humanity's biological and cultural success."[17]

Sustainability and knowledge of place intersect in settings as seemingly ordinary as the school lunch table. (See Chapter 6.) Nothing could be more basic to sustainability than food. But students, like others who are used to readily available food shipped thousands of kilometers, often do not know where their food comes from or how it is grown. They are unaware of the food system's susceptibility to disruption, whether from climate change, fossil fuel shortages, or economic upheaval. They do not understand the role of the food system in the local economy or the importance of creating resilient local food systems.

The California Thursdays program is one response. The Center for Ecoliteracy had chosen to work with Nutrition Services at the Oakland Unified School District (OUSD), a district with stark disparities in family income and academic achievement, in part because its then-superintendent saw clearly that improving school food is "part of the basic work we have to do in order to correct systemic injustice, pursue equity, and give our children the best future possible."[18]

OUSD's food service director committed to local procurement after research by a fifth-grade class revealed that the district was serving asparagus that had been shipped more than twenty-seven-thousand kilometers, although asparagus is grown within one-hundred kilometers of Oakland. Distributors often could not tell the district where their food originated. Recipes that feature local food and meet federal specifications had to be created. The district, like many others, had been offering heat-and-serve meals for so long that staff members needed training in cooking from scratch.[19]

California Thursdays emerged as a "bite-size" strategy in 2013–14, as part of the Center's Rethinking School Lunch Oakland program, beginning with one meal a month featuring fresh California ingredients. In a few months, it had expanded to one day a week. Other districts joined. By late 2016, the California Thursdays network had grown to seventy-four districts serving more than 318 million meals annually (about one-third of the school meals served in California). The network's buying power created incentives for distributors to introduce "California-grown" product lines. Students have written research papers tracing the environmental and economic impacts of local food, and lessons using school meals teach students about the ethnic and agricultural heritage of their home place.[20]

Globalizing Schooling for Sustainability

Increasingly, schooling for sustainability is an international movement. The 2005–14 period was the United Nations Decade of Education for Sustainable Development. The International School Grounds Alliance was created in 2011 by educators representing countries on six continents to promote, among other values, outdoor education, hands-on learning, and preservation of local ecology. Its activity guides, in English and Chinese, describe lessons and activities from organizations in seventeen countries.[21]

The most extensive international program, Eco-Schools, was launched by four European countries in 1994, in response to the 1992 "Earth Summit" in Rio de Janeiro, which had called education "critical for achieving environmental and ethical awareness, values and attitudes, skills and behavior consistent with sustainable development." Participation in Eco-Schools has increased from one hundred and thirty-nine schools in 1995 to more than forty-nine thousand schools in sixty-two countries, enrolling over 16.5 million students. Eco-Schools spread to the United States in 2009 under the auspices of the National Wildlife Federation, and U.S. membership has since expanded from 103 to 4,375 schools.[22]

Recognizing that schools' circumstances can vary widely, the program prescribes a general process but leaves many implementation details to national operators. After a school registers to join the program, it forms an eco-committee to assess the school's environmental impact and develop action plans, identifying priorities with respect to ten themes ranging from school grounds to energy, climate change, and global citizenship. Schools integrate action plans into the curriculum and develop eco-codes. After achieving a "high level of performance," usually after at least two years, schools may apply to be assessed by outside evaluators and awarded Green Flag status. Of the more than forty-nine thousand participating schools, more than sixteen thousand have been awarded Green Flag status. (See Figure 3–1).[23]

The program is noteworthy in requiring that efforts be student-led, with projects extending into the wider community. For example, at Ynysddu

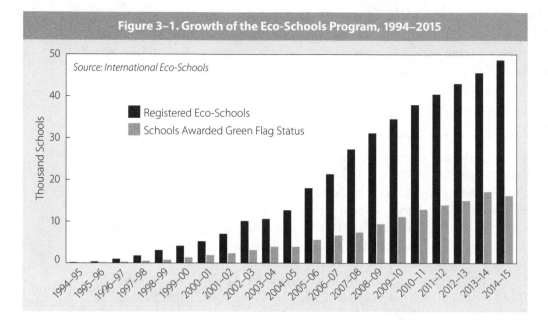

Figure 3–1. Growth of the Eco-Schools Program, 1994–2015

Primary School in Wales, students, staff, and the community set aside an area surrounding the school and planted fruit trees and perennial vegetables to be tended by the children and community members, with the produce made available to the community. Projects in Mauritius in the Eco-Schools Indian Ocean program include rainwater harvesting, soil stabilization, and sanitation. Local officials attest that the most successful projects are those that have built strong partnerships with governmental institutions, nongovernmental organizations, and the private sector.[24]

In the future for ecoliteracy education, a variety of collaborations between schools and government will be important. In keeping with ecological principles, a diversity of approaches seems most promising. For example, STRAW partners with county water management agencies, and California Thursdays is supported in part by the California Department of Food and Agriculture. Government support also will be important in efforts to update standardized curricular materials so that they better reflect climate change realities, as has been mandated in Portland, Oregon, through a district-wide "climate literacy" resolution. (See Box 3–2.)[25]

In some places, mandated curricula may play a role, especially if teachers and schools can receive help in meeting requirements. The Sustainable

Box 3–2. Teaching the Climate Crisis

According to a recent survey by the Pew Research Center, fewer than half of U.S. citizens surveyed (45 percent) believe that climate change is a "very serious problem." Compare that to India (76 percent), Brazil (86 percent), or Burkina Faso (79 percent). Interestingly, Pew notes that the "countries with high per capita levels of carbon emissions are less intensely concerned about climate change."

The relatively ho-hum attitude about climate change in the United States no doubt has many explanations. But one of these is the widespread doubt, expressed in mainstream textbooks, about the human origins of climate change and its severity. Writing in *Environmental Education Research*, Diego Román and K. C. Busch cite as culprits numerous sixth-grade science texts currently used in California schools, including *Focus on Earth Science*, which states: "Not all scientists agree about the causes of global warming. Some scientists think that the 0.7 Celsius degree rise in global temperatures over the past 120 years may be due in part to natural variations in climate."

Social studies texts are equally egregious. The widely used Holt McDougal text, *Modern World History*, includes three paragraphs on global warming. The second one begins: "Not all scientists agree with the theory of the greenhouse effect." Acknowledging that Earth's climate is "slowly warming," the global studies textbook tells students: "To combat this problem, the industrialized nations have called for limits on the release of greenhouse gases. In the past, developed nations were the worst polluters." The textbook then turns poor countries into eco-villains, noting that, "[s]o far, developing countries have resisted strict limits," even though the per capita greenhouse gas emissions of developed countries still far exceed those of any so-called developing countries.

The good news is that school districts and teachers are beginning to reject this approach. The most comprehensive "climate justice" policy of any U.S. school district was passed in Portland, Oregon, in May 2016. Initiated by Educating for Climate Justice, an organization of parents, teachers, students, and climate activists, the Portland board of education passed a resolution requiring the school district to "abandon the use of any adopted text material that is found to express doubt about the severity of the climate crisis or its roots in human activity." The resolution commits Portland to develop a plan to "address climate change and climate justice" in all of its seventy-eight schools.

School board members voted unanimously to affirm: "All Portland Public Schools students should develop confidence and passion when it comes to making a positive difference in society, and come to see themselves as activists and leaders for social and environmental justice—especially through seeing the diversity of people around the world who are fighting the root causes of climate change." At its 2016 representative assembly, the National Education Association, the largest teachers union in the United States with about 3 million members, used the Portland resolution as a model when it

continued on next page

Box 3–2. continued

passed a measure urging local affiliates to create and promote climate literacy resolutions in their own communities.

These policy initiatives have been sparked, in part, by imaginative classroom efforts to bring the climate crisis alive for students. In Portland, the public K–8 Sunnyside Environmental School has held weeklong teach-ins on climate change and energy issues, featuring role plays, simulations, and presentations from community groups. In March 2016, Sunnyside students traveled to the U.S. District Court in Eugene, Oregon, to witness the historic Our Children's Trust lawsuit hearing, in which twenty-one young plaintiffs are suing the federal government for failing to protect their constitutional rights to life, liberty, and the pursuit of happiness by neglecting to adequately address the climate crisis—an action that climate activists Bill McKibben and Naomi Klein called "the most important lawsuit on the planet right now."

In classroom activities, teachers have put fossil fuel companies on trial; introduced (through video and poetry) students to the people of Kiribati, whose island nation is being drowned by the rising ocean; played a game simulating the conflict between a profit-driven economy and climate sanity; and asked students to role play individuals around the world who suffer from but also exploit the climate crisis. In one activity, a teacher takes out a cigarette and lighter and casually asks the class whether they would mind if he smoked. When they object, he presses them on why and asserts his individual "freedom." He uses this to provoke students to compare the air they share in the classroom to the atmospheric commons.

The gulf between the enormity of the climate crisis and our schools' climate curriculum is huge. Increasingly, the classroom will be one of many battlefields where we decide how we will make sense of, and respond to, a warming planet.

—*Bill Bigelow, curriculum editor of Rethinking Schools and codirector of the Zinn Education Project*

Source: See endnote 25.

Schools Project emerged in Vermont because teachers sought assistance from Shelburne Farms to meet new state standards. Some of the Swedish Nature Schools represent another model. The *Naturskolan* in Lund is funded by the local municipality to coach and consult with teachers wanting to teach about science and sustainable development within the national curriculum. Sometimes, efforts emerge in spite of governments. Global Action Plan International introduced conservation programs in parts of the former Soviet Union via schools because they found that teachers and schools were more trusted than government authorities.[26]

The challenges of the coming decades are daunting. A key, says one environmental science teacher, is that students can be motivated to build capacities and envision responses, "if they're not being fed the idea that it's hopeless." Education for ecological literacy is, finally, about hope—about recognizing, in the words of the Biomimicry Group, that "life creates conditions conducive to life," and that we can find in nature the guidance and inspiration that we need to design and maintain healthy, sustainable communities.[27]

CHAPTER 4

Education for the Eighth Fire: Indigeneity and Native Ways of Learning

Melissa K. Nelson

How do educators prepare the next generation of adults and leaders in an era of radical climate disruption, ecological tipping points, economic volatility, and social and political inequity? In other words, should there be a new type of education for a dystopian future? Although the world has never been a stable, serene place, the changes to the Earth globally and the social disruptions happening in this second decade of the twenty-first century are unprecedented and alarming. The present epoch has been called the Anthropocene, a time of human domination of the Earth. This new epoch requires educators to completely rethink the purpose and goals of education, especially in terms of preparing young people for a viable and hopeful future.

Indigenous peoples, however, have centuries of experience with radical unwanted change from the many ongoing forms of colonialism that have been thrust upon them—western education being one of the most insidious and destructive. It is fair to say that most Indigenous communities and communities of color have experienced some form of oppression, racism, violence, and discrimination through western education. As a consequence, these students often are reluctant to fully trust and give their attention to a system that may have harmed them or their families, or may harm them again. This is one factor behind the high dropout rates for Native students and students of color, as well as for those who are brown, female, artistic, queer, disabled, or on a different learning spectrum. Education is in need of radical transformation for many communities in our growing, diverse world.[1]

Melissa K. Nelson is an Anishinaabe/Cree/Métis (Turtle Mountain Chippewa) ecologist, educator, and scholar-activist who serves as an associate professor of American Indian Studies at San Francisco State University and as president of the Cultural Conservancy.

Some scholars have noted that Native Americans already experience a "post-apocalyptic" reality, or "already inhabit what our ancestors would have called a dystopian future." Land dispossession, genocidal policies, relocation, containment on reservations, resource extraction, toxic pollution, institutional and environmental racism—these are realities for modern Native Americans and for many other Indigenous peoples around the globe. In many ways, these groups are already living in a dystopic reality, and young people feel that. Where is there room to find what many call a "geography of hope"?

The etymological root of the word education is *educare*, to "bring out, draw out." It is like drawing water out from a well. In this sense, education is primarily endogenous (from within) rather than exogenous (from outside/external). Of course, with the teacher/student relationship, education focuses on a subtle dance between internal and external learning. Many postcolonial, social justice educators refuse to accept what educational philosopher Paulo Freire

Melissa K. Nelson

Two participants in the Guardians of the Waters youth program plant corn at Indian Valley Organic Farm and Garden, Novato, California.

has termed the dehumanizing and oppressive "banking model" of western education, which thinks of knowledge like capital that can be deposited into empty students and later extracted like money.[2]

This banking model of education is also antithetical to Indigenous ways of learning and knowing. A primary difference is that Indigenous education often focuses on the *process of learning* as much, if not more, than the *content of knowing*, thereby focusing on knowledge as a living system, not as a commodity. Tewa educator Greg Cajete talks about this process as "coming-to-know." It is not just about what we know, but how we come to know it and why we know it. According to Native worldviews, gaining knowledge comes with great responsibility; therefore, educational programs should reflect and promote life-affirming cultural values and ethical actions.[3]

Given the approximately three hundred and seventy million Indigenous peoples on the planet and the five million American Indians/Alaskan Natives in the United States today (who represent hundreds of Indigenous nations, tribes, and cultures), there are innumerable types of Indigenous learning processes. These ways of knowing or "Indigenous Knowledge Systems" are based in distinct languages, epistemologies, and ontologies—ways of knowing and being in the world. These diverse worldviews inherently express themselves in different pedagogies, or ways of teaching.

Three Elements of Indigenous Learning and Knowing

Indigenous education is based on ways of learning and knowing that include elements that are quite distinct from western education. Three leading elements are learning spirits, embodied learning, and symbiotic and contextual learning. These are all interconnected and are linked directly to a participatory relationship and to an intimate understanding of the natural world that provides a sense of identity, belonging, and purpose. Additionally, these elements help reveal an inherent attachment to place and a sense of guardianship and stewardship of the natural world that surrounds a person and provides the nutrients of life. These educational practices are rooted in learning from place and gaining knowledge about how to live harmoniously within the limits of the ecological processes of local regions.

There are innumerable differences and variations on these fundamental and interconnected elements of Indigenous education, depending on place and culture. Yet, at their core, they demonstrate that humans have a holistic and sophisticated capacity for learning when they listen to the "learning spirits" of the land (and their ancestors), exercise their whole faculties, and practice their cultural values and environmental ethics on the lands that they live on and with their communities. Many of these Indigenous educational "methods" can be integrated into non-Indigenous educational programs to help all students be better connected to holistic learning practices and the natural world. This also will help prepare them for the dystopian future that they may soon join.

Learning Spirits

Chickasaw law professor James "Sakej" Youngblood Henderson and Mi-kmaw educator Marie Battiste discuss the concept of a learning spirit. These spirits are latent in the ecosystems of the land. Throughout a person's lifetime, he or she awakens to and remembers different learning spirits for different phases of his or her life. The idea is that every human, and perhaps even every life form, has the capacity to learn and grow: life equals learning. Henderson says that these learning spirits "are like cognitive gravity." They are strong attractors that lead us to the paths and experiences that we need at particular times, to nurture and bring forth our gifts and skills.[4]

There are learning spirits contained within each person from birth that awaken over time or come to the person at different periods of life. Perhaps they are inherent in the cellular tissue or DNA of any living organism, perhaps they come to people from ancestral spirits or other entities, or possibly a person's own unique personality wants to express these learning spirits as the person resonates with certain places and matures over time. Whatever it is, many tribes have this notion and understanding that one of the main roles that people have as human beings is to pay attention to and support the blossoming of their learning spirits. As parents, teachers, and elders, we need to nurture the learning spirits of young people especially. Whether explicit or not, many Indigenous education programs work to educate the whole human and to respect the seen and unseen communities that they come from, including their learning spirits.

Embodied Learning

Many Indigenous education programs focus on the whole body as an organism for learning: the mind, body, heart, and spirit. They include all of the major senses—sight, sound, taste, smell, and touch—as well as intuition and kinesthetic sensory awareness (or proprioception), a whole-body, nonmental physical awareness of self and space. The mind is the most obvious place to learn and teach as we have language, ideas, questions, and logical and translogical thought processes. Our minds are beautiful and complex organs that can be our greatest allies, but, as many have said, the mind is a good servant but not a good master. It needs the ethical connection of our learning spirits and other modes of knowing to balance out its reductionism and tendency to create binaries.

The Eurocentric overemphasis on the mind—specifically on the logical, linear, reductive, and shortsighted parts of the mind—has had grave consequences for humanity and the planet. The overemphasis on creating increasingly complex technology and using it without regard for its long-term consequences has enabled humans to destroy forests, mountains, and habitats and to drive numerous species to extinction. We are currently in the sixth great extinction crisis—all at the hands of humans. By 2050, 15–35 percent of the known species on Earth are expected to go extinct due to human activity. Even our own species, *Homo sapiens*, is at risk.[5]

Additionally, due to the unleashing of nuclear power and the release of massive amounts of stored carbon into the atmosphere, humans have forever changed both the geological strata of the Earth below and the atmosphere above. In addition to these impacts, the Eurocentric obsession with logical positivism, reductionism, Cartesian dualism, and abstraction has meant that humans have been forced to ignore and atrophy their other learning capacities, especially other ways of learning and knowing through the heart, body, and spirit.

The policies and practices of cultural genocide used by religious and government-run boarding schools in Australia, Canada, and the United States in the nineteenth and twentieth centuries have had long-term, devastating impacts on the health of Indigenous peoples. In addition, this type of education has harmed all peoples, regardless of ethnic background, as we all have other creative, imaginative, and emotional ways of knowing that have been disrupted, repressed, and atrophied. These transrational creative ways of learning need to be cultivated and nurtured for true, holistic education to be ignited and expressed.

Indigenous peoples learn through their bodies when they "learn by doing," as in dancing, wood carving, basket weaving, paddling a canoe, planting trees, or cleaning seeds. It engages the external and internal senses. There is a palpable physical learning and knowing that is expressed when engaged in these important cultural activities. The rhythms of movement and music and dance are excellent stimuli for embodied learning and memory. Sports, too, can be a great way to use the body as a site of learning: thus, the Haudenosaunee have a strong emphasis on lacrosse, and the Maya had a strong tradition of ball games. Many tribes use stick games as well as "gambling" games with dice and other objects to encourage young people to cultivate a strong physical awareness and embodied learning process. These games also teach about topics such as mathematics and geometry.

Additionally, because people's "gut feelings" often tell them things faster than the mind, it is important to listen to the body's wisdom and ways of knowing. Science is now confirming the notion of a "second brain" in the gut. People also have their emotions, their hearts, and their spirits, whether they subscribe to a spiritual worldview or not. Much research and work has been done on emotional intelligence and ecopsychology in the past twenty years. Educators and researchers are uncovering the significant role that emotional intelligence and place attachment have in overall learning and success in education and life. (See Chapter 8.)[6]

Symbiotic and Contextual Learning

Learning is an important individual endeavor, whether reading a book, writing an essay, pondering a significant question, or going on a "learning quest" (literally or metaphorically). But for much Indigenous-based education, learning is a collaborative, community process. It is about interrelationships and fostering reciprocity.

Indigenous education is about facilitating productive spaces for the transmission of knowledge among generations: from elders to youth and from youth to elders, as well as peer-to-peer. But Indigenous education does not just focus on learning from other humans, it also focuses on learning from cultural treasures, or "texts," such as regalia, pottery, baskets, masks, canoes, etc. These cultural treasures are made from natural materials and are imbued with great symbolic meaning and knowledge. These treasures are a perfect marriage of human creativity and the gifts of nature, the nonhuman relatives that give humans food, shelter, medicine, clothing, cultural items, and so on.

Melissa K. Nelson

A participant in the Guardians of the Waters youth program gathers tule reeds, Sonoma Mountain, California.

Learning from the land is a primary method of gaining knowledge. Many Indigenous worldviews see nature as the ultimate teacher. It is essential to learn from and be in reciprocal relationships with the more-than-human world: the local plants, animals, and elements that give us life, the sun, moon, wind, fire, soil. Potawatomie botanist Robin Kimmerer states, "This orientation to the world as an ongoing gift exchange between the human and the more-than-human world is foundational in Indigenous environmental philosophy." This direct, intimate connection to the natural world is the basis of Native sciences and is critical for fostering an emotional and spiritual bond with place.[7]

Implied within the concept of symbiosis—living together in community—there is an important emphasis on dialogue, exchange, and council. These exchanges are intergenerational, intertribal, intercultural, and often inter-species. Unlike normal classroom discussions, which can be very productive in terms of outcomes, this kind of communication usually is not based on achieving certain results (although it can be). Many Indigenous learning circles focus on a different type of learning and sharing. First, there is generally not a single, specific teacher, but all in the circle are considered teachers and learners. Second, everyone sits in a circle, making an even playing field (no one is higher or lower or privileged as special). Third, there is often no time limit on what one shares. It can be very short or very long, and there is no cross-talk or interruption, unless the speaker asks for it. Fourth, there is an emphasis on deep listening, where personal assumptions and biases are internally questioned and suspended so as to truly hear another's voice and meaning. Long spaces of silence are welcome and considered healthy.[8]

This Indigenous learning space, often called a learning lodge, invites deeper questions that generally are not asked and engenders a collective inquiry—a type of "group mind" or group meditation through words—that is palpably different from other learning spaces. In this circle, there is shared wonder and sometimes grief or frustration at the complexity or contradiction of certain questions or topics. Most importantly, judgment and competitive ego activity are strongly discouraged in this circle, whereas deepening the capacity to hold divergent views as equal to one's own is encouraged, thus making a safe and, many would say, ceremonial or even sacred learning space where one can share without fear of being wrong, reprimanded, scolded, or ignored. It is shameless education.

Indigenous Environmental Education: Linking All to the Natural World

Given these principles and elements, it should be clear that Indigenous education is inherently environmental education. It starts with a cosmological orientation to the sun, moon, and stars in relation to local geography (mountains and rivers) and ecology (plants and animals), which creates eco-cultural landscapes and sacred places. One cannot learn about California Indians, for example, without learning about their reliance on abalone and the Pacific Ocean or acorns and oak woodlands. One cannot learn about a birch bark canoe without learning about the Anishinaabeg peoples of the Great Lakes and their reliance on the water and their dependence on wild rice and fish for sustenance.

One cannot learn about the history of *any* place without understanding the First Peoples of the land and their unique cultural and environmental practices, as well as the impacts of conquest, and cultural resilience. Indigenous learning is always contextual, starting with exactly where you are—cosmologically, geographically, ecologically, culturally, and historically. This starts with honoring the local peoples and places where one learns. In many traditional protocols, one has to ask permission from the land, its spirits, and the Native peoples that dwell in that place to learn in a specific location. An offering generally is made in this process of asking for permission and support. The land has so much to teach us. As Wintun ethnobotanist and herbalist Sage LaPena has said: "[W]herever you go, there you are. There is no place on Earth that is not a place-based classroom."[9]

Guardians of the Waters Youth Program

Since 2013, the Cultural Conservancy, an Indigenous-rights community organization based in San Francisco, California, has been offering a Native youth summer internship program called Guardians of the Waters. Through this program, fifteen to twenty-five Indigenous youth (many of them urban and intertribal) are selected to participate in a six-week immersion experience, exploring Native water consciousness through a diversity of site visits, participating in canoe making, and learning about Native foodways through exposure to seeds, native plants, and traditional agriculture. Guest teachers for the program include artists, navigators, farmers, Knowledge Holders, and teachers that are associated with the flourishing Indigenous canoe and Indigenous food movements in the United States and around the world.[10]

Tribal canoes are sacred vessels that connect nature and culture for travel on water. Through human ingenuity—or "indigenuity," as Native studies scholar Daniel Wildcat calls it—Native communities transform trees and plant materials such as bark (and sometimes animal skins) into functional and beautiful canoes for water travel. These vessels have great historical and cultural significance to Indigenous communities who live by rivers, lakes, and oceans. Learning about ancestral canoe traditions, whether a Yurok dugout canoe, a Hawaiian double-hull canoe, or a Micmac bark canoe, connects one to ancestral waters and cultural traditions. Relearning canoe traditions such as paddle making and ocean navigation links people directly to the ecology of water— whether fresh water or salty, flowing or still, polluted or clean—and to their own personal waters (humans are 60 percent water). Learning how to carve paddles, build canoes, and navigate by the stars has inspired the youth in this program to reclaim their role as "Guardians of the Waters," where they become advocates for the sacredness of water and the need to protect it as a relative.[11]

Through talking circles and learning exchanges, the youth share their connection—or, more frequently, disconnection—to ancestral waters, thereby instigating a process of personal exploration and reconnection. By visiting springs, creeks, rivers, bays, and the ocean, they learn firsthand that the water cycle is both an ecological and spiritual system of renewal. By witnessing dammed and diverted waterways filled with trash and invasive species, they also learn about the need to rediscover dormant knowledge of water guardianship and how this is a part of Indigenous revitalization and environmental justice.[12]

A participant in the Guardians of the Waters youth program tests out a newly created tule reed boat, Occidental Arts and Ecology Center, California.

The youth program also emphasizes the importance of ancestral seeds and native foodways. It conveys the view that the future of life is based on healthy seeds, and, just as the waters need protection and renewal, so do the seeds and native foods that have sustained Indigenous nations, and all peoples, for thousands of years. Program participants visit the Indian Valley Organic Farm and Garden, a certified organic teaching garden at the College of Marin, to learn about organic agriculture, native seeds, ethnobotany, and Three Sisters Native agriculture (the joint planting of corn, squash, and beans). Ideally, they experience all aspects of growing food, from seed planting and farm tending (watering, weeding, composting, natural pest management) to harvesting and feasting, all with an eye toward maintaining carbon-rich, healthy organic soils. The act of "getting dirty" on a farm is a visceral way to literally reconnect to Mother Earth and to explore and learn about the magical alchemy of soil, water, and seeds—the basis of all life.[13]

Learning directly from the land and First Peoples, Guardians of the Waters youth gain critical understanding of the ecological and cultural contexts in which they reside. They also learn from the facilitators, peer teachers, Traditional Knowledge Holders, and each other, experiencing the "symbiosis" of community learning and sharing. By gathering and processing native plants such as tule and acorns, using chisels to carve redwood and cedar, and weaving baskets, as well as learning Hawaiian navigation chants and dances and Anishinaabe planting songs, youth engage in embodied ways of learning and knowing and are inspired to claim their roles as water and seed guardians.

Indigenizing the Ivory Tower

In 2012, Native educator Rose von Thater-Braan and Indigenous scholar-activist Melissa K. Nelson co-taught a variety of American Indian Studies' Native science classes at San Francisco State University. They introduced students to the concept of a learning lodge to explore Indigenous ways of learning, with a particular focus on exploring the concept of learning spirits, engaging in embodied forms of learning, and creating a safe context for holistic inquiry. The experience was transformational, leading many of the participating students to change their academic and career choices and to commit more deeply to a new type of learning. By incorporating artwork, sound, plants, and movement as valid ways of learning, students commented that they felt liberated from purely intellectual forms of inquiry. The instructors used these embodied methods to share concepts such as quantum time, plant pollination, and ocean dynamics.

During the learning lodges, students had a higher level of participation in dialogue, as they felt that a safe and ceremonial learning space had been created. For example, students who were usually hesitant to speak up in typical university classrooms felt freer to be present and visible and to express divergent or unique perspectives without fear or judgment. Students also commented that they experienced a deeper quality of listening and an altered sense of time. The students learned the importance of "the good mind," grounding and clearing themselves mentally, emotionally, and physically before engaging in a learning process where deep listening is required. This active presence, which Buddhists may call mindfulness, changes one's relationship to time and was new and refreshing for the students, helping them gain a greater sense of belonging and well-being.

In addition to cultivating these more internal Indigenous learning processes, the classes also focused on the importance of symbiotic and contextual learning: gaining knowledge directly from local landscapes and Native communities. A required part of the coursework was four intensive all-day Saturday field visits to different natural areas to learn about local ecosystems and cultural landscapes. Getting out of the classroom and into natural areas is key to Indigenous education. The class visited Indian Canyon, a beautiful, natural oak woodland canyon and living cultural heritage area of the Ohlone people; an urban watershed in the Presidio national park; an organic teaching farm where native foods are grown; and a shellmound or beach site along the Pacific Ocean.

During these field visits, the participants learned directly from the original peoples of the land—the Ohlone, Miwok, and other Native California Indian peoples. These Native Knowledge Holders, and Cultural Practitioners shared their traditional ecological knowledge on a variety of topics, depending on the place and theme. The students learned about the history of the land from both Indigenous oral narratives and conventional human and environmental histories. They interacted with native plants and learned about traditional foods and medicines. They also learned about invasive plant species such as eucalyptus and French broom and about local efforts to limit their growth, learning in the process about ecological restoration or "eco-cultural restoration": bringing back the health and well-being of Native communities, First Peoples, and the ecosystems that they rely on.

Indigeneity and the New People of the Eighth Fire

The Anishinaabeg peoples carry the Seventh Fire Prophecy, which is relevant for our precarious times. The prophecy states that Native peoples in North America are currently living in the Seventh Fire, a time of cultural recovery after the devastating impacts of colonization on Native minds, bodies, hearts, and spirits. Native people are mending the "split-head society" and healing the historical trauma of the boarding-school era and other oppressive impacts due to the policies of the U.S. and Canadian governments during the nineteenth and twentieth centuries. According to this prophecy, urban and rural, mixed-raced Native people will rediscover dormant traditional knowledge and practices, create new lifeways, and begin to weave together a new culture of modern indigeneity. These will be the New People of the Eighth Fire, devoted to kinship, peace, and reciprocity.[14]

But Indigenous peoples cannot do this alone. The "light-skinned" people need to make a decision whether they will continue down the path of exploitation, oppression, and destruction, which is characterized by a scorched-earth path; or whether they will choose to remember their own indigeneity to pursue a path of respect, reverence, and renewal, where they will join the New People and a green path of beauty and peace will prevail. Interestingly, Okanagan writer and Knowledge Holder Jeannette Armstrong has suggested that indigeneity is more about a lived environmental ethic based on ecoliteracy and collective governance than a bloodline. Likewise, Mohawk seed saver Rowen White encourages all peoples to "rehydrate the native seeds and wisdom in our own DNA." Wes Jackson, the famed agronomist and founder of the Land Institute, writes that humans must again "become native to place."[15]

Given the powerful statements and prophecies above, the possibility of reclaiming indigeneity appears to be a viable answer to the question of how to create a new type of education in order to prepare for or avoid a dystopian future. In this Anthropocene epoch, it seems urgent that humanity transform and evolve into an era of indigeneity and kincentrism. Yet how to do this without falling into the trap of White appropriation and exploitation of Native ways?[16]

The Kogi of Colombia call modern industrial humans "Little Brother," and they refer to themselves—in their role as Indigenous Knowledge Holders that remain deeply tied to the heartbeat of Mother Earth—as "Elder Brother." What will it take for Little Brother to listen? Is there still time, or is it too late? Regardless of whether Little Brother listens or not, Indigenous peoples are exercising their self-determination and educational rights, fueled by prophetic teachings, to renew Indigenous lifeways and teach them to younger generations to help light the Eighth Fire and create the New People. From the many examples of Indigenous education now taking place, it is clear that peoples of all walks of life are listening to these teachings, decolonizing their minds, and preparing to learn anew to create the New People for a green future.

CHAPTER 5

Pathway to Stewardship:
A Framework for Children and Youth

Jacob Rodenburg and Nicole Bell

As an environmental educator, it is difficult not to get discouraged. The news about the state of the environment is ever more sobering. Climate change, habitat destruction, species depletion, rising sea levels, pollution, and the list goes on. Teaching about these formidable challenges can seem daunting, overwhelming, and, at times, simply hopeless. And despite our best efforts, things just seem to be getting worse.[1]

Perhaps like a reversed telescope, environmental education is being looked at in the wrong way. Instead of dealing with reactions to problems and trying to solve environmental issues as they arise, it may be worthwhile to consider what sort of citizens we believe should populate the Earth. Or, as Simeon Ogonda, a youth development leader from Kenya, asks, "Many of us often wonder what kind of planet we're leaving behind for our children. But few ask the opposite: what kind of children are we leaving behind for our planet?" Raising environmentally engaged citizens requires more than just a few educators participating in this work. Rather, it is a collective responsibility: each of us has a stake in fostering the stewards of tomorrow.[2]

Increasingly, there are alarm signals that something is wrong with our children's mental and physical health. There are rising levels of anxiety, attention-deficit/hyperactivity disorder (ADHD), and antisocial behavior in children. A sedentary, indoor lifestyle—where the average child spends more than seven hours a day in front of a glowing screen and less than twenty minutes a day in active outdoor play—is leading to unprecedented rates of childhood obesity.

Jacob Rodenburg is the executive director of Camp Kawartha Outdoor Education Centre in Ontario, Canada, and a part-time teacher at Trent University in Peterborough, Ontario. **Nicole Bell** is Anishinaabe (bear clan) from Kitigan Zibi First Nation and an assistant professor and senior Indigenous adviser at Trent University's School of Education and Professional Learning.

Today's children may be the first in generations not to live as long as their parents.[3]

At the same time, there is mounting evidence that exposure to nature while growing up reduces stress, improves physical and mental health, stimulates creativity, builds self-esteem, and encourages cooperation, collaboration, and self-regulation. In his book *Last Child in the Woods*, Richard Louv posits that children need contact with nature (or, as he calls it, "vitamin N") as an essential part of a healthy childhood. The work of Joy Palmer, an environmental education researcher, found that regular exposure to nature is the single most important factor in fostering care and concern for the environment. But if direct contact with natural environments is critical in fostering stewardship, and children are spending more and more time indoors, then where will tomorrow's stewards come from?[4]

Charting the Path

Our children may well benefit from an environmental framework for education—centered on stewardship and anchored in Indigenous ways of knowing—that involves the entire community: parents, grandparents, educators, schools, organizations, community leaders, health professionals, municipal officials, and businesses. Such a framework is being developed based on a model from Ontario, Canada, called "The Pathway to Stewardship."[5]

Stewardship can be defined as a sense of connection to, caring about, and responsibility for each other and the natural world around us. It involves personal action to protect and enhance the health and well-being of both natural and human communities by providing children with the right tools and experiences at every age to know, love, respect, and protect the very life systems that sustain and nurture us all. Being a steward should not imply entitlement or power or dominion over the Earth. Rather, fostering stewardship means teaching children how to become engaged citizens of and for the Earth.

The Pathway to Stewardship model emerged out of a conversation between a group of community stakeholders in Ontario, including educators, professors, Indigenous leaders, public health officials, and conservationists. They wanted to find ways in which multiple sectors could coordinate their efforts in order to promote stewardship throughout all ages and stages of a child's development. The group began by conducting broad-based research into environmental education, Indigenous teachings, child development, and the factors promoting mental and physical health in children. They also interviewed more than seventy-five community leaders who expressed an interest in environmental issues, with the aim of exploring the formative experiences that these leaders had while growing up that helped shape their interest in the environment. The group felt that the findings emerging from both these interviews and the meta-research could provide a solid foundation for a workable stewardship framework for their community.[6]

From the research and interviews, various themes began to emerge that stem from the impacts of modern technology and culture on our relationship with the natural world. They suggest that repeated, rich experiences in the natural environment, developing a sense of place, and engaging in meaningful, age-appropriate action are all important aspects in creating an ethic of stewardship. These themes are consistent with the teachings of many of Canada's First Nations, including the Anishinaabe* (Ojibway) people who, for thousands of years, have lived in a region stretching from the Great Lakes westward to present-day Alberta and northward to Hudson's Bay. (See Box 5–1.) First Nations teachings are circular, holistic, and relevant to any grade and stage of a child's development. Stewardship has been deeply rooted in the traditional cultures of First Nations for millennia and offers insights into how to effectively raise caretakers/stewards of the environment.[7]

The Elements of Stewardship

Schools, organizations, and communities in Ontario, Canada, are working to foster future stewards by incorporating several key elements. These include: tending and caring, awe and wonder, a sense of place, interconnectedness, mentoring at all ages, time to explore and discover, and engaged action.

Tending and caring. A fundamental value in building a foundation for stewardship is the understanding that all living things—human and nonhuman—deserve to be treated with respect. In Indigenous worldviews, highly espoused values are love and humility, which create capacity and desire for harmony and well-being and the recognition that humans are a sacred part of creation. Cultivating love and humility inspires sensitivity toward others and a desire for good relations and balance with all of life.

At Edmison Heights Public School in Peterborough, Ontario, teacher Drew Monkman—with the help of the school council and student

---

* Anishinaabe is the word used by many Algonquin nations to name themselves in their language.

Box 5–1. Anishinaabe Teachings

For thousands of years, the Anishinaabe First Nations have respected, learned from, cared for, and made use of their natural environment. They teach us that the connection between ourselves and the land on which we live is not a conquest but a relationship—and, like cultivating any relationship, it requires our love, time, effort, and commitment.

Anishinaabe traditional teachings share that life should be lived according to the seven original or ancestral teachings embedded within an Anishinaabe philosophy of life. These original teachings are the guiding principles for how individuals are to treat each other and can be articulated in relation to how individuals are to treat the natural environment; they include love, honesty, respect, truth, bravery wisdom, and humility.

The Anishinaabe worldview includes the concept of the individual, growing through the stages of life (infant, child, youth, adult), who has a spirit, heart, mind, and body and who therefore connects, feels, thinks, and acts—which leads to respect, relationship, reciprocity, and responsibility (4 Rs) as the individual lives on the planet with all other living things. These foundational understandings are captured in the so-called medicine wheel framework. (See Figure 5–1.)

The traditional teachings of Indigenous people, writes Native author and activist Melissa K. Nelson, "are the literal and metaphorical instructions, passed on orally from generation to generation, for how to be a good human being living in reciprocal relation

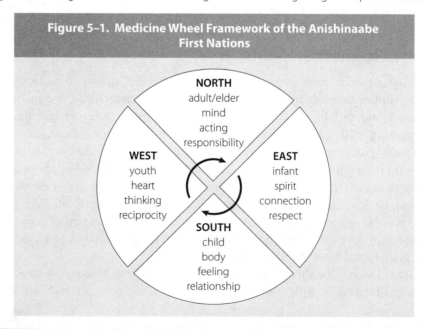

Figure 5–1. Medicine Wheel Framework of the Anishinaabe First Nations

Box 5–1. continued

with all of our seen and unseen relatives. They are natural laws, that, when ignored, have natural consequences."

The traditional teachings "contain wisdom and imply practices that may be essential if the ecosphere is to survive human practices of consumption and exploitation," observes political scientist Joyce Green. Attaining this place of wisdom begins with respect, or re-spect (meaning literally to "look back" or to "re-see") to recognize the interconnections among all living things. Relationship is practiced through a conception of personal and collective responsibility to each other, to the biosphere, and to a future in which one will not be present. Reciprocity is maintained in Anishinaabe tradition by the act of giving thanks and giving back. Anishinaabe teachings articulate the need to always give thanks for anything that one is given, thus recognizing that as humans we are dependent on the natural world for our survival.

Environmental education that espouses such values ensures responsible action to those who will be living seven generations forward. As Peter Kulchyski, a Native studies professor at the University of Manitoba, explains, "[i]f there is an Aboriginal environmentalism, . . . it stems from this kind of social structural fact: in these communities there is a knowledge that great-grandchildren will see the impact on the land of decisions made today." To live life in a sustainable way is to live respectfully in reciprocal relationships and to take responsibility, because what we do to the Earth, we do to ourselves.

Source: See endnote 7.

volunteers—established a schoolyard naturalized area, converting a former lawn into a wildflower meadow, a forest with wood-chipped trails, and a pollinator garden. To cultivate caring for the natural world, older grades teach younger grades by being "habitat helpers." Students have created field guides and nature scavenger hunts for the area. Older grades "adopt a tree" and keep a journal of each tree's progress throughout the school year.[8]

Awe and wonder. The engine of learning is curiosity. Curiosity, in turn, is fueled by a healthy sense of awe and wonder. As adults, we need to model a positive and healthy connection to our environment. We forget as adults how powerful language can be. If we want to cultivate a sense of wonder, we need to use the language of wonder. Words can inspire or discourage. Saying "put that down, don't touch that, it's dirty" sends a coded message to children that the outdoors is hazardous. From an Indigenous perspective, calling the Earth "dirty" is akin to calling one's mother dirty, since the Earth is considered to be Mother Earth. Allowing children the opportunity to respond creatively to the environment fosters love and respect. Using words like "wow, amazing" and "wonderful" encourage connection. For example, children at the Camp Kawartha Environment Centre, an outdoor education center in eastern Ontario, lie under the canopy of large oaks and create "poe*tree*," sharing inspiring words about the gifts of a forest.[9]

Sense of place. An important part of developing a sense of comfort and belonging is spending enough time outdoors in the same place to become deeply familiar and connected with it. For those who have developed a particular attachment to a place while growing up, that sense of place becomes part of their identity. For Indigenous people, this strong feeling of rootedness drives their identity as caretakers of the Earth. It is important to give children plenty of time to develop those deep attachments to place, whether that is a favorite park or a nearby green space. Overwhelmingly, the community leaders interviewed for the Pathway to Stewardship initiative cited special natural places that they grew to know and love as a very important part of their childhood.[10]

Interconnectedness. Children benefit from many opportunities to learn how their lives are connected to the lives of other people and other living things. We breathe the same air, drink the same water, and the food we eat contains nutrients that have been shared by many others for many years. This understanding reinforces the innate need to belong. Stewardship involves understanding that we belong to a community that extends far beyond our close friends and relatives. (See Box 5–2.) Our actions do not end in the present, and the consequences of what we do echo far into the future. The Indigenous worldview considers how the results of our actions will impact seven generations forward. At Camp Kawartha, children learn about sustainable living in action, how the materials used in building can be recycled, how energy consumption can be drastically reduced, and how we can include natural landscapes and corridors in our own backyards.[11]

Mentoring at all ages. Both research and discussions with community leaders revealed that having access to a caring mentor is critical in developing stewardship. In the early years, this is usually a close relative—a parent or grandparent who spends time with the child, exploring together and sharing the delights of discovery. As a child grows older, the mentor is often a teacher or a youth leader who becomes a trusted and admired role model. In Indigenous communities, the role of the Elder as teacher of the children ensured the passing on of traditional knowledge to the next generation. In a unique Eco-Mentorship Certificate Program offered by Trent University and

Box 5–2. Ubuntu and Ecological Citizenship

If education in the twenty-first century is to be relevant, then it has to respond to our two interconnected planetary crises: the ecological crisis and the rapid growth of technologies that potentially threaten life on a global scale. As technologies become increasingly intertwined with humans, it is difficult to determine what "being human" now is. Education can help affirm the human not as an isolated individual but as someone embedded in the processes of life. It is here that the concept of "Ubuntu" holds particular promise.

Ubuntu means humanness. In South Africa's Xhosa language, Ubuntu is derived from the aphorism, "Umuntu ngumuntu ngabanye Bantu," which suggests that a person's humanity is ideally expressed in relationship with others, and, in turn, individuality is truly expressed: "We are, therefore I am." Ubuntu suggests that our moral obligation is to care for others, because when they are harmed, we are harmed. This obligation extends to all of life, since everything in the cosmos is related: when I harm nature, then I am harmed.

An education informed by Ubuntu focuses on developing a subject that is ecological rather than atomized. The philosophy of Ubuntu has been integrated in various forms in thousands of South African eco-schools, where children come to experience the oneness of self, school, community, the global community, and the environment. The eco-schools program, coordinated by the Wildlife and Environment Society of South Africa, aims to create environmental awareness and action in schools and the local community.

Vuzamanzi Primary School, in Cape Town's informal settlement of Khayelitsha, works with community members to develop projects such as community gardens, which provide food for the undernourished and a habitat for pollinators. The school's garden project, Ekasi Project Green, aims to move learning beyond the classroom by broadening the way that learners and community members think about food security. The classroom is not a confined space, isolated from the environment; rather, children learn in local environments so as to learn to care for them. In eco-schools, children spend time close to the land so that they learn that nature is not something that we possess, control, or inhabit, but, rather, it inhabits us.

Where Ubuntu is lived, every child belongs to the whole community and is cared for by the community. Perhaps, over time, this philosophy can help create a more sustainable relationship between people, other species, and the Earth.

—*Lesley Le Grange, Distinguished Professor of Curriculum Studies at Stellenbosch University, South Africa*

Source: See endnote 11.

Camp Kawartha, student teachers acquire the skills, strategies, and knowledge to help them deliver environmental programming in their future classrooms.[12]

Time to explore and discover. Another recurring recommendation, both in research and in feedback from community leaders, points to the benefits of limiting screen time: television, computers, and cell phones. Too much screen time limits physical activity, impairs social and creative development, and serves to disconnect children from their natural surroundings. Community leaders recalled the "free-range" time to explore nearby nature that they were provided as children by their caregivers. Time to play, romp, and discover fosters initiative and independence, stimulates creativity, and promotes resiliency. Medicine wheel teachings of the Anishinaabe/Ojibway people share that time to explore nature with all the senses is essential to building respect and a positive relationship with the natural world. In a Forest Friday program for homeschoolers in Peterborough, Ontario, children spend the entire day—no matter what the weather—following tracks, exploring wetlands, and making forts.[13]

Engaged action. Well-meaning educators often believe that caring arises out of knowledge. The formula goes something like this: if children know more, then they will care more. There is no question that fostering knowledge about environmental issues is important. However, it turns out that teaching about environmental issues is more effective if the issue is of personal relevance and if children are provided with meaningful, age-appropriate action to improve the issue at a local level.

Everyone, no matter their age or ability, can do something positive for the environment. Tending a garden, raising butterflies, caring for a natural area, and reducing our energy consumption are just some of the simple ways that we empower our youth to make a positive impact. From an Indigenous perspective, acting with responsibility means responding with our abilities (response-ability). The idea of "agency" is key: kids can solve a problem provided they are given the right tools and strategies for their age. Every positive action leads to a sense of hope, and every bit of hope is empowering. As children grow older, they can begin to explore the idea of sustainable living: reducing their carbon footprint, investigating alternatives to fossil fuels, and learning about product lifecycles and social justice. In *Connecting the Dots: Key Strategies That Transform Learning for Environmental Education, Citizenship, and Sustainability*, authors Stan Kozak and Susan Elliott offer concrete steps that educators can implement to "prepare our young people to take their place as informed, engaged citizens."[14]

How to Nurture Stewards

It is important to recognize that children of different ages respond to the environment in markedly different ways. Well-meaning educators may want to talk to small children about the imminent dangers of climate change and the effects of global warming, but small children simply do not have the cognitive faculties to process such large and multidimensional issues. Instead, like ever-widening fields of self, children first discover their bodies, their senses, and the environment immediately around them. As they become older, they recognize cause and effect, action and reaction. They discover empathy and compassion. Older still, and they recognize that they are embedded in a community of people and other living things. As teenagers, they begin to be ready to understand larger, more complex issues facing their community and beyond. As mature teenagers, they can take on issues of social and environmental justice.

The Stewardship Framework articulates a variety of stewardship principles and suggests ways to implement these, derived from research and interviews aimed at children from early childhood to the teenage years. (See Table

A fourth-grade class plants marsh grasses along the shoreline while studying the ecosystem of the Elizabeth River in Norfolk, Virginia.

5–1.) Nurturing stewards is a proactive undertaking. Building on a sense of wonder and awe, educators can start by modeling empathy and respect for all life. At each stage, children need opportunities to develop their spirit, heart, mind, and body. As children begin to learn about how the world functions, they understand the impacts that people can have and they explore solutions to challenges within their community. As youth develop leadership skills by participating in local action, they develop confidence, and a sense of agency and belonging. Engaged stewards arise when we teach our children to know, love, understand, and protect the very land they stand upon.

This is a call for educators, parents, community leaders, and youth groups to coordinate their efforts so that they may take collective responsibility for fostering stewardship. Every community has its own environmental challenges, as well as its own resources and opportunities for environmental education. Despite best intentions, environmental education is often delivered in an ad hoc and siloed manner by individual schools and/or organizations. At times, efforts are duplicated, and key stewardship developmental opportunities are

Table 5–1. Fostering Stewardship

For Young Children (Ages 3 to 6)

Core Stewardship Principle	Stewardship Opportunity
A time for deepening relationships and understanding.	Choose an outdoor place to explore and play in. Visit regularly. Provide loose parts for kids to manipulate (sticks, stones, tree slices).
Reinforce and expand the developing sense of empathy.	Plant, tend, and harvest something that can be eaten. Raise butterflies, care for an animal.
Celebrate seasons.	Find simple ways to recognize and enjoy the change of each season.
Cultivate sensory awareness of nearby nature.	Identify natural sounds and smells. Explore micro-environments (peek under rocks/logs, create a mini trail).
Encourage the idea of "neighbor-wood"—that our community consists of other living things as well as humans and built structures.	Get to know plants, birds, and five insects living in your area. Create a mural that depicts the characters of your "neighborwood."
Offer a creative response to time spent outside.	Develop art projects using natural materials. Create a story or a play about the characters in your "neighborwood."

For Middle Childhood (Ages 7 to 12)

Core Stewardship Principle	Stewardship Opportunity
Develop more-complex outdoor skills.	Try non-motorized outdoor activities, such as hiking, survival skills (shelter building, fire making, foraging wild edibles), orienteering, birding, and astronomy. Spend at least seven hours a week practicing these skills.
Explore human impacts on the environment. Develop leadership and decision-making skills by planning and implementing a simple community-based project.	Create a small naturalized area. Manage a school recycling or composting project. Plan a small stream/river cleanup project. Make a poster or video to educate your community about your project. Research and write about the history of the piece of land you occupy.
Expand understanding of the relationships between living things and their habitats.	Explore biodiversity in a nearby natural area. Conduct a small-scale biophysical inventory, finding at least ten species each of plants, animals, and insects. Explain three ways that this ecosystem helps the environment. Get involved in citizen science projects: monitor bird, butterfly, and amphibian populations. Monitor ecosystem health by conducting basic water and soil tests.
Expand understanding of sustainable lifestyles.	Be an energy detective. Find out what kind of energy is used for heating, cooling, lights, and appliances at home or school. What renewable energy systems can you observe in your region? Design an energy-efficient home that is healthy for both people and the planet. Think about using natural materials, passive solar design, rainwater harvesting, renewable energy, and innovative ways to treat human waste.

Table 5–1. continued	

For Older Children (Ages 13 and older)

Core Stewardship Principle	Stewardship Opportunity
Expand skill and confidence in outdoor awareness, responsibility, and survival.	Research the meaning of sustainable harvest. How can the environment provide our needs without being damaged by human impact? Learn how to find your way in a natural area using a map, compass, and/or GPS. Learn how to recognize at least two constellations in the night sky in each season. Learn how to tell the four directions using clues in the sky.
Deepen understanding of how modern lifestyles affect the environment. Expand leadership and problem-solving skills by seeking solutions to ecological imbalances.	Calculate your ecological footprint. Research how your country's lifestyle consumes global resources, and how this compares with other countries. What does sustainability mean? Set a goal to reduce your ecological footprint for a month and assess how successful you are. Get your family and school involved, too.
Expand abilities to understand and empathize with others while exploring and responding to local social and environmental issues.	Find an organization that is making a difference in your community. Volunteer. Teach someone younger than you an outdoor skill. Find someone to tell you how your area has changed over the years. Find a local hero who is working to protect the environment. Arrange for them to speak at your school. Volunteer in a natural area to help with trail maintenance, ecological restoration, or control of invasive species. Help with a community tree-planting project. Participate in planning, planting, maintenance, and monitoring. Do you think it was a successful project? Would you make any changes in future projects?
Learn about social and environmental justice.	Find an issue of local concern that you feel strongly about. What problem needs to be solved? How does this issue align with global issues? Get involved. Learn simple action skills: how to make a presentation, how to write a convincing letter, how to organize an event. Learn how to listen and try to understand multiple points of view. Find a mentor who can help you learn and do more to solve this problem.
Express your feelings about your local environment.	Create a story, poem, visual art piece, or play that captures your feelings about the land you occupy. Write a letter to your ancestors. What would you say is worth protecting for your children and for their children?

missed. One way to move forward is by forming a stakeholders group and developing a collaborative approach among community organizations—including schools, early childhood programs, youth leaders, parent councils, municipalities, and faith groups—to ensure that every child has access to key stewardship opportunities throughout their development. Then, encourage other communities to do the same. In the end, it takes the heart and conviction of a village to raise a steward.

CHAPTER 6

Growing a New School Food Culture

Luis González Reyes

Around the world, the commitment to a fair, healthy, and sustainable food model within the sphere of formal education represents a cultural transformation, not just for students but for the broader educational community and for society as a whole. Getting there is not easy, however. Truly transforming the relationship that schools have with food—and ensuring that food is a vector for lasting societal change—is a multidirectional process of teaching and learning that involves a broad range of stakeholders. This includes students and teachers, cooks and cafeteria monitors, food suppliers and intermediaries, families and neighborhoods, as well as numerous other actors from the communities in which schools are located.

Food as an Educational Vector

Food serves as an important educational vector for at least four reasons:

First, it provides a valuable framework for understanding the wide-ranging environmental and social crises that the world now faces. The industrial agrifood system consumes large amounts of land, water, and energy; is responsible for widespread environmental degradation; has contributed to the systematic destruction of rural lifestyles; and is a leading contributor to climate change. Under this system, a handful of large multinational corporations, driven by profit, determine the type, quality, and price of foods, as well as how and where they are produced. Yet, despite the promises of the industrial model, hunger

Luis González Reyes is an educator and author who is responsible for the "ecosocialization" of schools run by the FUHEM Foundation in Madrid, Spain, including coordinating the ecosocial curriculum, the ecological school cafeteria, and the blog Tiempo de Actuar (tiempodeactuar.es).

and malnutrition persist. Meanwhile, the spread of high-calorie, meat-based diets has broad implications for public health, contributing to the rising incidence of obesity, diabetes, and cancer worldwide.[1]

In contrast, a food system based on balanced diets and on agroecology—an approach to growing food that is based on ecologically sustainable and socially just cultivation practices—could play a central role in restoring the local and global environments, while also providing livelihoods and healthier food.[2]

Second, food is an important educational vector because it provides a means by which to address many other important issues in schooling. For example, food is a popular area among educators for working on psychomotor development, for strengthening the senses, and for teaching the rules of coexistence. Food also makes it possible to approach areas of knowledge such as ecology, landscape modification, the global system of production and consumption, and the relationships among different productive sectors.

Third, in the coming decades, people will likely need to participate more directly in the food system. The current multiple crises—energy, climate, environmental, economic, cultural, and political—represent a major turning point for civilization. As the use of fossil fuels declines over time, the re-ruralization of societies will be inevitable. If a basic aim of schools is to help students understand and cope with the world in which they live, then it will be important to prepare our children for a future that is very different from what they are familiar with today. This means not only giving students the tools they need to understand and shape the world, but also letting them become agents of social change, so that they can help society transition toward a democratic future that meets people's needs without plundering the environment.[3]

Finally, food is a good educational vector because it is highly important in our daily lives. It is a critical determinant of people's health and has an influence on everything from hunger and obesity to hormonal imbalances and cancers. Food is continuously present in our lives and keeps us alive. It is a central focus of conversation and a key component of our social and family lives. Given food's centrality to human existence, we cannot waste its potential as an educational tool.

Modifying the Ingredients in a School Cafeteria Changes Everything

Although the school cafeteria is by no means the heart of the educational process, it is the primary location of food within an educational setting. And key to school food is the specific ingredients that go into the meals that are served. School meal plans can play an important role in orienting students and staff toward food choices that are healthier for both people and the planet. This includes offering meals that contain lower amounts of fat, sugar, and animal protein (meat, dairy, and eggs) and higher amounts of fresh, organic produce and unrefined (whole) grains. To support local communities, cafeterias can strive to use ingredients that are in-season and locally sourced.

By introducing organic ingredients, schools can provide tastier food on the table, although students may at first be unaccustomed to the stronger flavors of some organic vegetables. In temperate regions, where many foods cannot be grown locally year-round, seasonal produce options for much of the school year might include winter vegetables such as pumpkin, leek, spinach, Swiss chard, and beetroot, as well as a high percentage of sprouts. These vegetables tend not to be the most popular foods (particularly among young people), however, and opting for greater variety in the dining room through organic and seasonal offerings could lead to student resistance and, ultimately, to wasted food.

A cafeteria worker prepares fresh vegetable cups for the National School Lunch Program in the kitchen of Washington-Lee High School in Arlington, Virginia.

U.S. Department of Agriculture

School kitchens would need to adapt to using the changed ingredients. In most cases, local produce does not arrive in washed, peeled, and chopped form, and more time is needed to prepare it, increasing the workload of cafeteria workers. The use of new ingredients also affects how meals are designed and cooked. Opting for seasonal foods and dishes that rely on minimal animal protein requires some degree of culinary retraining, as many school cooks are accustomed to using specific produce, such as peppers or canned tomatoes, as the basis for their meals year-round. In regions where two dishes are served in the school meal, cooks may struggle to prepare culturally acceptable second courses that lack animal protein.

Changes to school meals also affect teachers and cafeteria monitors, as they must be aware of the new food culture and be trained in conveying the changes to students and staff. As families notice the changes, this can result in cultural clashes over, for example, the reduction in animal protein. For cafeteria managers, logistical challenges may arise because of the need to contract with a larger number of suppliers of organic and local ingredients. Diverse local farmers would be responsible for supplying different components of the meal (sometimes through an organized distribution network), as opposed to the common practice today of contracting with a single large distributor that specializes in institutional bulk procurement of ingredients that are often difficult to trace and verify.

Changing the ingredients in school meals is a process that involves raising awareness, discussion, and training within the entire educational community. But schools around the world are taking up the challenge, sometimes with support from local governments and nongovernmental groups. The city of Malmö, Sweden, engaged in a multistage, participatory negotiation process involving cooks, teachers, health staff, and decision makers to develop its policy of providing a daily vegetarian option in all public schools and offering vegetarian-only meals one day a week. In Spain, the schools run by FUHEM Foundation—an organization that promotes social justice and environmental sustainability through educational and ecosocial activities—work with broader educational communities to promote healthy meals in Madrid, and Menjadors Ecologics has had similar success with this in the region of Catalonia. In the United States, Georgia Organics works with teachers, parents, and cafeteria staff to connect organic food from Georgia farms to Georgia families, and in Illinois, the nonprofit Seven Generations Ahead includes parents in food education to encourage healthy eating.[4]

A leading challenge in transforming school food is the cost. Paradoxically, agroindustrial food production is cheaper than its agroecological counterpart. Schools can use four basic strategies to overcome this challenge: 1) using seasonal produce, which can reduce the price of raw materials in addition to being more environmentally friendly; 2) negotiating contracts with suppliers for the entire school year and in coordination with other schools, to allow for larger volumes and better prices; 3) minimizing intermediaries and using shorter marketing channels to reduce costs; and 4) adjusting the raw materials required.

By using these four strategies, most schools should be able to reduce the costs of providing local, organic, and/or seasonal non-animal food products. Ecomenja in Catalonia offers students ingredients that are 95 percent organic and 75 percent Catalan every single day. In the United States, California's Sausalito Marin City School District is the first district in the country to provide meals that are 100 percent organic, free of genetically modified organisms (GMOs), and 90 percent locally produced. The National Farm to School Network helps schools districts around the United States strengthen their connections with local farmers.[5]

In places where low-income populations are prevalent and/or financing is lacking, additional strategies may be needed. One healthy strategy is to reduce the consumption of animal protein, which can help greatly to reduce costs. Another option is for schools that rely on an outside vendor for cafeteria management to switch to less-costly direct management, which would enable any savings to be reinvested in efforts to improve the quality of ingredients. At the Gómez Moreno school in Granada, Spain, local families manage the cafeteria, enabling greater control over the ingredients used and generating additional educational opportunities within the school. In the U.S. city of Detroit, Michigan, doing away with an outside management company enabled the school district to double its food budget and improve its meal choices. District cafeterias now serve more fresh fruits and vegetables, participate in Meatless Mondays, and are committed to introducing students to a new raw vegetable every week. In addition, 22 percent of the food served comes from farms around the state.[6]

In the transition to new lunchroom ingredients, government support can be decisive. In Brazil, the government has stipulated that 70 percent of federal lunch funds must be used to purchase foods that are natural or not highly processed, and 30 percent of lunch funds must be spent on food produced by family farmers. Across Brazil, participatory councils made up of parents, students, and government representatives are tasked with monitoring school lunch programs to ensure that they meet these requirements, although enforcement is still a challenge. Since 2002, Rome, Italy, under the leadership of the city council, has been introducing organic and local food into school cafeterias; it also has plans to boost purchases from cooperatives and to reuse food waste. With the final cost nearing $5.87 per meal, the city provides assistance to families that have fewer resources.[7]

Many schools serve not only lunch, but also breakfast and other meals, depending on the type of school and on family needs. Schools also may supply food for student birthday celebrations, staff appreciation parties, and other events, many of which involve high consumption of sugary drinks, high-sodium snacks, and other junk food. A good strategy would be to apply any changes that are made in the school cafeteria to these other activities as well, as Catalonia's Menjadors Ecologics has done.[8]

Creating Meaning Within the Broader Educational Community

In most cases, it is not the school cafeteria, but rather the home, that is the center of child nutrition. It therefore makes sense to extend any meal transformation initiatives beyond the school walls. In Spain, FUHEM has supported the creation of school-coordinated "consumer groups" in which teachers, administration, service staff, and families use their combined purchasing power and the school's existing connections to support local food, rather than buying items sold by large agrifood companies and shipped from across the world. The consumer groups put school employees and families in direct contact with local agroecological producers—with minimal or no use of intermediaries—enabling them to take home the same quality of produce that is being consumed at the school, but at a price similar to that of agro-industrial foods.[9]

Such initiatives give rise to other benefits. Most importantly, the schools offering these types of programs take on more meaning for families because they link critical discourses with coherent practices. They break down the walls that separate the school from the neighborhood, making it possible for what happens inside to take place outside as well. School-coordinated efforts to engage community members also represent an egalitarian meeting space between families and educational professionals, which strengthens affinities, and they create a space where families can take center stage and make a contribution to the school. Finally, in the case of the consumer groups, these initiatives are a useful tool for family conciliation because parents are able to pick up their groceries at the same time as their children. Ultimately, such activities make the school not only an educational center, but also a food center, with all that this entails.[10]

For the families and individuals that are involved in FUHEM's consumer groups, the new practices often lead to a change in values, thereby providing a powerful key to social change. Food shopping becomes more personalized, as the buyers have a clearer sense of who grows and distributes the food that they eat. There is greater understanding of food-related information that previously was unknown (such as that peas come in pods, or that there are different seasons for produce). Finally, there is direct participation in the management of something as basic as food (such as in the choice of suppliers, shopping criteria, or organizational models), and an activity that was once restricted to the family sphere becomes a collective enterprise.[11]

Meaning also can be created when undernourished students attend schools that provide sufficient, nutritious food. For example, school garden projects provide opportunities to teach students about cultivating crops and eating in a more balanced manner. The students eat what they grow, and the projects create meaning by providing other educational opportunities and spurring an exploration of values. Examples of successful school garden projects from around the world include Cidades Sem Fome in São Paulo, Brazil; the School Garden Project in Beijing, China; the Healthy Garden and Kitchen Program in Lima, Peru (which also uses school orchards to work with students with developmental disabilities); the Edible Garden City in Singapore; and Mbuyuni Garden in Dar es Salaam, Tanzania. When the school gardens belong not just to the school but to the wider community, this can open even more opportunities.[12]

At the Heart of School Learning

Another key space to introduce food education at school is the classroom itself, which is at the core of student learning and where more-formal knowledge acquisition takes place. Learning about food—ideally through a combination of classroom instruction and hands-on experience, such as cultivating a school garden—is as important a subject as mathematics and language, and should be assessed in a similar way.

Luis González Reyes

A cooking class for a FUHEM Foundation cooking team.

How do we transfer the activities of the cafeteria to the classroom? The easiest way is probably by starting from areas of conflict. As noted earlier, changing the ingredients used in the school kitchen implies a series of transformations that are accompanied by conflicts—from unfamiliar tastes to new ways of cooking. Moreover, these conflicts are not distant ones: they are experienced in the first person during the meal itself, which is of great importance to everyone. This provides a unique opportunity to provide meaning to what is being experienced in the cafeteria, through a wide range of activities. Educators of

FUHEM, for example, use the existence of conflicts to facilitate greater understanding of food in their workshops with children.

Conversely, how do we transfer the activities of the classroom to the cafeteria? A first step is to acknowledge that a school cafeteria is in itself a "classroom." It is always educational—from exposing student palates to new tastes to developing and testing social norms—although usually this role goes unnoticed. Creating change in the cafeteria opens the door to the transformation of food into a conscious, meaningful, educational act. Cafeteria educators play an important role, and their involvement in broader food education projects is crucial. The Edible Schoolyard Project, based in Berkeley, California, offers trainings for food service professionals as well as teachers, administrators, and advocates to help them implement programs in the kitchen, garden, and lunchroom that bring academic subjects to life.[13]

What specific content needs to be addressed in schools? For food education, key areas of focus include the negative social and environmental impacts of industrial agriculture, as well as the benefits and potentials of agroecology and its specific practices. Students should learn about the workings of the agroecological economy, from cultivation (including specific foods to grow) to marketing, preparation, and distribution. Running an "ecological cafeteria," or cooking or selling the food itself, could be the subject of school study. In New York, the nonprofit organization GrowNYC coordinates a project called "Learn It, Grow It, Eat It" that teaches low-income youth about food cultivation and commercialization through activities such as cooking lessons, tending community gardens, teaching younger children in the garden, and running a weekly farm stand.[14]

Schools that lack formal curricula on food education can introduce these subjects. The United Kingdom, for example, has added cooking classes to the official school curriculum. But this is not enough. An interdisciplinary approach to food education is necessary because it reaches all students and places food-related competencies at the heart of the learning process, giving them greater meaning. Such a cross-cutting approach also enables in-depth and continuous work on these issues across diverse subjects and courses, enabling learning to link multiple perspectives (a skill that is more important than ever). Moreover, interdisciplinary learning makes it possible to address food-related content without undermining regulated studies.[15]

There are multiple ways to facilitate the inclusion of food-related competencies across the curriculum. These range from "hidden curricula"—such as the types of images, stories, or examples used in classes and the messages that they transmit—to more explicit introduction of food education into school exercises, which goes beyond simply providing background to primary learning, and requires working on it, too. An even more powerful approach is to implement projects that are inspired by current-day issues linked to the food system. Such projects could culminate in real-life experiences, such as the student-driven creation of a consumer group for food purchasing.

Perhaps most critical is the need to bring an agroecological focus (when relevant) to the treatment of all academic disciplines, especially in the natural and social sciences. In an innovative effort to link formal curricula with content related to healthy eating, the Curriculum of Cuisine project in Portland, Oregon, works with teachers in participating schools to bring professional chefs into the classroom five to seven times a semester. The visiting chefs might, for example, join a language arts class and work with students to connect their reading and writing to culinary experiences. The Edible Schoolyard Project and FUHEM also integrate their lessons into standard academic subjects.[16]

The cross-cutting inclusion of food-related competencies in schools is an onerous task. It requires considerable effort in awareness raising, training, and the creation of new programs and materials. The development of classroom-ready resources is critical, as few teachers have the skills, motivation, and time to do this for themselves. It also is important that the materials are designed for multidirectional teaching-learning, so that they serve as a learning tool not just for students, but also for teachers and families. The South African organization SEED works in this direction by authoring textbooks with teachers, creating programs that help teachers grow and use outdoor learning spaces, and partnering with under-resourced schools to integrate environmental education into teaching practice. But there is still much that can be done.[17]

An important step before even introducing this type of food education into the classroom is to eliminate those elements of the existing curriculum that run counter to the goals of achieving a more democratic, environmentally conscious, and sustainable civilization. With regard to food, this includes ongoing praise for GMOs and for the Green Revolution (and its embrace of toxic fertilizers and pesticides); the lack of criticism of multinational food companies and commodity-related financial markets; the mostly pejorative view of rural societies and peoples; and the idea that only large corporations can (and do) feed the world.

Also important is how the new food-related content is conveyed. If the aim is to encourage outcomes such as democratic deepening, cooperation, equality in differences, and responsibility for one's actions, then schools need to embrace an appropriate method for doing this. Instruction cannot be based solely on the one-way transmission of knowledge from teachers to students, but rather it should allow for the joint creation of knowledge within the broader educational community (although teachers still play a central role in transmitting and organizing this knowledge). Such a method also provides for democratic management of the classroom and the school. Cooperative and project learning are key, as is "dialogic" learning, or the use of dialogue to consider and evaluate the diverse viewpoints present within the community. Such teaching practices help to accelerate learning in an inclusive manner, as has been demonstrated by the INCLUD-ED project in Europe.[18]

Another central element is to acknowledge that people learn not only rationally, but also through their emotions. This is why an experiential focus, such as the transformation of the school cafeteria, is vital. Because it engages the five senses, it has greater potential to effect long-term behavioral change.

Going Beyond the Mainstream, and Looking to the Future

Most of the examples presented above relate to experiences in or around urban schools; however, other alternatives are worth mentioning briefly. Some of the most important peasant movements in the world, such as the Zapatistas in Mexico and the Landless Workers Movement (MST) in Brazil, have created their own schools. In these rural communities, the food circuit—from production to consumption, and integrating waste management—is something that enters into daily life, which is why food is a naturally integrated vector for education in both movements. In these schools, students learn to cultivate and eat the foods that they plant, making the productive and educational fields essentially the same. In addition, the curricula that these movements have created for their schools naturally integrate food alongside other areas of knowledge, such as history or mathematics. All this is done from the paradigm of agroecology. Thus, the ways to create meaning are even more powerful in these rural cases than in urban school settings, as they come out more naturally.[19]

From cities in Europe to rural Mexico, schools around the world are beginning to provide experiences that teach students to grow and eat organic and healthy food; that enable students and staff to eat these foods in school cafeterias; and that integrate food education and experiences into the curriculum, using materials developed by the educational community. Through these initiatives, these schools have, and are, building a higher meaning for the community and are shaping the growers and eaters of today and tomorrow.

CHAPTER 7

The Centrality of Character Education for Creating and Sustaining a Just World

Marvin W. Berkowitz

When Tim Crutchley and Kristen Pelster were hired as the new principal and assistant principal (respectively) at Ridgewood Middle School in Arnold, Missouri, the school was in a shambles. Serving a mostly poor rural and suburban population, the building was covered in graffiti, the grounds were rotting and rusting, student behavior was unacceptable (it was the only school in the district with a police presence), and academic achievement was abysmal (only one in four students met state standards for communication arts, and less than 7 percent met standards for math). The students knew that the school did not care about them and behaved accordingly.[1]

This school failure was the product of years of neglect and an unofficial district policy to allow failing teachers in other district schools to be transferred to Ridgewood, resulting in two-thirds of the staff being educators who should not have been teaching. The new superintendent, Diana Bourisaw, quickly diagnosed this problem, hired the administrative team of Crutchley and Pelster, and charged them with the task of essentially "cleaning up Dodge City." Fortunately, both administrators had graduated from the Leadership Academy in Character Education (LACE) in St. Louis and understood that the school climate—the quality and character of school life—was the first hurdle that needed to be tackled, which meant starting with authentically caring about students.[2]

This met with great resistance from the existing staff, but Crutchley and Pelster persevered. They helped clean up the physical plant. They modeled student-centered teaching methods and behavior management for the instructors. They drove to truant students' homes and dragged them out of

Marvin W. Berkowitz is the Sanford N. McDonnell Professor of Character Education and codirector of the Center for Character and Citizenship at the University of Missouri-St. Louis.

bed and to school. They personally staffed an hour-and-a-half daily study hall for students who did not hand in homework assignments, because they understood that students cannot learn if they do not do the work, and because they believed that students at this age should not be able to opt out of learning. They had open, frank discussions with staff, and they encouraged burned-out and incompetent staff to leave. And they created structures for students to own, run, and co-create the school. They created a leadership class that ran new student orientation and special events, and they delivered character lessons in advisories. They allowed teachers to create new classes and structures for students who were not succeeding within traditional structures.[3]

After three years of remarkable wisdom and dedication (and a commitment of hours way beyond what could reasonably be asked of them), Crutchley and Pelster replaced two-thirds of the staff with educators who shared their vision of a great school, allowed the staff and students to have profound control of the school, built a school that truly centered on kids and what was in their best interest, drastically reduced misbehavior by students, and achieved impressive success in academics. As a result, that small minority of students who were meeting state standards ballooned to about 70 percent in both communication arts and mathematics.

In 2006, Ridgewood Middle School was recognized as a "National School of Character" by Character.org, a Washington, D.C.-based nonprofit that provides leadership and advocacy for character in schools and communities worldwide. To earn this distinction, Ridgewood was evaluated for how well it implemented each of Character.org's *Eleven Principles of Effective Character Education*, which include "providing students with opportunities for moral action," "fostering students' self-motivation," and "engaging families and community members as partners in the character-building effort."[4]

Why Character Education?

What Ridgewood Middle School accomplished is not atypical, except for the speed with which they did it. Good character education is good education. When done comprehensively and well, it leads to a caring and fair school climate, prosocial and responsible student behavior, increased academic achievement, and development of character in youth. Society's future depends upon the character of its youth and how that will manifest when they become adult citizens. Intelligent, comprehensive, effective character education will contribute greatly to the positive future that our world needs.

But what is character education? Unfortunately, that is more difficult to answer than it might seem. One thing character education is *not* is optional. *How* schools attempt to nurture the development of character is an option, but *whether* they affect character development is not. Schools (and the people who populate them) always shape character, for better or for worse, and whether intentionally or not. Hence, character education is ubiquitous, but also quite variable. The real question is how to have a positive impact on the character of our youth, by intentionally and systematically applying evidence-based practices.

Character is a broad concept that includes all the psychological characteristics of an individual that both motivate and enable him or her to do the right thing. This includes values, identity, moral emotions (such as compassion, empathy, guilt, and shame), the capacity to reason critically about social and moral issues, and so on. Most important is "moral character," the part of one's character that supports one's ability and inclination to do what is right. Other parts of character have to do with excellence ("performance character"), regardless of whether one excels at doing good or evil; with one's functioning as a citizen ("civic character"), such as participation in the political process; and with one's capacity to learn, think, and reason ("intellectual character").

These are somewhat overlapping, as performance character enables one to be effective in the moral realm, and there are parts of intellectual and civic character that have specific moral aspects to them (searching for the common good, respecting truth in intellectual inquiry, etc.). When aiming toward the larger goals of sustainability and making a better world for all, particular focus is needed on the moral aspects of character, including those parts of civic and intellectual character that are about morality, and performance character as a support for moral character.

Applying developmental psychology to the broad global demands of sustainability is challenging. To value and enact sustainable behaviors, one has to be able to think in the long term. It is much more difficult for children and adolescents to do this than it is for adults. This does not mean that children cannot care about the future; it is just hard for them to envision it, logically think about it, plan for it, monitor long-term change, etc. A second challenge is the tendency of people to favor consequences first to the self, and then to others who are close to them. Sustainability is about much broader impact than oneself and one's friends and relatives, and often it can conflict with what would be best for these interests. That choice takes mature moral reasoning and wisdom, which again do not come easily or early for most people.

So What Works?

When done well—meaning comprehensively, authentically, and via evidence-based practices and principles—character education is very effective. A 2005 analysis of the existing school-based research on character education found that implementing it most commonly results in improvements in the capacity to critically reason about morality and social issues, increases in prosocial behavior and attitudes, better problem-solving skills, less drug use and violent behavior, better school behavior in general, more knowledge about and healthier attitudes toward risky behaviors, greater emotional competency, and improved academic achievement. Ridgewood Middle School, like many schools that make character education the authentic centerpiece of school reform and success, demonstrates these results clearly.[5]

Such findings have been shown repeatedly in research, including in large-scale statistical analyses of hundreds of separate studies. A 2011 meta-analysis of social and emotional learning program evaluation studies found a consistent pattern of such programs increasing academic achievement. (See Chapter 8.) These findings were echoed in a set of studies of hundreds of schools in Missouri implementing "The CHARACTERplus Way," where the authors found 33 percent more students achieving state standards in communication arts and nearly 50 percent more achieving state standards in mathematics than in schools not implementing character education. Character education schools also showed significant decreases in discipline referrals, particularly around moral issues such as justice and fairness. There also were significant differences in school climate favoring the character education schools.[6]

Research has found that six foundational principles, which can be termed "PRIMED," help guide effective character education. The following paragraphs provide examples and specific strategies within each.[7]

P Is for Prioritization

The first principle is that for character education to be effective, it has to be an *authentic* priority in the school. It cannot merely be an add-on or "silo-ed" part of the school. It has to drive the train, ideally by having the principal or head of school be its champion and leader. The Center for Character and Citizenship has been running the LACE academy in St. Louis (and other places) for nearly two decades. Experience has shown that when a school sends someone who does not have broad decision-making authority—such as an assistant principal, counselor, or teacher—to learn how to lead a comprehensive school reform designed to foster character development, the outcomes are less effective, as the task of character education usually is not a true priority of the school.[8]

Rob Lescher, the principal of Busch Middle School in the St. Louis Public Schools, graduated from LACE and decided that character education would be the foundation for building a great school in a struggling and unaccredited urban school district. He championed it in all staff meetings, in staff evalu-

Students at Busch Middle School of Character celebrate the school being named a "Missouri School of Character" in 2015.

ations, and in his choices of new initiatives. Slowly, he transformed the staff and the school culture by never taking his leadership foot off the character education pedal. As a humble learner, he asked for advice, professional development, and mentorship. In 2016, Busch was named a National School of Character based on its climate, implementation, and student outcome data.

A second set of prioritization strategies centers on rhetoric. At Evergreen Secondary School in Singapore, for example, teachers are told that they are "character coaches" first and academic teachers second. In Colorado, principal Charles Elbot led Slavens School, a kindergarten through grade eight (K–8) school in Denver, to excellence in character education in part by working with stakeholders to craft a shared language around it—"The High Road"—which served as the organizing frame for character excellence in the school. Slavens began its character education journey in 1997 ranked twenty-seventh for academic achievement of eighty elementary schools in the Denver Public School district. Four years later (in 2001), it was recognized by Character.org as a National School of Character and ranked first in the district.[9]

A third prioritization strategy is the emphasis on professional development for character education. Julie Frugo, head of school at Premier Charter School (a K-8 urban charter in St. Louis), led her school to excellence and national recognition in part by being what she calls a professional growth leader. She created a culture of professional growth, modeled it, and invested in the development of her staff through in-service training, visits by national experts, and an internal process of learning among staff.

A final strategy for prioritizing character education is the allocation of limited resources. Pat McEvoy, the former principal at Maplewood-Richmond Heights High School and currently principal at Bayless High School (both in the St. Louis region), has said that "if I have to buy less footballs to pay for a character education project, I will do so." When he took over Maplewood-Richmond Heights, it was infested with gang activity, violence was common, and 10 percent of junior and senior girls were pregnant. By investing in character education, he drove out the gangs, ended the violence, and reduced the pregnancy rate to zero.

R Is for Relationships

Positive relationships are foundational to good schools, both academically and for character development. However, many educators assume that if the school (or classroom) is a good one, relationships will naturally develop. And they will—for some, but many will be left out. Relationships can be understood as central to the school climate, which the National School Climate Center defines broadly as the "patterns of students', parents', and school personnel's experience of school life." Therefore, character education must permeate the school community and all of its stakeholders. This includes students, teachers, and administrators, but also the wider community of support staff, parents, and community members such as local government, shop owners, and law enforcement.

Those latter relationships likely will not develop unless they are intentionally targeted. Schools with strong character education programs tend to have a systematic way to connect support staff to parents, to connect students to support staff, to connect teachers to community members, and to develop a strong, interlocking set of relationships. Former principal Amy Johnston recognized this at Francis Howell Middle School in Weldon Spring, Missouri, when she said, "If we don't make time in our master schedule for relationships, they won't happen." To create the necessary time, she created a cross-grade, looped advisory program in which students would have sustained and close relationships with at least one teacher and with students in all grades. She also linked support staff to the advisories.[10]

Using peer-interactive educational methods is a core strategy that supports both academic achievement and character development. Cooperative learning is among the most studied of these methods. The Developmental Studies Center (now the Center for the Collaborative Classroom), based in Alameda, California, used cooperative learning to help craft the very successful character education program called the Child Development Project, and offers an excellent guide to cooperative learning in their book *Blueprints for a Collaborative Classroom.*[11]

Karen Smith, the former principal of Mark Twain Elementary in Brentwood, Missouri, building on the work of Fairbrook Elementary in Dayton, Ohio, created a "family program" with weekly meetings of groups of two students from each grade focused on character issues, led by fifth-grade students. At Mark Twain, she asked every adult who worked in the building (herself as principal, her secretary Marie, the cook, etc.) to lead a family, helping to expand the number of adult mentors and role models for students while being inclusive, respectful, and empowering of all adults working in the school. Under her leadership, the achievement gap plummeted and Mark Twain was recognized as both a Blue Ribbon School and a National School of Character.

Students of various ages work collaboratively at Mark Twain Elementary School in Brentwood, Missouri.

Mark Twain Elementary School

I Is for Intrinsic Motivation

Ultimately, the goal of character education is to shape the nature of the child; that is, to help the child become more moral and effective at navigating the world in ways that add to the world—to be *value-added*. What we do not want is people who do the right thing only when others are watching or when there is a payoff for doing so. Far too many schools load on such extrinsic incentives, even to the point of giving money or parking spots to students who achieve good grades or meet other behavioral goals. All too often, schools rely on tangible rewards and public recognition ceremonies as their primary means of promoting character and managing student behavior.

Such extrinsic strategies are fraught with peril and can actually reduce the internalization of the desired values. One elementary school in the Kansas City Public Schools worked for a year to promote an ethic of service in its elementary grade students, only to discover that when there was no tangible reward, the students would not even throw a surprise retirement party for the most beloved member of the staff. Their own conclusion is that "the students never internalized the value of service. They just did it for the rewards."

Character is about motivation along with capacity and knowledge. Doing the right thing because it is a valued good is necessary for character to be complete. One way to foster the internalization of values is through rethinking one's behavior management (discipline) strategies. Developmental discipline and restorative practices are related ways of managing behavior respectfully, collaboratively, and productively; that is, in ways that nurture the long-term positive growth of the child rather than merely trying to stop the specific undesirable behavior immediately.[12]

Another common strategy is for schools to provide opportunities for students to serve others in meaningful ways, either through service activities or by integrating them into the academic curriculum through service learning. When Barbara Lewis taught at Jackson Elementary in Salt Lake City, Utah, her students studied hazardous waste disposal and, when they discovered that a disposal site was within blocks of the school, they worked with the Utah legislature to create a state Superfund site to clean it up. Ron Berger, when teaching in a rural Massachusetts elementary school, led his students to test and clean up local water supplies and create a safe pathway for migrating amphibians, a method that he now helps spread worldwide through Expeditionary Learning Schools.[13]

M Is for Modeling

When people are asked to reflect on their own character and what led to it, they almost always invoke significant role models, mostly their parents. Yet schools rarely strategically leverage this key element of effective character education. When they do, it is mostly by stating that the adults in the school are or should be positive role models, yet the schools rarely have a strategy for making this happen. One way to implement this is to make character modeling an explicit part of the adult culture and an intentional point of discussion among the adults in the school. Staff at Amy Johnston's school, Francis Howell Middle, would say one's "character crown is slipping" as a way to address when adults failed to be models of the character they wanted to see in their students. Adults would say it about themselves or about their coworkers as a reminder of the importance of modeling character.

Another strategy is to focus on the academic study of role models, as is done in the Giraffe Project. Giraffe, a K–12 curriculum, has three main parts. First, the implementers collect and create biographies of Giraffes (people who stick their necks out for others) around the world, and students study them. Second, students search out Giraffes in their own communities, and study and honor them. Lastly, students become Giraffes by finding ways to serve others.[14]

E Is for Empowerment

Ideally, schools should respect each member of the educational community through what can be called a "pedagogy of empowerment." Yet schools all around the world tend to be authoritarian and hierarchical, which is ironic in democratic societies that need to socialize each generation to responsibly shepherd the democratic societies that are bequeathed to them. This is one place where the overlap between areas of character is most apparent—for example, the overlap between moral and civic character. The best ways to instill empowerment relate to sharing power and democratizing school systems. The Center for the Collaborative Classroom's Caring School Community program highlights the regular use of class meetings to empower students to make decisions, plan events, and solve problems. This program has been shown to promote character and a positive school climate and has been lauded by the U.S. Department of Education for both drug and violence prevention.[15]

Another empowerment strategy is for principals to share leadership with others, such as having student advisory committees or transforming faculty meetings into class meeting formats. When Tim Crutchley left Ridgewood Middle School to work in the district office, Kristen Pelster was promoted to principal. When students came to her to suggest that there was a cheating problem (even though neither she nor the teachers thought there was one), she told the students, "I don't see a cheating problem, and I asked the teachers and they don't either. But I believe you. So what are you going to do about it?" The students took the challenge and created an academic integrity program. None of the teachers nor students knew that when academic integrity programs are student-owned and led, they are more effective; they were just being true to a pedagogy of empowerment.

And D Is for Developmental Perspective

Students, and their education, can be seen through many lenses. Having a developmental perspective leads to choosing strategies that are designed specifically to support the long-term growth of students. A simple, common, and straightforward strategy is to teach about character—to promote the study of character. The Jubilee Centre in Birmingham, in the United Kingdom, has offered a set of British core virtues that they call the Knightly Virtues, including humility, honesty, justice, service, and courage. This project uses literature to teach the virtues, and it leads to better understanding of the virtues and to improved behavior.[16]

Another strategy, and one that strongly aligns with the research on parenting, is to set high expectations (both for academic achievement and character development), but with ample supports (scaffolding) to provide the resources needed to have a chance at meeting those expectations. This was central to what Ron Berger called "educating for an ethic of excellence." One strategy was to teach students to do multiple drafts of work and to provide each other with constructive critical feedback. Another strategy is to teach and enact personal goal-setting around character. The Inspiring Purpose program in Scotland, which builds on the John Templeton Foundation's Laws of Life Essay Contest, asks students to choose a virtue of importance to them, to study it and exemplars of it, and to make it a personal priority by creating a graphic portfolio about it and their personal journey.[17]

Inspiring Purpose/ www.inspiringpurpose.org.uk

Students at Ikusasalethu High School in South Africa display their Inspiring Purpose posters.

Where to Begin?

As ever, initiatives rely on leadership, priorities, and resources. If one authentically wants to use character education to foster the positive development of the next generation in order to make a better world, then the priority is there. It is critical that key leaders share that priority. In Singapore, for example, the past and current Ministers of Education (as well as the Prime Minister) strongly supported character education as critical to national flourishing, never missing a chance to publicly proclaim it and providing impressive resources, including a significant branch of the Ministry of Education dedicated exclusively to character and citizenship.

Finding resources is not difficult; however, discerning those that are more effective and evidence-based is. For those who want to adopt an existing program, U.S. national organizations such as Character.org (formerly the Character Education Partnership), the National School Climate Center, Expeditionary Learning Schools, and the Collaborative for Academic, Social, and Emotional Learning (CASEL) all have evidence-based frameworks for effective implementation. At a national level, it is wise to invest in dedicated staff and resources and to rely on expertise, although this unfortunately remains limited, and on research. It also is wise to balance a culturally specific approach with avoiding reinventing the wheel.[18]

Ultimately, if we want to support the development of character in students, we need to know and implement research-supported strategies. Those strategies cluster under the six categories of the "PRIMED for Character" model: prioritizing character education as a central purpose of the school; being strategic and intentional about nurturing healthy relationships among all stakeholders; using practices that lead to the internalization of values and intrinsic motivation to do good in the world; modeling the character we want to see in students; sharing power through a pedagogy of empowerment; and strategically creating the conditions that lead to positive development, especially over the long term.[19]

A more sustainable, just, and compassionate world will only happen if there are more people able and motivated to steer the world in that direction. This is precisely the definition of character: "characteristics that motivate and enable one to act as a competent moral agent." The students at Ridgewood Middle School were effectively written off until Diana Bourisaw, as the new superintendent, and Tim Crutchley and Kristen Pelster, as the new school leaders, brought in character education to allow those students to see that they were valued and of value. When their voices were liberated and their beautiful minds were allowed to help craft a great school, they could reach their potential as learners, as people, and as shepherds for the future of their communities and the world.

And it was sustainable. Nearly two decades after starting the character education journey with abysmal academic outcomes and student behavior, through three very different principals, Ridgewood, still with the same under-resourced student population, was ranked thirteenth out of four hundred and thirty-four middle schools in Missouri in state academic achievement scores.

Character education ultimately is a process of motivating and equipping all children to more strongly repair the world. But people of character do not grow on trees and cannot be manufactured in an assembly line. They must be slowly and carefully nurtured, and it does indeed take a village to raise a child of character. Of course, parenting is at the center of that village, but schools are critical, too.

CHAPTER 8

Social and Emotional Learning for a Challenging Century

Pamela Barker and Amy McConnell Franklin

What does it take to ensure that students are present to learn? What does it take for them to care enough about the world and each other to commit to collective efforts to address one of the major challenges of our time: climate change? It is no longer enough to simply teach reading, writing, and arithmetic in a one-size-fits-all approach that is useful for finding jobs in industrial societies. In addition to these cognitive skills, our children must learn the necessary social and emotional skills to thrive in an information-rich and climate shifting society where knowledge, innovation, collaboration, and adaptation are key. Considering the looming threat of climate change, there is no time to waste.

Daniel, a young resident of the city of New Orleans, faced immense social and emotional burdens in coping with the devastating impacts of Hurricane Katrina, which struck the U.S. Gulf Coast in late August 2005. For the next five years, he and his family experienced twelve post-disaster moves around Louisiana and other U.S. states. Each move required a change of schools, leaving friends and supportive relationships, forging new relationships with classmates and teachers, and navigating new environments. Daniel faced all of this with limited social and financial resources. At new schools, he was bullied. At home, he took care of his baby sister and helped his mother manage the post-disaster bureaucracy so that the family could obtain "adequate" food and shelter.[1]

Hurricane Katrina and its fallout placed extraordinary demands on Daniel's awareness, in particular the skills that are required to rapidly read emotional

Pamela Barker is an educational consultant in social and emotional learning and restorative practices in Fort Collins, Colorado. **Amy McConnell Franklin** is the director of social and emotional learning and mindfulness at United World College Thailand.

data (his own and that of others) to stay safe. It demanded his quick response to cultural norms, as well as having to read cues from others and to regulate his own behaviors in order to stay focused while navigating the chaos of life at school and at home. It demanded his clear intention: to have goals for himself and his family as well as seemingly endless patience and empathy. It demanded that Daniel make a choice: to stay optimistic in the face of constant and complex challenges, and to remain motivated even in the face of uncertainty and hopelessness.[2]

Like Daniel, today's youth will face acute and chronic pressures due to climate change: from extreme weather events and public health crises to generalized increased stress, anxiety, depression, trauma, social network disruption, and growing conflict over limited resources. As a changing climate places immense social and emotional demands on youth, educators must rally to support them. Yet most educational systems are not preparing either students or teachers with the necessary aptitudes that they will need to act on behalf of themselves and others and to engage effectively with the circumstances. It is not surprising that many react to climate change with apathy, pessimism, or paralysis.

Social and emotional learning strategies can play a significant role in transforming these maladaptive reactions into adaptive skills. It is essential that efforts to prepare youth for the political, social, economic, and emotional turbulence of a warming planet be systematically woven into all aspects of their education (from pre-kindergarten to university), at home, and in their everyday experiences. Incorporating systematic social and emotional learning programming can help to prepare students for an uncertain future.

What Is Social and Emotional Learning?

Emotional intelligence, a building block of social and emotional competence, allows a person to use emotion to enhance reasoning and decision making. Social and emotional learning is a process that promotes cognitive, affective, and behavioral education. As people learn to connect with one another and with the natural world in respectful ways, they more often choose to align their thoughts, feelings, and actions in caring and compassionate ways. This increases their ability to take perspective and motivates them to navigate complexity while accounting for the internal experiences of self and others and for the innate legitimacy of the natural world.

According to the Collaborative for Academic, Social, and Emotional Learning (CASEL)—an international leader in research, practice, and policy in this field—social and emotional learning is "the process through which children and adults acquire and effectively apply the knowledge, attitudes, and skills necessary to understand and manage emotions, set and achieve positive goals, feel and show empathy for others, establish and maintain positive relationships, and make responsible decisions." CASEL's model includes the social and emotional competencies of self-awareness, social awareness, relationship skills, decision-making skills, and self-management. Many U.S.-based and international programs that teach these competencies employ other, related, terms such as global citizenship; ethics, peace, and civic education; and emotional intelligence.[3]

A robust and growing body of research indicates that social and emotional learning is a set of skills that is teachable and learnable through explicit instruction and that become more permanent with practice. A 2011 meta-analysis of two hundred and thirteen school-based studies involving more than two hundred and seventy thousand students found that participants in social and emotional learning showed improvements in social and emotional skills, attitudes, and behavior; academic performance; and student-teacher relationships. They also exhibited fewer conduct problems. Long-term outcomes included increased high school graduation rates, college and career readiness, better relationships, engaged citizenship, and reduced criminality, substance abuse, and mental illness. Effective implementation of social and emotional learning also enhances school readiness, school climate and learning, empathy, compassion, civic responsibility, mental and physical well-being, and resilience.[4]

Research indicates that training in social and emotional skills generally has a greater impact than training in cognitive skills on improving social behavior and labor market outcomes. The economic case is strong: a 2015 review of six prominent social and emotional learning programs (five in the United States and one in Sweden) found that the economic gains realized through savings related to social and emotional learning outcomes—such as increased high school completion and career readiness and decreased substance abuse and criminality—can outweigh the costs substantially. The interaction of both cognitive and social/emotional skills can further improve children's life outcomes. These outcomes represent what is most critical for youth to engage proactively with a climate-shifted future and a rapidly changing world. (See Box 8–1.)[5]

Integrating Social and Emotional Learning in Education

Students who feel safe and included remain engaged, are comfortable taking risks and putting forth effort when challenged, are able to navigate their

Box 8–1. Social and Emotional Learning and Climate Change

Research on climate change indicates that the two key strategies to address the effects of climate change—mitigation and adaptation—require strong social and emotional skills. Thomas Doherty and Susan Clayton found that adaptive behavior is optimal when people have high emotional literacy and engagement and can solve problems creatively and collaboratively. Janet Swim and colleagues found that individual factors of behavior, belief, and emotional responses influence whether or not humans cope well with climate change and if they will behave in ways that mitigate it. Tom Crompton and Tim Kasser note that three social and emotional skills—empathy, intrinsic motivation, and aware-ness of interconnectedness with others and with nature—are key attributes that are correlated with pro-environmental choices, which is particularly relevant to mitigating climate change.

Schools, with their broad reach, offer ideal environments in which to build these capac-ities, and teacher training can provide access to effective content and teaching practices. Developments in the fields of emotional intelligence and social and emotional learning can aid educators in integrating social and emotional skills into schools, helping children to navigate the emotional challenges that come with climate change.

Source: See endnote 5.

emotions to stay focused and motivated, and recognize when they are on or off task. They know when they are contributing to the group and when they are distracting, are able to recognize the causes and consequences of their feelings and moods, and are able to govern their own emotions appropriately. They are ready and able to learn and to work ethically and effectively.

Katie Parsanko-Malone, a high school science teacher in Fort Collins, Col-orado, was frustrated with her students' low engagement. She knew it was not so much rooted in what or how she taught, but in the disruptions of their insecure home lives and in their lack of skills and support to cope adequately. Many of her students experienced ongoing trauma and unusual responsibili-ties, were poor, and faced housing, employment, and food insecurity—similar characteristics to populations around the world that are most vulnerable to climate change. Understandably, these students were not available to learn.[6]

To restore her passion for teaching and her students' passion for learning, Parsanko-Malone intentionally integrated social and emotional learning with the instructional content that she taught in the classroom. This provided the tipping point in her class, pushing her students toward both engagement and real learning. She explained in an interview:

Once I began to see their trauma as the primary barrier to their academic success, I saw models of resilience throughout the environmental systems curriculum that I teach. From succession and evolution to symbiotic relationships, parallels quickly began to emerge. Ecology was a natural link to bridge their internal, emotional interactions with those of small and large communities. As we examined factors that influenced and changed climate on a global scale, students were asked to examine the factors that influence our classroom climate. As we recognized the consequences of a change to our classroom climate on individuals and communities, we were building the awareness and empathy that allows us to think critically as members of a global community.[7]

Parsanko-Malone's approach helps her students engage with learning and develop skills to generate solutions, rather than flounder in fear and anxiety or ignorance and denial. They are learning the science of their own emotions, as well as the social science of their communities. She says they are able to see relationships that they had not recognized before and are building skills that may increase their willingness and ability to engage in activities and habits that positively affect both society and the planet. This combination allows them to see themselves not as separate from the natural world, which leads to apathy and disregard, but as connected, with an enhanced sense of interdependence.[8]

Taking Social and Emotional Learning to Scale

Increased efforts to systematically integrate social and emotional learning are needed. However, the challenges of making this learning mainstream across the global educational landscape are significant. Insufficiently trained teachers often receive inadequate systemic and strategic support in this area from administrators, and both may struggle because they have not yet developed social and emotional competence themselves. Similar to other mandated or recommended initiatives, social and emotional learning is often implemented with good will yet poor understanding and commitment on the part of the educational community. Furthermore, and despite evidence to the contrary, such learning is perceived to be cost prohibitive and to take precious time away from teaching academic content and preparing for high-stakes testing.

Overcoming these challenges requires attention to essential stages of scaling programs through proper dissemination, adoption, implementation, evaluation, and support for sustainability. Figure 8–1 provides a framework

Figure 8–1. Components of Systematic Programming in Social and Emotional Learning

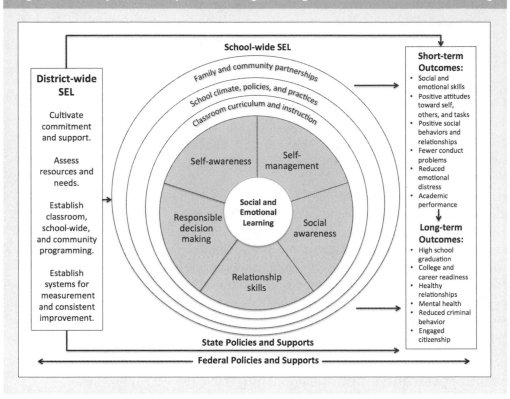

highlighting the components of effective implementation of social and emotional learning in the United States, as well as summarizing student outcomes. Collaboration by the architects of program implementation at each stage builds relationships and trust, thereby creating engagement and influencing adoption with accuracy in other communities.[9]

CASEL has established the "SELect" designation to rate evidence-based social and emotional programs and shares best practice guidelines for district and school teams on how to select and implement programs. To be designated as SELect, a program must promote students' social and emotional competence by providing opportunities for practice, as well as multiyear programming; deliver high-quality training and support throughout implementation to ensure success; and be supported by at least one well-designed study documenting positive impacts on student behavior and/or academic performance.

CASEL helps organizations choose SELect programs based on three key principles: 1) the school and district teams—not individuals—lead the selection of the program, engaging diverse stakeholders; 2) ongoing school and district planning, programming, and evaluation leads to more effective implementation; and 3) consideration of local contextual factors, including student characteristics and existing programs, is essential. These criteria align with best practices in implementation science for human service programs.[10]

RULER, a SELect program developed by researchers at the Yale Center for Emotional Intelligence, meets these criteria. It is grounded in emotional intelligence and ecological systems theories. RULER is an acronym for the underlying framework for the knowledge and skills of Recognizing, Understanding, Labeling, Expressing, and Regulating emotions—skills of social and emotional learning. The implementation goal for RULER is to foster emotional intelligence in children and adults by forming a learning community that values safe emotional expression and that infuses academic learning with social and emotional learning, while creating multiple vectors for spreading social and emotional competence. RULER has been adopted in eight large U.S. school districts, including Bridgeport Public Schools in Connecticut, Colorado Springs Academy District 20 in Colorado, and Seattle Public Schools District in Washington.[11]

In the first year of program implementation, RULER's four sequential "anchors"—the *Charter, Mood Meter, Meta-Moment,* and *Blueprint*—are taught, practiced, and integrated by all stakeholders, including classroom teachers, school staff (custodial, administrative, transportation), students, and families. The *Charter* is an agreement among stakeholders describing how they want to feel when working together, specific behaviors required to create and sustain those feelings, and strategies for handling conflict and undesired feelings. The *Mood Meter* is a self-awareness tool for students to develop emotional literacy—including a nuanced emotional vocabulary and a way of navigating through feelings—and for setting goals and strategizing around emotions. Educators also can use it to guide, maintain, or modify best mood states for learning tasks. The *Meta-Moment* is a self-regulation tool that students, parents, and educators can use to recognize common patterns, to choose to act consciously versus react, and to strategize a "best self." Finally, the *Blueprint* is a problem-solving and empathy-building tool for emotionally laden past, present, and future challenges. It asks a student, "What happened," "How did I (other person) feel," "What caused me (other person) to feel this way," and "What could I do differently next time?" In the second year of RULER program implementation, the *Feeling Words Curriculum* is introduced and provides integration of emotional vocabulary into existing lessons.[12]

Another avenue to foster social and emotional learning is mindfulness, an ancient practice for cultivating attention skills and emotional balance. Educational research reveals that mindfulness awareness practices promote social and emotional competence and also can reduce stress. Mindfulness is relatively easy to implement in the classroom and is widely accepted by students, administrators, and teachers. A study of mindfulness in thirty Chicago public schools is following two thousand kindergarten through second grade students—primarily minority students from low-income families—over four years to determine if mindfulness training will help reduce the achievement gap of children living in poverty. The students participate in ten to twelve minutes of daily activity focused on breathing, physical sensations, or emotions, which has been shown to help control hyperarousal so that students can be more available to learn and to choose adaptive responses. Teachers in the study report increased instruction time gained through fewer behavioral issues, greater focus, and quicker refocus and recovery after transitions and upsets. Mindfulness also has proven to be a useful teaching tool at United World College Thailand. (See Box 8–2.)[13]

Regardless of efforts to prevent it, conflict in schools happens. An increasingly popular social and emotional learning approach to addressing conflict in educational environments is "restorative practices," a model that brings together victims and offenders to explore the conflict: what harm was caused, who is responsible, and how it can be repaired. Notably, restorative practices are strength, not deficit, focused. They facilitate the development of social and emotional skills of empathy, inclusivity, relationship building, and problem solving. Many schools in Australia, Canada, the United Kingdom, and the United States have successfully incorporated restorative practices to reduce suspension and expulsion. These schools have reduced bullying, violence, and teacher stress sharply and have increased academic learning and parental involvement.[14]

Restorative practices turn conflict into a generative experience that keeps kids in school who otherwise may drop out. Restorative practices also are used simply to build connection through "connection circles." Eric Rasmussen, a science teacher at Silver Creek High School in Longmont, Colorado, uses connection circles to build empathy and respect among his students and to facilitate discussion of academic content. For example, in his Earth science class, before starting a unit on climate change, he asks: "If someone had a different viewpoint than you, what is a strategy you could use to work with them?"[15]

Box 8–2. Well-being Studies in Thailand

The taproot of United World College Thailand lies deep in the practice of mindfulness and social and emotional learning. Founded in 2006 as the Phuket International Academy, the school community joined the United World College (UWC) movement in 2016. It is a kindergarten through grade twelve (K–12) school with students representing more than eighty-six countries. The mission of the UWC movement is "to make education a force to unite people, nations, and cultures for peace and a sustainable future."

Recognizing that how we treat one another is as important as how well we read or do math, all teachers, administrators, and many parents are trained and mentored in social and emotional concepts and skills and mindful meditation practices. All K–12 students attend "Wellbeing," a curriculum that allocates three hours and twenty minutes per week to cultivate intrinsic motivation, empathy, and a sense of interdependence and interconnectedness. Wellbeing includes:

- 20 minutes per day for Time In: mindful movement, a stillness practice, and reflection
- 50 minutes per week for Council Meetings: a community-building practice of authentic expression and empathic, non-judgmental listening and receptivity
- 50 minutes per week for direct instruction on developmentally appropriate components of personal and social well-being, including personal and cultural identity, academic honesty, emotional and social intelligence skills, responsible engagement with social media, sex education, cultural competency, and health and substance use.

The educational goal of UWC Thailand is "to realize our highest human potential, cultivate genuine happiness, and take mindful and compassionate action." This develops more self-aware and socially aware citizens who have a well-developed sense of self-efficacy and are able to more often choose to align their actions with a conscious and compassionate intention for the greater good. Developing these social and emotional capacities in the school's stakeholders nourishes the UWC mission.

Source: See endnote 13.

Barrowford Primary School in the United Kingdom uses restorative practices to build relationships and increase learning among its students. With this approach, the school's educators strive to promote respect, responsibility, repair, and reintegration. They address conflict by encouraging students to constructively engage: listening to each other's perspectives, being accountable to one's actions, and offering amends for harm done. Through this, they have developed a vital learning community.[16]

Ideally, school-based efforts disseminate social and emotional learning skills and understanding community-wide through extracurricular programming, parent education, and collaboration with community partners. In Loveland, Colorado, instructors with the Changing Leads program at Hearts and Horses Therapeutic Riding Center help struggling students in the community achieve greater balance in their lives through carefully designed activities aimed at developing their social and emotional skills in the experiential setting of the riding arena. (See Box 8–3).[17]

Prioritizing Social and Emotional Learning

To prioritize the development of those qualities that are needed to navigate turbulence—such as resilience, emotional regulation, collaborative capacity, civic-mindedness, empathy, and mental well-being—CASEL recommends freestanding educational standards for social and emotional learning in the United States. All fifty U.S. states have social and emotional learning standards for preschool, and four states have developed comprehensive standards for K–12. The large-scale Collaborative Districts Initiative aims to expand and improve programming in social and emotional learning by providing school districts with ongoing support that is critical to overcoming challenges. This support includes funding for planning, implementation, and monitoring to establish coordinated and sustainable, evidence-based programming and to achieve district outcome targets for social and emotional learning. To date, the Initiative supports eight of the two hundred largest school districts in the United States and is a model for best practices in social and emotional learning. External evaluations of participating districts have shown consistent improvements in school culture and climate, as well as student outcomes.[18]

Many international governments are leading efforts toward large-scale programming in social and emotional learning. In 2015, the Organisation for Economic Co-operation and Development noted that twenty-six of its thirty-four member countries at the time had stated social and emotional skill objectives. Many international programs align with the CASEL framework and are focused on individual as well as global citizenship to address twenty-first century challenges. Others are informed by cultural ideals or regional philosophy through stories of reciprocity and kinship, humility, morality, generosity, altruism, and interdependence, indicating essential ways of acting to promote living peacefully and supporting others in community. International collaboration among researchers, policy makers, and educators is critical in

Box 8–3. Changing Leads: Social and Emotional Growth Through Equine-Facilitated Learning

Tamara Merritt, special programs director at Hearts and Horses Therapeutic Riding Center in Loveland, Colorado, and educational psychologist Pamela Barker teach middle-school students who struggle socially and academically in school and in life. Their program, aptly named Changing Leads, draws from the idea that a horse is in balance when it is on the correct lead. Often, humans need to "change their lead" in order to balance the weight of stressors in their lives. Horses and humans both strive for balance, and Changing Leads can help students achieve this balance through systematic development of social and emotional skills in an experiential setting

As innately social creatures, horses mirror what humans are feeling and can facilitate powerful personal exploration and transformation of behavior. When students enter the natural, mostly silent world of the horse, their verbal and non-verbal behavior is reflected back to them. They experience a "felt sense" of connection, trust, empathy, confidence, accountability, safety, and peace. This creates the ideal environment for social and emotional skill building.

Each day begins with an hour participating in the "horse hook," a classroom curriculum designed to enhance students' subsequent interactions with the horses by front-loading the social and emotional skills that they will practice in the arena. Students interact with each other and with their adult equine-learning volunteers, sharing stories and activities to build social and emotional skills. Students then participate in ground and mounted work in the arena for two hours, doing activities that call on and strengthen those skills through interaction with their equine partners. For example, students learn to ride "boot to boot" (side-by-side, keeping their horses' bodies and their own in perfect parallel) and learn to regulate their horses' movement and their own bodies. Students also learn to calm their horses and themselves through centered posture, focused intention, and presence, and to trust their classmates to lead them blindfolded through an obstacle course representing a chosen challenge (for example, graduating from high school or avoiding school drama) while leading their horses.

Teachers who accompany the students to the ranch learn alongside the students, so that they, too, gain the skills and can incorporate the lessons and language learned into the school and classroom environment. This adult participation adds power to the transference of students' skills to school.

Changing Leads's outcomes include increased school attendance, school engagement, increased self-confidence, self-awareness, improved trust, greater emotional vocabulary, and heightened empathy among students. Participants also show significant increases in the life-outcomes measurements of self-efficacy, personal achievement, and relationship quality.

Source: See endnote 17.

developing a shared framework (including terminology and concepts) to promote global mainstreaming of social and emotional learning.[19]

Universal Social and Emotional Learning

A century ago, the idea that all children around the globe had the right and ability to learn to read and write was deemed naive. The average literacy rate globally is now nearly 90 percent. While social and emotional skills historically were fostered through a variety of venues, modernity requires that this responsibility fall on formal school systems. Due to the diligence of educators, legislators, and academics, the necessity of all children developing social and emotional skills through social and emotional learning is an attainable goal. No time should be wasted in making it happen. There is strong evidence that social and emotional learning will contribute to the capacity for creative and compassionate problem solving and to the development of the mutually responsible relationships that are required for mitigating and adapting to the effects of climate change and other sustainability challenges. As resources become scarce, shelter more insecure, inequality more widespread, and our global community more at risk, these skills and our empathy for each other will be tested.

Recalling Daniel in the wake of Hurricane Katrina, as an eighteen-year-old high school sophomore with aspirations for college but insufficient grades and limited resources to get him there, he did his best to adapt yet still faced inordinate social and emotional burdens from the storm. If the schools that he moved in and out of had been prepared with social and emotional skills training and had operated within a culture of social and emotional competence, perhaps they could have better supported Daniel and others facing similar upheaval. As it stood, far too many youth like Daniel faced these challenges without the support of school-based social and emotional training, with many suffering enduring wounds in the wake of the storm.[20]

There are thousands of children like Daniel who must now, or in the future, adapt to devastating changes that a warming planet has created. Simply leaving the development of social and emotional skills to chance versus deliberately teaching them to all children—particularly those living in communities on the front lines who are most vulnerable and those who may emerge as leaders and policy makers—puts all of our futures at risk. The ripple effect of teaching social and emotional learning universally can create a surge of compassionate, inclusive action that will build the individual and collective resilience to thrive in turbulent times.

CHAPTER 9

Prioritizing Play

David Whitebread

In an August 2016 article in *The Atlantic*, contributing writer Tim Walker describes his family's experience visiting one of the few remaining "adventure playgrounds" in New York City, located on a small urban lot on Governor's Island. "It looks like a dumpster playground . . . like some slum," his wife observed about the site, which had opened a few months earlier and is modeled after a junkyard. Walker goes on to describe "an area that looked like—to my eyes, at least—an Occupy Wall Street campground, with shoddy constructions of plywood, wooden pallets, and blue tarp . . . littered with car tires, plastic crates, orange cones, and a sea of unidentifiable debris."[1]

Walker, an American teacher based in Finland, and his wife and two preschool-aged children had been visiting New York for the summer and were keen to find out what was happening at the playground, billed as a place where kids can explore and tinker using real tools, such as hammers and handsaws, and where parents are not supposed to intervene. The Walkers had come from a country where children still experience a childhood full of adventurous outdoor play, and their experience in a large American city, with dramatically fewer such opportunities, captures, in microcosm, what is happening to children all over the world.[2]

This concern may seem trivial. Children's play is often regarded as something that they enjoy because they are not yet mature, and something that they will grow out of as they become serious, responsible adults. The demise of play may be regarded as unfortunate, but not something with any serious consequences. However, there is considerable evidence to the contrary. In recent

David Whitebread is the director of the Play in Education, Development and Learning (PEDAL) Research Centre at the University of Cambridge.

years, an increasingly rich and rigorous body of research has demonstrated that play is a fundamental human characteristic that supports our unique qualities as creative problem solvers, as innovators, and as a highly adaptive species. Arguably, our culture, our science, and our technological achievements all arise, at least in part, from our playfulness. By limiting the playful opportunities that are available to our children in the modern world, there is a grave risk that we are reducing the development of the very skills and capabilities that they will need the most in order to confront the challenges of the twenty-first century and beyond.

The threats and barriers to nurturing children's playfulness, both in their homes and in their communities, are many. One culprit is the many hours that young children now spend in school focused on academic, rather than play-based, learning. Recent psychological evidence is beginning to reveal the mechanisms through which play supports the development of a person's intellectual abilities and emotional well-being. Studies show that the erosion of play has consequences for our children's mental health and for the development of their abilities as learners. Fortunately, there are many ways in which children's play can be supported, resulting in beneficial outcomes for both the individual and society.

Where Has Play Gone?

Over the last two generations, there has been a dramatic cultural change—in the United Kingdom, the United States, and many other urbanized and technologically advanced countries—in the life experiences of our children. The generation of "baby boomers" born just after World War II experienced a childhood in which they played out in the street or in local fields and parks, in groups of children often of very mixed ages. This relative freedom and experience of free play and "unstructured" time has largely disappeared for many children across the world. As documented in the report *The Importance of Play*, around half of all children worldwide now live in urban settings, where they experience greatly reduced opportunities for free play—particularly free play outdoors and in natural environments. Parental concerns about traffic, "stranger danger," germs and disease, and so on, have resulted in children's home lives that are heavily structured and supervised to an extent that would have been regarded as bizarre only a couple of generations ago.[3]

Alongside this loss of outdoor free play opportunities at home and in communities, there is increasing concern about children's dwindling time for free and creative play while they are in school. In the name of "raising standards" and "increased rigor," young children in nursery and primary schools in the United Kingdom, the United States, and many other countries are being subjected to a curriculum that emphasizes learning facts and basic skills at the expense of developing the higher-order capabilities and dispositions that support children in becoming powerful, confident, and enthusiastic learners. This is driven by a view among politicians and policy makers that the aim of education is to give their country an economic edge, which can best be measured by scores on standardized tests that cover a narrow range of literacy and numerical skills.[4]

Many commentators have rightly expressed concern that, at the start of the twenty-first century—when, more than ever, we need to be educating our children to be creative problem solvers, innovators, and confident, adaptable learners—this is a highly unhelpful situation. Among the many other concerns is the loss of teaching approaches and learning opportunities grounded in activities

Sand play at a Montessori kindergarten in Shah Alam, Malaysia.

that are playful. In the United Kingdom, opportunities for young children to learn through playful activities are now almost or completely absent from Year 1 (when the children are five-to-six years of age) onward in many primary schools, and these opportunities are under increasing threat in classes for four- to five-year-olds and even preschoolers. This damaging trend in educational provision, which Finnish educationalist Pasi Sahlberg has termed the "GERM" (Global Educational Reform Movement), has infected educational systems across the world and manifests itself in a range of ways. The symptoms of this disease of "marketizing" education, Sahlberg argues, include providing parents with a choice of schools (thus positioning them as consumers in relation to the educational system), increasingly high-stakes standardized testing, and draconian school accountability systems.[5]

One particularly damaging idea, as illustrated in the United Kingdom example, is that "earlier is better." The policy makers and politicians who espouse this view assert that the lack of what is now commonly referred to as "school readiness" in children from relatively disadvantaged home backgrounds can be remedied through the earlier introduction of formal schooling. This has resulted in moves in numerous countries across the world to lower the age at which children are required to start school. Over the last two to three decades, even in countries where the official school starting age has remained constant, a rapidly growing percentage of young children is attending increasingly formal preschool options.[6]

Play and Mental Health

For those of us who work with young children and conduct research about their development, one of the most disturbing trends as we enter the twenty-first century is the sudden and rapid increase in mental health problems for this age group. Recent figures collated by the U.K. National Health Service's statistical wing, NHS Digital, are deeply worrying. According to data covering 60 percent of mental health trusts in England, 235,189 young people under the age of eighteen are currently in need of specialist mental health care. While the teenage or adolescent years have long been recognized as an emotionally challenging period, shockingly, these latest figures include 11,849 children under the age of five and 53,659 children between the ages of six and ten. Among the conditions suffered by these children are increases in clinical-level anxiety, depression, and self-harming. Childline, a U.K. charity that provides counseling support to children in difficulties via telephone, email, and social media, reported 19,481 contacts in 2015–16 from young people who were considering suicide, twice the figure recorded just five years previously.[7]

Of course, a range of possible economic, social, and cultural trends may be underlying this disturbing picture of increased stress among children living in the modern world, including those living in relatively affluent, technologically advanced countries such as the United Kingdom and the United States. Many commentators and researchers in this area, however, have suggested that the loss of unstructured, unsupervised opportunities for children to play freely with their friends and families and to learn playfully in school is likely a significant element in this deteriorating situation.

Numerous studies by anthropologists and developmental psychologists have linked the loss of free play opportunities to the rise in children's mental health problems and to difficulties that arise when they start their schooling. American anthropologist Peter Gray has documented the sharp decline in free play over the past half century in the United States and other developed nations and has explicitly linked it to a parallel increase in anxiety, depression, suicide, feelings of helplessness, and narcissism in children, adolescents, and young adults. Based on these compellingly matched trends, Gray mounts a strong argument that the decline in play has contributed to the rise in the psychopathology of young people. He indicates, in line with evidence from developmental psychology, that play experiences provide important opportunities for children to develop so-called self-regulation capabilities, including abilities related to making decisions, solving problems, exerting self-control, following rules, and regulating their emotions. Through all of these effects, Gray argues, play promotes mental health.[8]

Numerous studies in developmental psychology have demonstrated a strong relationship between playfulness—particularly in relation to pretend or imaginative play—and emotional well-being in children. Other studies have demonstrated the circumstances in which children in "maltreating" families display less socially competent behaviors and less self-initiated play compared to children in typical, emotionally supportive households. Significant links have also been demonstrated between high rates of stress in mothers, symptoms of attention-deficit/hyperactivity disorder (ADHD) in their zero- to four-year-old children, and a disinterest in play being shown by these children. A range of studies has indicated the positive impact of play therapy with children in these kinds of emotionally deprived environments.[9]

Play and Learning

The movement across Europe, and in many countries worldwide, to lower the school starting age might be influenced to some extent by increasing acknowledgement of the fact that high-quality early education can have a positive impact. There also is clear evidence of strong economic drivers behind this trend, such as the push to get women back to work as soon as possible after childbirth. The huge irony, however, is that the move toward earlier schooling, rather than incorporating the play-based approaches that are fundamental to the achievement of high quality, has shifted in the direction of more-formal instruction. This flies in the face of a considerable body of international evidence, all of which points in the opposite direction.

So far, there is no serious evidence that putting children into formal schooling earlier improves the quality of their education or their achievements as learners. At the same time, two significant strands of evidence indicate that a longer period of informal playful learning—and a later start to formal education—is advantageous. This evidence includes: 1) studies relating to the age at which children start formal schooling, and the consequences of this for the length of time that children spend in informal preschool, play-based educational settings; and 2) evidence from a wide range of disciplines relating to the benefits of play experiences for children's development as learners.

Three exemplar studies illustrate the findings of the first strand of research, looking directly at the consequences of an earlier versus delayed transition to formal schooling. First, a longitudinal study of three thousand children, funded by the U.K. Department of Education, showed that high-quality, play-based preschool education made a significant difference to academic learning and well-being through the primary school years, and that an extended period (three years) of such schooling was particularly beneficial for children from disadvantaged households. Second, studies in New Zealand have demonstrated that the early introduction of formal learning approaches to literacy does not improve children's reading development in the long run and may even be damaging. Finally, a more recent international study that re-analyzed data from the reading portion of the 2006 Programme for International Student Assessment (PISA)—a standardized testing protocol used in many countries to measure "core competencies" in mathematics, science, and reading—found no significant association between reading achievement and the age of school entry.[10]

Seattle Parks

Play at a Seattle Parks and Recreation preschool program.

With regard to the second strand of evidence—showing the advantages of play experiences in early development—studies in the areas of evolutionary anthropology, neuroscience, psychology, and education have investigated children's play in both educational contexts and in domestic areas of their life. One U.S. study explored individual differences in naturally occurring "object play" (play involving the exploration and manipulation of objects in ways that use curiosity) among young children in kindergartens and found that children who engaged most frequently in this type of play over a school year demonstrated increased problem-solving abilities. A pair of related studies of slightly older children in the United Kingdom revealed a similar effect in the area of writing: when supported by opportunities to engage in pretend or imaginary play, children's writing showed enhancements in both technical aspects and in the quality of the ideas developed.[11]

Within the domestic context, studies on both sides of the Atlantic have indicated the importance of unstructured free play and of play between children and their parents. In the United States, a longitudinal study from 2014 showed that the more unstructured time that children had while out of school—including both unsupervised indoor and outdoor free play and family excursions to the seaside, museums, etc.—the better their skills were on undertaking a learning task in school. In a 2015 study in the United Kingdom, the strongest predictor of "school readiness" and of language and cognitive development among children at the point of starting school was an item completed by their teachers indicating that the child "talks about fun activities at home."[12]

How Does Play Support Learning and Development?

Work in developmental psychology is beginning to identify the two crucial processes that underpin the relationship between play and learning. First, a number of types of play—most significantly pretend or imaginary play—involve children in increasingly sophisticated processes of symbolic representation, which are fundamental to language, literacy, numeracy, artistic expression, visual media, and the many other ways in which humans represent meaning. In symbolic play—such as when a child makes up a silly word or rhyme, draws a picture of herself as "Skipping Girl" wearing a t-shirt with "SG" on it, or plays "spooky music" sounds on her family's piano—each of these underlying systems are themselves also the object of playful activity. A body of evidence has shown that a playful approach to language learning, as opposed to formal instruction, offers the most powerful support for the early development of phonological and literacy skills.[13]

Second, through all kinds of physical, constructional, and social play, children develop their skills of intellectual and emotional self-regulation—that is, they learn to be aware of and to control their own physical, mental, and behavioral activity. Children who attend preschools that are based predominantly on models emphasizing play rather than academic outcomes have been found to achieve higher scores on measures of self-regulation. In a series of comprehensive studies of educational factors that have an impact on learning, self-regulatory abilities have been shown to be much stronger predictors of academic achievement and emotional well-being than any other developing abilities, including general intelligence. Educational interventions that support children's self-regulation have emerged as being the most effective in supporting children's learning. In a major longitudinal U.S. study, a key element of self-regulation—attention span-persistence—in four-year-olds greatly predicted math and reading achievement at age twenty-one and the odds of completing college by age twenty-five.[14]

In other bodies of research, studies have examined numerous characteristics of play that make this type of activity a particularly powerful context for learning. For example, studies have highlighted the way that children challenge themselves and incorporate a spontaneous purposefulness in their play. Such studies have begun to elucidate the ways in which children develop increasingly sophisticated strategies for solving problems and for understanding causal relations between physical events. More than thirty years ago, a study exploring the ubiquitous and spontaneous attempts by eighteen- to forty-two-month-old children to stack cups in the correct order documented the increasingly sophisticated strategies that they developed and used to correct their errors. A more recent study has shown that within exploratory play, where there is ambiguous information about a causal relation, preschoolers spontaneously select and design interventions that effectively provide them with more information.[15]

The Role of Adults in Supporting Children's Play

A range of very recent studies has begun to explore a variety of ways in which adults can support children's play. Studies of parents of five-month- to two-year-old children have shown that, with babies and toddlers, playful parents spontaneously modify their behavior to stimulate and cue playful behaviors. This early cueing of play for young children by adults appears to have similar characteristics to the ways in which adults modify their language during "motherese" when speaking with young children. Other studies, however, have demonstrated that young children's learning of playful behaviors from their adult caregivers is not mere imitation. Adult playfulness stimulates children to produce both observed and novel play activities.[16]

The role of the adult in supporting children's learning through play within educational settings has begun to be more actively and scientifically investigated in recent years as well. Within this work, the notion of "guided play" has emerged, with various studies showing that if the teacher takes the role of co-player, rather than directing the play, this can support learning more strongly than either free play or direct instruction.[17]

Prioritizing Play

There is a strong case, then, that supporting playful opportunities for our children and attempting to halt or reverse the decline in these opportunities is of vital importance to the well-being of both children themselves and society as a whole. While there clearly are grounds for concern, some hopeful signs are emerging, allowing for a cautious optimism. Clear and growing indications of a resurgence of interest in children's play are evident within the research community, within the activities of philanthropic and charitable social enterprise organizations, in some policy and governmental arenas, and, perhaps most heartening of all, in a plethora of small initiatives, all over the world, to make local communities and schools more child-friendly and playful. A few shining examples of these developments stand out.

Within the scientific research community, there is a resurgence in work attempting to understand the many unanswered questions in relation to human play behavior. Several recent volumes attest to the number of young researchers worldwide who are beginning to take an interest in play as an important topic for inquiry. A similar resurgence in interest is evident among nongovernmental and philanthropic organizations, as indicated by the activities of the Denmark-based LEGO Foundation. Among the many areas of its work are support for new play research centers, as well as projects worldwide that provide enhanced play-based preschool opportunities for children in disadvantaged communities. The LEGO Foundation recently sponsored the Play, Learning and Narrative Skills (PLaNS) project at the University of Cambridge, demonstrating the potential of a playful approach to supporting primary school children's development as writers. The foundation also is the founding donor of the Play in Education, Development and Learning (PEDAL) Research Centre at Cambridge, established to conduct rigorous scientific research to address the many important questions relating to the role of play and playfulness in human development.[18]

Even at the policy level, in many parts of the world, play-based learning is being recognized as a key component in high-quality early childhood education, as endorsed in the United Nations Sustainable Development Goals for

2015–30 and in many national and regional policy documents. But perhaps the most uplifting examples of the worldwide recognition of the importance of children's play are initiatives within local communities and schools. The resurgence of "Play Streets" in the United Kingdom, the United States, and other urbanized parts of the world is one shining example. Here, with the agreement of local authorities, residential streets and areas are blocked off to traffic on a regular basis so that children and their families can play outside with their neighbors. Within schooling, the development of materials-based outdoor play facilities in the Anji district of China is a marvellous example of supporting children's curiosity, creativity, and love of adventure through play.[19]

Play has often been dismissed in the popular mind as essentially trivial. It has been seen as something that young children do, but that serves no purpose and that they eventually grow out of. On the contrary, play has critical importance as the engine of human creativity and adaptability. For the good of our children and our societies, we need to begin to develop policies, within our communities and our schooling systems, that support and nurture children's natural and adaptive playfulness. It is incumbent upon all of us—researchers, educationalists, philanthropists, politicians, and parents—that we start making decisions, large and small, that prioritize play.

CHAPTER 10

Looking the Monster in the Eye: Drawing Comics for Sustainability

Marilyn Mehlmann with Esbjörn Jorsäter, Alexander Mehlmann, and Olena Pometun

Esbjörn Jorsäter, a Swedish comics teacher, invited the children and grandchildren of refugees and other immigrants in Sweden to tell the life story of a parent or grandparent. The narrative was to be in the form of a "heroic journey," a classic story form where the protagonist awakens to a challenge, confronts a monster, and returns to acclamation (as depicted in popular films such as *Star Wars* or *Harry Potter*). The social impact of the project was dramatic: from viewing their parents as something of an embarrassment—all too common a perspective among the children of immigrants—many students came to see them as heroes.

In India, another teacher, Shankar Musafir, had managed to arrange for a study visit by students from Pakistan—in itself a heroic journey, given the tensions between the two countries. He invited both the visiting and host students to draw comics of their expectations prior to meeting one another, and then to continue to sketch together during the visit. The exercise proved highly effective at bringing out fears and prejudices in a nonthreatening, even humorous, way. For example, the Pakistanis imagined that the Indian girls would all show up in "seductive" saris, whereas the Indians imagined that the Pakistani girls would be heavily veiled. Surprise: when they finally met, they were all wearing t-shirts and jeans.[1]

Marilyn Mehlmann is a psychologist and management consultant focused on sustainable development and methods development and serves as head of development and training at Global Action Plan International (GAP). **Esbjörn Jorsäter** teaches the use and practice of comics as an educational tool to teachers and youth leaders worldwide and is codesigner of the Drawing for Life program and book. **Alexander Mehlmann** is network coordinator at GAP and has long experience in project management, including the Drawing for Life project. **Olena Pometun** is a professor at the National Academy of Pedagogical Science in Kiev, Ukraine, and introduced the use of comics into lessons for sustainable development in Ukrainian public schools.

Why Comics?

The idea of using comics as an educational medium is hardly new. In Japan, informational *manga*—a term that refers to all kinds of cartooning, comics, and animation—has been used widely to influence people both young and old since the Second World War. In the 1970s, the country saw a publishing boom for educational manga, including the successful use of comics to facilitate public discussion of sensitive topics such as sexuality. The educational role of comics in Japan continues today: in 2013, the Asahi Glass Foundation developed the environmentally focused Gring and Woodin series to raise awareness of climate change and other issues among students, teachers, and parents. (See Box 10–1.)[2]

As an educational medium, comics—as well as other cultural manifestations such as theater and film—are still used primarily as a means to impart knowledge or to raise awareness about issues (today's "monsters") that are defined by educators. This is no criticism of this important informational role. In ancient Egypt, a picture book that provided instructions on growing and harvesting food served its population well for more than a thousand years. More recently, the United Nations Educational, Scientific and Cultural Organization (UNESCO) released the youth version of its 2016 *Global Education Monitoring Report*—an annual publication that monitors progress toward global education targets—in comic form. While this is an excellent strategy for engaging readers, however, it is not sufficient in an educational context.[3]

Education for sustainable development is no ordinary instructional challenge. To be successful, it needs to encompass transformation and engaged action, which, in turn, presuppose the engagement and empowerment of students. The experiences in Sweden and India described above were particularly effective in achieving transformational outcomes—helping to develop awareness and empathy—because they required students to design and draw the comics themselves, rather than just read them. Comics can play a powerful role in teaching, in engaging students with often abstract concepts related to sustainability, and in enabling them to face and tackle difficult questions.

The Educational Challenge

For sustainability education to be effective, simply transferring knowledge is not enough. Students need to be actively engaged and provided an opportunity to be teachers as well as students, helping to achieve the quest for a sustainable world. Students need the opportunity to identify their own "monsters" and to find ways to defeat them. This, in turn, makes unusual demands on teachers, who need, as sustainability education expert Charles Hopkins has noted, "competencies [that] go beyond factual knowledge and include skills, values, perceptions and action skills." In other words, teachers need to see themselves as learners as well. The big personal monster for many sustainability educators is that of not knowing: allowing oneself to admit and to explain that no one can say what a truly "sustainable" society will look like. We can educate—but for a future that we cannot imagine unless in a shared consciousness with others.[4]

This critical question is thrown into particular relief when dealing with art. As authors Susana Gonçalves and Suzanne Majhanovich write in their anthology *Art and Intercultural Dialogue*: "Art initiates, fosters, and protects diversity. . . . Imagination, creativity, innovation, and problem solving are intertwined in the process of art creation. These ingredients are at the same time the manifestation of diversity and the result of interaction, dialogue, and cultural influence." The book explores different forms of artistic expression— from the visual arts to literature and cinema—asking questions such as, "How can art contribute to sustain or promote social cohesion in neighborhoods, in the groups and community, and in the larger society?"[5]

Four Elements of Using Comics for Sustainability Education

To engage people in action for sustainability, the transfer of knowledge is not enough. Although enjoying art produced by others may be informational and inspiring, engaging oneself in the actual creation and co-creation of works of art has far greater potential as an educational tool. In the case of comics, realizing the potential of education for sustainable development through this artistic medium requires a focus on four principal elements:

- How to narrate a heroic journey,
- How to understand "sustainable development" in a personal context,
- How to imagine desired futures, and
- How to draw comics.

Narrating a Heroic Journey

Storytelling is a well-established educational technique, especially when topics span a variety of subjects that normally are separated. For young children,

Box 10–1. Comics: An Effective Way to Educate About the Environment

We sometimes assume that the more people learn about climate change and its effects, the more they will share concern about the crisis and turn that concern into action. But this is not necessarily the case. Knowledge has only a limited effect on shaping concern, and, in most cases, it has an even more limited effect on generating action. Sometimes political beliefs override scientific findings about global warming. Moreover, studies show that public apathy about climate change is not necessarily a result of poor comprehension of the science: in some cases, high science literacy overlaps with attitudes showing less concern about climate change.

Considering this, is there still an opportunity to raise people's risk perceptions about climate change? According to one study, increased knowledge about climate change among young people is positively correlated with acceptance of human-caused global warming, regardless of the individual's worldview. Thus, finding ways to cultivate early understandings of climate change must be prioritized. This is why Japan's Asahi Glass Foundation developed an environmental comic series, Gring and Woodin, to educate children and adolescents about global warming and other challenges.

The Foundation chose the comic medium because it is a powerful communication

Box 10–1. continued

tool that is eloquent and flexible enough to promote understandings about climate change and nature conservation. The series—available online and targeted at students, teachers, and parents—depicts the adventures of an environmental hero, Gring the rabbit, and his companion Woodin, who is a tiny tree-like creature but who also plays the role

of an environmental doomsday clock that gauges people's concern about the environment. The surrealism of the characters and their adventures—as they travel through various ecosystems helping to save the Water Kingdom from pollution, aided by a variety of strange creatures and individuals—keeps readers engaged and keeps the overall tone light. Yet each comic also teaches readers about environmental issues (from recycling to climate change) and regularly reminds readers of human interconnectedness with and dependence on nature.

In combination with other forms of early science education, comics, games, and various engaging storytelling efforts can play an important role in helping to teach children about—and, more importantly, to love—the environment.

—*Tetsuro Yasuda, Secretary General, The Asahi Glass Foundation*
Source: See endnote 2.

retelling stories has proved to be an effective tool for both reinforcing knowledge and improving literacy. This seems to apply, in particular, to oral storytelling: children who tell a story verbally are both more prolific with details and tend to have a stronger storyline than they do when writing a story down. But when students are taught to craft their own stories and also to document them in comic form, new benefits open up: they learn to work together, to engage with questions of ethics, and to make explicit their tacit knowledge about the challenges and opportunities facing humanity. In other words, they become innovators and teachers as well as learners.[6]

One traditional route to teaching storytelling is via myth: a universal story that leaves room for local legends and personal stories. An enduring myth that is especially applicable to a period of rapid transition—such as that facing the planet today—is that of the "heroic journey." When Esbjörn Jorsäter designed the comics program for children of immigrants in Sweden, he drew on this tradition to create a structure that could be easily visualized and taught. (See Figure 10–1.)[7]

Figure 10–1. The Hero's Journey

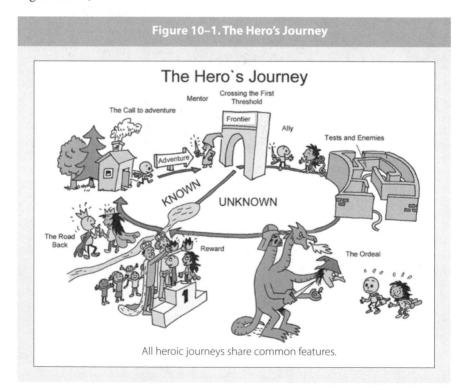

All heroic journeys share common features.

Central to the myth of the heroic journey is the threat of a monster—a threat that, for students of sustainability, can easily be assimilated and transformed. "We've seen the monsters of unsustainability. . . war, corruption, social erosion, ecological collapse," observed comic-art students in Kosovo, a territory that endured widespread ethnic cleansing and environmental devastation during the 1990s. Storytelling through comics offers an opportunity to learn—and teach—the art of looking these monsters in the eye without losing heart. It enables students to identify and describe their own monsters, rather than simply adopting those proposed by teachers or textbooks.[8]

Understanding "Sustainable Development" in a Personal Context

Even if transferring knowledge—from teacher or book to pupil—is not enough to create lasting outcomes, some degree of knowledge transfer is still called for. The heroic journey is, of necessity, a leap into the unknown— but *something* needs to be known, or there will be no direction. But what knowledge, how much, and how? How to impart "enough" knowledge about sustainability (and unsustainability) to awaken the imagination and critical faculties of students without implying answers—or inducing despair?

For this purpose, most of the available roadmaps to sustainable development are of dubious service, including the prominent United Nations Sustainable Development Goals (SDGs), which set aspirational global targets in areas such as poverty alleviation, sustainable cities, and clean water and sanitation. Although the seventeen SDGs are powerful in themselves, they are not easy to translate to a personal context. To appeal more readily to the individual, a recently developed comics program for schools, *Drawing for Life*, starts with the simpler concept of the ecological footprint—not because it is more accurate or "better," but because it translates more directly into personal experience, risks, and opportunities. The course material includes a short questionnaire that enables students to make a rough calculation of their own footprint. This has proved to be the "open sesame" to animated discussions and deep questioning.[9]

Imagining Desired Futures

If the ecological footprint tells something about the current situation—helping to assess the scale of impacts today—the next step on the journey for students is to envisage a desired future. A possible starting point is an assumption (by no means given, but with some research to support it) that most people would prefer to live in a more sustainable society—one where people support each other and where resources are managed sustainably and distributed equitably.[10]

From such a perspective, it makes sense to focus on "what we want"—on visions of desirable futures—and to chart a path (individual or collective) from here to there. While there are many methods and techniques for eliciting futures, including predictive modeling and scenario building, few are easily adaptable for the classroom. When using comics as a tool for sustainability education, one reasonably quick and easy approach to imagining desired futures is that of "utopia," a nonexistent but typically idyllic place.

The team that developed *Drawing for Life* built on a utopian concept that was originally developed for the Swedish National Museum of Natural History for an exhibit on outer space. The museum's aim was to "inspire in visitors the sense of awe at the beauty and fragility of our planet as reported by space travellers." The resulting interactive material was presented as a humorous "travel guide" to Planet Earth (termed "Planet TellUs"). When this material was later adapted for the comics program, it was transformed into a role play, based on the knowledge that students had acquired while working with ecological footprints. The envisioned utopia was the planet Pondera—a planet far from Earth where people live in harmony with themselves, with each other, and with all other species. Students were given the following exercise:

"Interplanetary hero" comic drawn by students at a 2012 workshop in Kosovo.

Imagine you are a native of the planet Pondera, which is totally sustainable. You will become a youth ambassador to planet TellUs, where they have some problems. Explain to them *your* story of how Pondera came to be so sustainable, in the form of a heroic journey. Do it in pictures, because speech communication may be difficult.

This exercise—using comics to write and draw about the journey toward sustainability on Pondera—was originally used as a standalone discussion starter in youth education programs in Sweden. It has since spread to other countries: for example, it serves as an activity at "sustainability" summer camps for students in Ukraine. It also has been (and is being) used as part of the *Drawing for Life* comics program by several hundred teachers and youth leaders in Albania, Belarus, Kosovo, and Macedonia, as well as Ukraine.[11]

Drawing Comics

A final focus for a comics program in sustainable development is teaching students how to draw comics: for example, using facial expression to communicate emotions, portraying bodies and dynamic movement, applying perspective and composition, and creating frames and panels to carry readers through the story. While it may seem as if teaching these creative skills should be the first requirement on the list, the beauty of comic art is its forgiving nature. Drawing comics does not require a high level of artistic or technical skill, and it can be taught reasonably quickly to teachers and youth leaders as well as students. The step-by-step educational approach developed by Jorsäter and incorporated into the *Drawing for Life* manual can help make teaching the actual drawing techniques considerably easier.[12]

Comics in Every Classroom

By taking inspiration from teachers of language, literature, and art; from artistic and creative students; and from other guides, schools can readily integrate comics programs into their curricula. Based on anecdotal evidence from many of the teachers and youth leaders who have used comics as an educational tool for sustainable development, participants frequently report two common experiences: "Aha" moments—sudden insights into connections between facts, phenomena, problems, and solutions—and laughter. Enjoying a lesson is, of course, important, but beyond that, laughter in such situations often is a signifier of emerging creativity and insights. This is a part of the transformative process that is demanded by education for sustainability.[13]

Reviewing the process and outcomes with teachers and youth leaders from several countries, *Drawing for Life* has demonstrated the transformative power of the comics approach. It also has made visible some of the major professional challenges facing any sustainability educator, including:

- Learning, and teaching, the art of "looking the monster in the eye" without losing heart;

- Imparting the skills needed for teamwork and cooperation as students work together to create narratives, storyboards, and finished comic projects;
- Engaging students in identifying their own monsters and finding ways to overcome them;
- Enabling students to envisage desirable futures in which they play an active role; and
- Maintaining a focus on sustainable development and still leaving room for the inspiration and experience of every student.

Is this a road open to anyone and everyone? Is every student—and teacher—capable of cultural creativity? In the case of creating comics, students frequently work in teams, and it often becomes apparent that one team member is particularly gifted at storytelling, another at design, another at final drawing and inking, etc. This, in turn, reinforces the vital point underscored by educational expert Charles Hopkins that all sustainable development is, of necessity, based on teamwork and cooperation: "Collaborative learning is essential for addressing the common challenges with which citizens are confronted, in their own communities and countries and globally."[14]

One of the participants in a comics project in Kosovo in 2012 displays his work.

Esbjörn Jorsäter

Comics—particularly the active creation of them—have demonstrated an educational potential far beyond the mere reading and enjoyment of them. The same case may be made for other art forms, such as drama: learning is not only reinforced, but also engendered through the process of creating. The challenge for humankind, collectively, on the next stage of our sustainability journey is to turn this insight to good account: to make ourselves not "only" teachers of and for sustainable development, but also teachers that are engaged in conveying active life skills, by whatever means come to hand.[15]

A student comic about cleaning up plastic pollution in Ukraine.

CHAPTER 11

Deeper Learning and the Future of Education

Dennis McGrath and Monica M. Martinez

When eighth-grade students at King Middle School in Portland, Maine, began a curricular unit on electricity, they were given two guiding questions: "How do we capture and use nature's energy?" and "How can you change your energy consumption to improve the world?" Teachers used these questions to integrate their course offerings and to provide a context for every assignment, classroom activity, field work experience, and project that the students did. The students quickly discovered that to understand how best to capture and use nature's energy, they first had to master the concepts of energy and energy transfer presented in science class. They then applied what they were learning through an assignment to conduct an energy audit on their own homes, and they learned in math class how to measure energy consumption as well as how to calculate their own carbon footprints.[1]

While exploring ways to reduce their energy use, the students learned in science class about alternative forms of energy and worked in teams in technology education class to design and build generators that were more energy-efficient. In English class, the students developed their research and writing skills as they learned how to report on the results of their home energy audit. The class also practiced writing a persuasive essay advocating for or against the use of wind power in their state, while developing a land-use proposal in social studies.[2]

Throughout all of these activities, the two guiding questions unified learning by helping the students identify key ideas, see commonalities, identify

Dennis McGrath is a professor of sociology at the Community College of Philadelphia. **Monica M. Martinez** is deeper learning senior fellow at the Hewlett Foundation. This chapter is based on the book *Deeper Learning: How Eight Innovative Public Schools Are Transforming Education in the Twenty-First Century* (The New Press: 2014).

relationships across subjects, develop data analysis skills, and explain their research. The students became more deeply engaged in learning by doing field work, and they developed collaboration skills as they worked in groups to build and test the generator. Equally importantly, throughout the curricular unit, the students learned concepts and skills that were important to their families and communities and that they could apply in grappling with future challenges in their lives.[3]

Three students of King Middle School in Portland, Maine, explain their civil rights project to then-Education Secretary Arne Duncan.

As it becomes more typical for students worldwide to attain at least a secondary education (and, increasingly, to pursue additional schooling), young people today are more likely than their older counterparts to achieve higher levels of education. However, most educational systems fail to adequately prepare students for the striking challenges and accelerating pace of change that they will face this century. As academic innovation advocate Ken Robinson has put it: "[C]urrent systems of education . . . were developed to meet the needs of a former age. Reform is not enough: they need to be transformed." Education must prepare young people with the skills that they need to confront the overarching challenge of climate change and the connected disasters that flow from it.[4]

Deeper Learning

Robinson echoes the call by many to move away from rote learning—which encourages passive rule-following in students—and toward educational experiences, such as the one described above, that promote critical thinking, creativity, and innovation. Many of those who are engaged in reimagining models of teaching and learning are turning to the concept of "deeper learning," an umbrella term for the skills and knowledge that students must possess to succeed in twenty-first century jobs and civic life. Deeper learning provides a vision of what high school graduates should be able to know and, most of all, do. It identifies a set of competencies that students must master in order to develop a deep understanding of academic content and to apply their knowledge to problems in the classroom, on the job, and in their communities.[5]

When students complete their secondary education, they should be able to:

- *Draw upon their knowledge in an academic discipline for the purpose of transferring that knowledge to other situations or tasks* because they have a strong foundation in subjects like reading, writing, math, and science, but most importantly because they understand the key principles and procedures within these subjects, rather than merely recalling facts.
- *Think critically and solve complex problems across disciplines* based on an ability to critically synthesize and analyze data and information; frame questions; recognize patterns, trends, and relationships; identify and solve problems; and assess or evaluate the effectiveness of the proposed solutions.
- *Work collaboratively with others* to complete tasks, produce shared work, and identify and develop solutions to complex problems.
- *Communicate complex concepts to others through multiple modes of expression—written, oral, and visual—*in a logical, purposeful, and meaningful way. This requires students to be able to clearly organize their findings, thoughts, and presentations.
- *Monitor and direct their own learning*—or know how to learn. Students will not only need to be cognizant of what they know, but also be able to identify the obstacles or barriers to their success and then determine and deploy the strategies to address these challenges. This requires a very high level of self-awareness as well as a tremendous amount of initiative, independence, intellectual flexibility, and concern for quality.
- *See themselves as academic achievers and expect to succeed in their learning pursuits.* Students need to develop trust in their own capacity and competence and to feel a strong sense of efficacy, resulting in their desire to engage in productive academic behaviors and to persevere even in the face of difficulty.

How Schools Are Developing These Six Competencies

Schools can promote these "deeper learning" competencies in their students in a variety of ways. Although schools worldwide are highly diverse, they share many common features, including offering a picture of what a transformed approach to education should look like. In deeper learning schools, the school's leader and its teachers share an educational vision. They purposefully design learning experiences to be aligned to their vision of what students should know and be able to do. Teachers have a very different approach to their role and understand that, for students to be successful, they should graduate with these six competencies and should be able to "learn how to learn," regardless of the future that they are inheriting.

Drawing on and Applying Academic Knowledge

For students to draw upon knowledge in a discipline and to apply it to other situations and tasks, they need multiple opportunities to understand findings, theories, and techniques by applying them to a range of challenging tasks and situations. In American History classes at Science Leadership Academy in Philadelphia, Pennsylvania, eleventh graders apply their knowledge of the legislative process to a citizen lobbying project. Students work together in small groups to identify and research an issue of concern to them in their city, such as housing, poverty, or services for young people; attend a city council meeting; interview key stakeholders and organizations; and ultimately present their ideas to the Philadelphia city council and urge support for specific legislation.[6]

Thinking Critically and Solving Complex Problems

For students to think critically and solve complex problems, they need frequent and sustained opportunities to conduct research, analyze, and synthesize information and evidence (both historical and contemporary) to formulate accurate hypotheses, well-reasoned arguments, and coherent explanations in support of a hypothesis. Teachers regularly have students use what they learn in their courses to solve local problems, such as when students at King Middle School in Maine worked with a local science center to identify invasive species in Casco Bay and then presented their findings to the city council and other local decision makers.[7]

Some teachers develop curricula that offer students a sustained opportunity to engage with critical global problems. Students at High Tech High in San Diego, California, helped a conservation agency gather evidence against wildlife poachers in Africa by analyzing the DNA of meat sold in markets. The biotech teacher had done field research in Africa and wanted to engage his students in applying their knowledge to a global issue. A consortium of life sciences companies had provided lab equipment for doing DNA typing, so the teacher decided to focus on the problem of poaching. He arranged for Tanzanian colleagues to send meat samples, and the students, working in pairs, tested the samples to isolate the DNA and identify the source of the meat. Once the students mastered the procedure of crude cell extraction, their project was to find ways to do it more cheaply and efficiently, with the goal of enabling conservationists to use the improved method to test for illegally poached meat in African street markets. At the end of the project, the teacher brought several of the students to Tanzania to conduct a bushmeat identification workshop for local wildlife protection officials.[8]

Working Collaboratively

Students need frequent and sustained opportunities to work in groups and to collaborate with each other, with teachers, and with experts. Group work has to be carefully scaffolded to ensure that students will be able to collaborate productively. This means helping students learn how to organize people, knowledge, and resources to achieve a goal (such as solving a problem or creating a product), as well as how to provide and receive feedback and incorporate multiple points of view.

Group work may start off as a small part of a classroom activity to help students learn how to draw upon each other's knowledge and perspective to solve problems, such as by having students work in groups to assess and explain the validity of a mathematical formula; critique each other's writing or help each other study for an exam; or discuss their perspective on a piece of literature that they read. As students advance in their collaboration skills, they may regularly work in groups supported by social media and other technology platforms.

Teachers also can design larger collaborative efforts. At Casco Bay High School in Portland, Maine, the "Junior Journey" expedition is the most intense learning experience of the students' four years of study, requiring intensive coordination and collaboration among teachers and students as they learn new skills and demonstrate high levels of teamwork. In 2008, the centerpiece of the expedition was a weeklong class trip to West Virginia, combining a service learning experience and a documentary project to interview people in the community. To prepare for the trip, students studied the history and economy of the Appalachian region and learned about its culture through music and literature. The teachers brought in local experts to teach the students how to conduct interviews and how to create a film documentary. Students deepened their collaborative skills as they formulated the questions that they would ask and allocated the tasks needed for the interview and filming process. When the class returned home, each student wrote a full oral history of his or her interviewee, and the students collaborated to produce fifteen three-minute documentaries that were presented at a formal event in the city.[9]

Communicating Complex Concepts

For students to be able to communicate complex concepts effectively, they need frequent and sustained opportunities to express themselves in writing and speaking assignments where they have to convey meaning, explain important concepts, present data and conclusions to an audience, and listen attentively to the feedback that they are offered. These assignments call on students to practice multiple forms of writing, including essays, journal reflections, technical reports, poems, and short stories. Students also make oral presentations to peers, the entire school community, or audiences outside of the school that may involve producing a video, documentary, or podcast; designing an infographic; or developing a presentation using PowerPoint or Prezi.

At MC² STEM High School in Cleveland, Ohio, students engaged in multiple opportunities to communicate as part of a cross-disciplinary project on the Age of Enlightenment. In English class, they wrote poems and short plays; in math class, they used software to compose an original piece of music; and in physics and engineering classes, they collaborated on designing a circuit board and an audio speaker that was capable of transmitting sound. The students' communication skills were further developed when they presented their work at the Rock and Roll Hall of Fame, located next to their Science Center and open to the public. Each student group set up a stand in the Hall's main auditorium, and the ninth graders described their work on building their speakers to family and friends as well as to general museum visitors. Each student group also presented the poems, plays, and music that they had written. Knowing that they would be presenting to a real audience motivated the students to deepen their skills and prepare for the events.[10]

Being Self-Directed and Learning to Learn

For students to be able to direct their own learning, teachers have to be less prescriptive and controlling and need to build in frequent and sustained opportunities for students to make their own choices and to reflect on their work and on the decisions they make during the learning process. To build student choices, both large and small, into every class, teachers can provide opportunities for students to choose the topic that they will debate, the problem that they will solve, the product that they will create, or, as at Avalon School in St. Paul, Minnesota, the individual projects that students will undertake to meet their state graduation standards. At many schools, teachers use journal writing to encourage student reflection.[11]

A signature practice of many schools is the student-led conference. Unlike the traditional teacher-parent conference that is common in U.S. schools, this conference is led by the student, not the teacher, as a way for students to take and show responsibility for their own learning. The student and teachers prepare for the conference by going through the student's work from each subject and deciding which of the sample papers, assignments, or tests best reflect the student's strengths and weaknesses.

A teacher looks on as fourth-grade students at Lisby Elementary School in Aberdeen, Maryland, explain their "Bad Hair Day Fixer" prototype, developed as part of the school's STEM Superstar program.

Developing an Academic Mindset

For students to develop an academic mindset, they need frequent and sustained opportunities to identify their interests, understand the standards of high-quality work, and see themselves as capable of doing this work. These experiences help students to internalize a clear model of their capabilities and to improve their ability to monitor and direct their own learning. At Impact Academy of Arts and Technology in Hayward, California, promotion to eleventh grade requires that each tenth-grade student complete and defend a Benchmark Portfolio before a panel of teachers to demonstrate their readiness. To graduate from the school, students must demonstrate readiness through the College Success Portfolio, consisting of four pieces of their work and a cover letter reflecting on what they have learned. Students use the Portfolios to show that they have learned their academic content, are able to apply the school's core competencies (inquiry, analysis, research, creative expression in core content areas, and leadership), and are able to reflect on what they have learned.[12]

The Future of Deeper Learning

There is increasing recognition of the need to transform current educational systems so that they promote critical thinking, problem solving, collaboration, and communication. The key to that transformation is reconceiving the role of teachers and revising the relationship between teachers and students: displacing the teacher as primarily a dispenser of knowledge and encouraging students to take responsibility for their own learning. As educational reform experts Michael Fullan and Maria Langworthy explain, the driving force will be new teaching practices that offer "a new model of learning partnerships between and among students and teachers, aiming toward deep learning goals and enabled by pervasive digital access." The goal is to enable students to be self-directed, independent learners that are capable of managing their own learning.[13]

Worldwide, examples abound of efforts to promote these aims—both by national education ministries and by institutional collaborations and public-private partnerships. Fullan and Langworthy describe innovative efforts in Australia, Canada, Singapore, the United Kingdom, and elsewhere to promote engagement in learning by connecting it to students' real lives and aspirations. Many of these efforts do not require extensive resources. At a small rural school in Ireland, teachers engaged students by connecting the curriculum to their own community and history. A yearlong, schoolwide learning project focused on the more than two hundred ringforts—circular earthen mound dwellings built in the Bronze Age—that are present in the area. Students studied the history of ringforts, including what people ate and wore when they lived in them; visited the sites with an archaeologist; mapped the forts using mathematical models; built their own version of a ringfort in their school; wrote a script and acted it out in their own film; and traveled to heritage meetings across Ireland to present what they learned.[14]

Fullan and Langworthy also offer examples of schools that are helping students "learn how to learn." A common strategy is peer tutoring to help students master the learning process by examining their own work and incorporating feedback from others. The Learning Communities Project in Mexico incorporates peer tutoring in rural schools throughout the country. Through tutoring networks, the students teach each other, using the curriculum as a map and with each student "teacher" developing his or her own area of expertise. More than six thousand schools across Mexico have joined this movement. Data from the first four thousand schools to participate indicate that these institutions have greatly increased the proportion of students scoring at "good" and "excellent" levels on national standardized tests.[15]

Many efforts to promote deeper learning recognize that the key is changing what goes on in the classroom. European Schoolnet is a network of thirty

European Ministries of Education that is focused on promoting innovation in teaching and learning. Schoolnet supports a wide variety of innovations including classrooms of the future, a Living School Lab to scale up best practices, and numerous teacher training programs. One Schoolnet project, Innovative Technologies for Engaging Classrooms (iTEC), involved more than 2,500 classrooms in using technology to promote student collaboration and inquiry-based learning. Teacher-developed projects ranged from a virtual museum in Spain to students building and using a three-dimensional printer in Austria.[16]

In the United States, the Common Core State Standards that have been implemented in a number of states have strengthened the curriculum that students follow. Common Core emphasizes that college and career readiness is the primary goal of school and needs to be reflected in classroom teaching. Higher-order thinking, communication skills, and conceptual understanding are given priority over rote learning and fact-based content knowledge. For example, the Common Core standards in mathematics require students to solve nonroutine problems and to reason by using logic and evidence, rather than simply memorizing formulas.[17]

There also has been a broad movement toward so-called personalized learning. This approach puts students at the center and redesigns teaching and learning so that:

- Instruction is aligned to rigorous college- and career-ready standards and to the social and emotional skills that students need to be successful in college and in a career.
- Instruction is customized, allowing each student to design learning experiences aligned to his or her interests.
- The pace of instruction is varied based on individual student needs, allowing students to accelerate or to take additional time based on their level of mastery.
- Data from formative assessments and student feedback is used to differentiate instruction and to provide robust supports and interventions so that every student remains on track to graduation.
- Students and parents have access to clear, transferable learning objectives and assessment results.[18]

The innovative approach, known as XQ: The Super School Project, aims to remake high schools in the United States by creating a new model of learning that better prepares young people for the modern world. In a push to promote "reimagining education," the project awarded $10 million each to ten schools in 2016. The project overlaps with the aims of deeper learning because it calls for the reinvention of learning to develop original thinkers and lifelong learners who are proficient in core academic subjects and are capable of collaborating with others to solve complex problems. A similar U.S. initiative, Next Generation Learning Challenges, is supported by a broad collaboration of national foundations and educational organizations and aims to apply current research on learning to new school designs by providing challenge grants to practitioners who want to redesign their schools. Like the XQ project, Next Generation Learning Challenges looks to identify and support breakthrough models of schools that are student-centered and mastery-based, and that creatively use technology to promote learning.[19]

While these myriad efforts differ somewhat in their initial focus and emphasis, they share some common concerns and strategies. They seek to:

- Empower students as learners so they can direct their own learning.
- Contextualize knowledge so that it is coherent and not disjointed, isolated, or taught in a vacuum, as a way to help students acquire understanding of content knowledge.
- Connect learning to real problems and issues and experiences, in order to make learning meaningful for students.
- Extend learning beyond the school to provide students with access to experts, authentic learning experiences, opportunities to contribute to their communities, and access to extended networks of support and learning. (See Box 11–1.)
- Inspire students by personalizing and customizing learning experiences. Teachers establish strong relationships with students to help them find and pursue their own learning passions.[20]

The future of deeper learning lies in a two-pronged effort. One is reimagining what teaching and learning should look like in the modern age. The other, related effort is reconfiguring the relationship between teachers and students so that students can develop the skills, creativity, and self-direction that they will need to confront climate change, sustainability, and the other global challenges they will likely face in their lifetimes.

Box 11–1. Mapping Is Learning

"If I were asked to name the most needed of all reforms in the spirit of education," John Dewey wrote back in 1893, "I should say: 'Cease conceiving of education as mere preparation for later life, and make it the full meaning of the present life.'" The Turkish nonprofit organization TEMA (Turkish Foundation for Combating Soil Erosion, for Reforestation and the Protection of Natural Habitats) offers an example of this type of meaningful, real-world education by mobilizing citizens to identify water threats via a participatory mapping exercise.

Joe Bryan, a geography professor at the University of Colorado, Boulder, notes that maps act like user interfaces that increase our understanding of nature and the world. Beneath the dots and lines lie vast amounts of information and knowledge. By helping to map local and regional threats, participants in Turkey's mapping exercise learn more about water quality and the environment, helping them to better understand the interactions between humans and nature. The map itself also serves as a learning tool that can be used for educational purposes.

In this citizen-science project, titled "The Map of Threats to Water Resources in Turkey," volunteers from across the country work with TEMA representatives to observe, compile, and analyze threats in their provinces related to water quality, water quantity, and access to water. Volunteers from diverse backgrounds use a standardized data collection form to send information to the project coordinators—without having to create an online account, thus minimizing the barriers between citizen volunteers and TEMA coordinators. The coordinators then map these threats using online and interactive open-access tools.

During the first wave of data collection in early 2016, the project mapped forty-six threats to water quality (including river and lake pollution and declines in biodiversity), twenty-nine threats to water quantity (including rivers and lakes drying up and groundwater depletion), and six cases related to access to safe drinking water. The map was published on World Water Day together with an assessment report, and journalists, activists, and teachers are using it for education, awareness raising, and advocacy. TEMA aims to publish annual updates to boost understanding of the evolution of such threats over time.

The involvement of ordinary citizens in identifying water threats helps, first, to promote advocacy and stewardship. Second, it helps to change the perception that Turkey has abundant and unspoiled water resources, by presenting in a visual manner the many threats around the country. It is hoped that the map will motivate citizens to take action, perhaps in the form of increased dialogue between the government and civil society. Participatory mapping also provides an opportunity to get more of Turkey's citizens out into nature—no small benefit in an era when people are spending more and more time indoors.

—*Ali Değer Özbakır, TEMA Foundation, Turkey*

—*Cem İskender Aydın, Université Versailles Saint Quentin-en-Yvelines, France*

Source: See endnote 20.

CHAPTER 12

All Systems Go! Developing a Generation of "Systems-Smart" Kids

Linda Booth Sweeney

Try this: Find a young person between the ages of four and twenty-four. Show them a picture of a cow and ask, "If you cut a cow in half, do you get two cows?" Even the four-year-old will shout, "No way!" Children understand that a cow has certain parts—hearts, lungs, legs, brain, and more—that belong together and have to be arranged in a certain way for the cow to live. You cannot have the tail in the front and the nose in the back.[1]

As adults, it is easy to miss this simple truth: a cow is a complex, living system, in the same way that the human body, a family, a classroom, a community, an organization, or an ocean is. A system is composed of parts and processes that interact over time—often in closed-loop patterns of cause and effect—to serve some purpose or function. Living systems, unlike a collection or "heap of stuff," share similar characteristics. In systems, it matters how the parts are arranged. That is why a cow cannot have the tail in the front and the nose in the back. And why a stomach does not work on its own, and the body does not work without a stomach. And systems often are connected to or nested within other systems (for instance, a person may be nested within a family, school, ecosystem, community, and nation).

Making the Shift: Systems as Context

Sounds simple, right? But here is the challenge: much of today's education remains focused on discrete disciplines—for example, math, science, and English. Science is taught in one class. The bell rings. The student moves on

Linda Booth Sweeney is an American systems educator, author, and cofounder of the Society for Organizational Learning Education Partnership.

to math and then, perhaps, to English—and never the twain shall meet. Such a fragmented approach reinforces the notion that knowledge is made up of many unrelated parts, leaving students well-trained to cope with obstacle-type or technical-based problems but less prepared to explore and understand complex systems issues. (See Box 12–1.) In medicine, for example, obstacle-type problems are those that can be clearly targeted and fixed, such as a broken arm or an acute disease, like appendicitis. A systems approach is more effective for chronic and complex diseases, such as diabetes, where the interaction of factors—lifestyle, family history, environment, etc.—also plays a role.

Issues such as climate change, economic breakdowns, food insecurity, biodiversity loss, and escalating conflict are matters not only of science, but also of geography, economics, philosophy, and history. They cut across several disciplines and are best understood when these domains are addressed together. Students and adults must be able to see such important issues as

Box 12–1. Teaching Big Systems Ideas

The following core systems ideas should be taught—where developmentally appropriate—across the grade levels:

- There are such things as systems. A system, unlike a collection or a "heap of stuff," is composed of parts and processes that interact over time to serve some purpose or function.
- Systems, as a whole, can exhibit properties and behaviors that are different from those of their parts.
- Simple systems can work in predictable ways; dynamic systems exhibit more complex and unpredictable behaviors.
- Although different, complex systems share similar patterns of behaviors such as "escalation," "boom and bust," and "limits to growth." These often are called "systems archetypes" or "kernel structures."
- Making systems visible, using tools such as connection circles, behavior-over-time graphs, causal-loop diagrams, stock-and-flow models, and computer simulations, helps students visualize, understand, and test ideas that are applicable throughout science, engineering, history, literature, and more.
- Constructing a simplified model of a system by defining a boundary in which it operates enables students to better understand and predict system behaviors.
- Change occurs through closed loops of cause and effect called feedback loops. There are two types of feedback loops: reinforcing (positive) and balancing (negative), which either amplify or control change.

systems— elements interacting and affecting one another. In the case of climate change, a systems view shows the link between politics, policy (for example, legislation related to carbon emissions and deforestation), the natural sciences (particularly forests, which help stabilize the climate by absorbing heat-trapping emissions from factories and vehicles), and a person's own consumption habits. Without a systems view, the complexity can be daunting, and the result is often policy resistance or, worse yet, polarization and political paralysis.

When We Don't See Systems

It is no surprise, then, that we often find ourselves tinkering with complex systems without understanding the system—how the key parts of the system are interconnected, or how the interactions of all the parts produce the often puzzling or confounding behavior of the system. Consider, for example, road-building programs that are meant to reduce congestion but that end up increasing traffic, delays, and pollution. Or pesticides that are meant to kill crop-damaging insects but that also kill the "good" insects that are controlling the population of "bad bugs."

We get into trouble when the desire to make the problem go away outweighs the desire to understand the problem or opportunity. In the 1920s, local ranchers in the U.S. states of Montana, Wyoming, and Idaho called for the removal of wolves from Yellowstone National Park. Over the next seventy years, wildlife biologists and park rangers watched as the integrity, or "wholeness," of the Yellowstone ecosystem unraveled. Without a top predator such as the wolf, the populations of deer and elk grew unchecked, which led to overgrazing of shrubs and plants, decreasing the nesting areas for migrating birds, changing the habitat for beaver colonies, and so on.[2]

Professor John Sterman at the Massachusetts Institute of Technology argues that these examples of "policy resistance" arise because we do not understand the full range of feedbacks operating in the system. Moreover, in all of the above examples—increasing traffic congestion, the rise of pesticide-resistant bugs, unchecked population growth—these behaviors are not the result of a single event but are produced by the interaction among the elements or individual factors within a system. Or, as the old adage goes, "the whole is greater than the sum of its parts." In systems terms, we call these behaviors "emergent properties."[3]

Decision-making research shows that when adults are faced with dynamically complex systems—containing emergent properties, multiple feedback processes, time delays, nonlinearities, and accumulations—performance is systematically biased and suboptimal. When the scope of the research is broadened to children, studies show that both young people and adults find it difficult to trace causality beyond simple one-way connections, and most do not spontaneously "close the loop" when feedback exists. Said simply, we tend to think in straight lines—"Problem" leads to "Solution"—when, in reality, closed loops of cause and effect exist.[4]

In one study, ten- and eleven-year-old students were asked to describe the relationship between wolves and rabbits. Most participants immediately responded that wolves eat rabbits, end of story. With additional questioning, the students were able to identify the mutual interdependence between rabbits and wolves, commonly called a predator-prey relationship: a relationship that is not one-way or linear, but rather is made up of a closed loop of cause and effect, with births and deaths of one species affecting the population of the other. (See Figure 12–1.)[5]

In addition, we tend to think in terms of immediate outcomes over longer-term effects. Economists have even coined a term—"discount rate"—to designate the extent to which we reduce our concern for future consequences in comparison with current events. If you have a high discount rate, you ignore information about consequences that will manifest themselves more than a few years into the future. This perspective is particularly evident, and damaging, among elected politicians, but we all suffer from it. If people got cancer immediately from smoking, few people would become addicted to cigarettes. But, in this case, the consequence lies decades in the future, so most smokers ignore it and opt for the immediate pleasure of smoking. High discount rates can act against sustainable

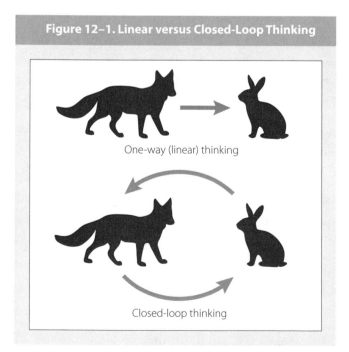

Figure 12–1. Linear versus Closed-Loop Thinking

One-way (linear) thinking

Closed-loop thinking

development efforts that aim to "meet the needs of the present without compromising the ability of future generations to meet their own needs."[6]

Learning to Think About Systems

Enter the age of the Anthropocene, the term that many scientists are now using to describe the significant global impact that humans are having on Earth's air, water, and land—or, more specifically, on its interacting systems, including our atmosphere, hydrosphere, biosphere, geosphere, and cryosphere.[7]

How can education—whether in school, on a farm, in a lab, or at the kitchen table—enable the next generations to live sustainably and navigate the radical changes that they are inheriting in this human-dominated epoch? We do not need a specialized degree to answer this question. Common sense tells us that to understand human impact on Earth's systems, we need to understand *systems*. The question then becomes: how can young people develop systems-based skills and "habits of mind"—for example, discerning what is a system and what is not; looking for recurring patterns of relationships across subjects and situations; making systems visible through maps and models; anticipating how the functioning of a living system will change if a part or a process is changed (assuming that nothing stands in isolation); and looking for causes and consequences in a slew of interconnected systems, including families, schools, local economies, the environment, and more?

The good news is that systems education is happening in schools, nature centers, community meeting rooms, board rooms, and even on playgrounds. Perhaps the best way to look at the emerging state of systems education is through the lens of a constellation. Such a view reveals the brilliance of multiple points of light, including: the new face of Earth system science, the rise of "Education for Sustainability," pioneering systems-based curricula, the teacher as systems thinker, innovative out-of-school learning and application opportunities, and the growing demand for corporate and nonprofit "systems leadership." Here, we discuss several of these stellar exemplars, pointing the way toward more-comprehensive systems-based education for all.

The New Face of Earth System Science

Historically, Earth science as a discipline has taken a back seat in U.S. schools. The more-talented college-track students often pass over Earth science classes, which are perceived to be less rigorous than the Nobel Prize-worthy subjects of physics, biology, and chemistry. But this picture is slowly changing, thanks, in part, to a growing recognition that an understanding of Earth's interconnected systems is critical to our planet's future.

According to the 2007 report *Revolutionizing Earth System Science Education for the 21st Century*, many U.S. states are reshaping their Earth science curricula, standards, and high-school science graduation requirements. These states are "revolutionizing their approaches to Earth science education by moving towards a '21st century' view of Earth science, with an increased focus on Earth as a system, use of new visualization technologies that reveal Earth's processes in powerful ways, recognition of Earth science as a cutting-edge domain of vital importance for our future, and responsiveness to the national call for workforce development."[8]

Since the formation of Earth system science as a discrete yet integrative field of study in the 1980s, the importance of this discipline was reinforced by the creation in 2000 of the Earth System Science Education Alliance (ESSEA). With additional support from the National Aeronautics and Space Administration and the National Science Foundation, ESSEA offers online programs for primary, and secondary, school teachers that are designed to improve the quality of geoscience instruction and to promote an understanding of Earth's interrelated systems. Worldwide, there also is growing interest in teaching "Big History," a systems-based perspective on the origins of Earth and on how humans have come to dramatically transform the planet. (See Box 12–2.)[9]

The Emergence of Education for Sustainability

Spurred on by the United Nations Decade of Education for Sustainable Development (2005–14), schools, school systems, universities, and education departments in many U.S. states are now incorporating "Education for Sustainability" (EfS), also called "Education for Sustainable Development" (ESD), into curricula and instruction. There is a developing consensus both in the United States and globally that systems thinking is an essential element of sustainability education. Both EfS and ESD use systems thinking to empower "students to make decisions that balance the need to preserve healthy ecosystems with the need to promote vibrant economies and equitable social systems for all generations to come." U.S.-based organizations such as the Center for Ecoliteracy, the Center for Green Schools, the Cloud Institute for Sustainability Education, Facing the Future, and Shelburne Farms, as well as global institutions such as the United Nations Educational, Scientific and Cultural

Box 12–2. Big History Teaches Systems Thinking and Transforms Worldviews

"Big History," a discipline first developed by historian David Christian, teaches the "origin story" of how humans came to be, based on our best science. A Big History course typically describes Big Bang cosmology, the creation of stars, and the dispersal of new chemical elements from dying stars, enabling the creation of planets. It covers the conditions for the emergence of life on Earth and eventually of our own species.

Many Big History courses identify the distinctiveness of our species as being our capacity for "collective learning," the ability to share ideas so efficiently that the information learned by individuals accumulates in our collective memory from one generation to the next. This creates a level of technological creativity that no other species has been able to match, enabling us to transform our biosphere. Big History also looks ahead, pondering whether humanity—and perhaps other intelligent species throughout the universe—end up growing in power faster than they grow in wisdom, never making it beyond "civilizational adolescence" to maturity, where they find balance with their planetary system and with themselves.

Big History is now taught in a broad range of high schools in Australia, the Netherlands, Scotland, South Korea, and the United States, among others. Its growth in popularity has been aided by the Bill Gates-sponsored Big History Project, which has developed a rich set of teaching resources that are freely available to high school teachers worldwide. The number of high schools with ongoing support by the Big History Project has increased from just a handful in 2012 to well over one thousand in mid-2016. The course is offered in a growing number of colleges as well, and even as an introductory lecture at the Presidio Graduate School (known for its MBA in Sustainable Management). Dominican University in San Rafael, California, has pioneered an undergraduate first semester "Introduction to Big History" course, followed by a broad array of liberal arts courses taught through the lens of Big History.

Learning Big History can transform a student's vision of humanity, which can lead the student to embrace sustainability values and behaviors. Students learn to think across multiple time scales. Their writing skills improve markedly as they learn to think more critically, using evidence to support their points. Perhaps most importantly, learning Big History teaches students to empathize with other peoples' perspectives.

—Dwight E. Collins, Professor Emeritus, Presidio Graduate School

—Russell M. Genet, Research Scholar in Residence, California Polytechnic Institute

—David Christian, Professor of History and Director of the Big History Institute, Macquarie University, Sydney, Australia

Source: See endnote 10.

Organization (UNESCO), the International Union for Conservation of Nature (IUCN), and the Swedish International Centre of Education for Sustainable Development share a common goal: for students to apply systems thinking to problem solving and decision making on the road to a sustainable future.[10]

These and numerous other groups provide professional development and coaching to improve teaching and learning for sustainability, working with faculty around the world to develop curricula, lessons, and projects that educate for sustainability. In the United States, at least twelve states and dozens of school districts have embraced policies that promote and support EfS, including California, Colorado, Maryland, Massachusetts, New Jersey, New York, Pennsylvania, Vermont, Washington, and Wisconsin. Globally, the picture is even brighter. In many countries, including Australia, Brazil, Canada, Japan, the Netherlands, New Zealand, Sweden, and Switzerland, ESD is housed within ministries of education.[11]

In the state of New York, for example, the Putnam/Northern Westchester Boards of Cooperative Educational Services Curriculum Center has developed an EfS curriculum that provides multidisciplinary web-based materials to address the question: How are we all going to live well within the means of nature? The curriculum challenges students to think about issues that affect their future—all within the context of their existing curricula in math, language arts, science, social studies, and the arts. The program is being integrated in kindergarten through grade twelve (K–12) classrooms in the region.[12]

K–12 Curriculum: Integration and Innovation

Educational standards can act as beacons for schools. Teachers and parents use them to make clear what students are expected to learn in each grade and subject. It is only in the last decade that "understanding systems" has begun to show up in state educational standards. The most recent U.S. science standards feature "patterns," "cause and effect," and "systems and system models," among others, as cross-cutting concepts. Although intended primarily to cut across science disciplines, these concepts also enable students and teachers to recognize recurrent features of complex systems across disciplinary boundaries.[13]

How might this look in a classroom? Imagine that you are a student in a fourth-grade science class. Your teacher tells you that ecosystems, like all dynamic systems, are made up of many cause-and-effect interrelationships. Walking through a national park, you may not be able to see how wolves and rabbits interact, but you learn that these two species are tightly connected in a predator-prey relationship and exist within a closed loop of cause and effect: any change in the circumstances of one species will ultimately influence the other, and vice versa.

You create a graph of the changing wolf population. The general pattern of behavior is one of oscillation, rather than of continued growth or decline. (See Figure 12–2.) Your teacher then asks, "Where else do you see this kind of up-and-down pattern of behavior?" Someone shouts, "Our thermostat!" Another says, "Hungry, eat, not hungry; hungry, eat, not hungry . . . ," and another asks "supply and demand?" The teacher explains: "This type of causal loop, known as a balancing feedback loop, returns a system (such as your household cooling system, your

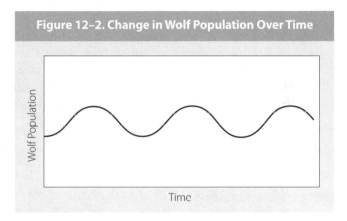

Figure 12–2. Change in Wolf Population Over Time

Wolf Population

Time

body, or an ecosystem) back to a state of equilibrium. By its very nature, balancing feedback works to self-regulate and self-correct systems, to bring them toward some goal or desired state and keep them there."

Similar cause-and-effect relationships can be introduced in history class. When presenting a unit on U.S. westward expansion, the teacher asks, "What set of interrelationships led to improved agricultural land, drawing settlers out west?" Students map out one set of causal connections, including increased railroad access and new technologies and inventions. They show how innovations such as the mechanical reaper and the steel plow helped convert open range into farmland, which enabled more settlers to work the land, which subsequently drove demand for further technologies that could increase the production and distribution of crops. As even more settlers headed west in search of farmland, driven in part by the promise of the Homestead Act, the amount of territory that was improved for agricultural use grew. In a case of reinforcing feedback, one change amplified another. The teacher then asks: "Where else do you see a similar pattern of behavior?" Students suggest the spread of a rumor or a virus. They are right.[14]

These students are now cued up to recognize these cause-and-effect patterns through analogy—or, more accurately, through "homologies": recurring structural similarities that surface in a wide variety of systems, from ecosystems and families to global politics and the Internet. As writer and cultural anthropologist Mary Catherine Bateson observes:

> Central to the study of ecosystems is the study of circular processes of self-regulation and self-correction. If children grasp the concept of self-regulating systems, they can apply it to systems of all kinds, including the functioning of their own bodies or families, schools or neighborhoods. When an abstract pattern has been recognized in a single memorable example, the possibility of multiple analogy is created. In this sense, ecology offers tools for thinking about why it is unwise to experiment with addictive drugs, about the course of family quarrels, or about damage done by racism.[15]

A compelling example of learning design that bridges disciplinary boundaries while inspiring innovative learning environments could be found in the classroom of Frank Draper. Draper was a longtime science teacher at Catalina Foothills High School in Tucson, Arizona, where he taught an honors, advanced field-science course for seniors. Over the course of his twenty-year teaching career, systems thinking concepts and tools transformed his science curriculum. As Draper notes: "Science, as it is generally taught in our country, is mostly a series of facts unrelated to each other in terms of dynamics and relationships."[16]

In his class, Draper integrated the sciences, combining anatomy, physiology, evolution, biogeography, ecology, geology, chemistry, and more into a unified understanding of how the natural world works. At any one time in his classroom, four labs would occur simultaneously, each involving one-quarter of the class and exploring similar self-regulating patterns found in different "systems": a cooling coffee cup, the thermodynamics of animal temperature, the feedback relationship driving cumulonimbus cloud buildup, and the impact of thermodynamics on plate tectonics. Students would spend a full class period on each of the four labs and then, on the fifth day, spend a portion of the class outdoors, trekking in the wild desert surrounding the campus. Classroom learning is, in Draper's view, ultimately about a "better understanding of the real world outside the classroom."[17]

Does this approach work? By the end of their high school experience, Draper says, the students who attended his class were able to explain the world "not as a series of discrete events, but as a rich, interconnected structure." He adds, "I have seen it so often: a systems-thinking worldview helps concepts make so much sense because they are retained better."[18]

Educators report that the use of visual tools such as causal-loop diagrams and simulations helps to level the playing field, enabling students with different language skills to clearly communicate and make their thinking visible. According to the Pittsburgh-based Waters Foundation, a nonprofit organization dedicated to systems thinking education, "Systems thinking engages even the most reluctant students with a mix of visual, verbal and kinesthetic strategies, offering an 'in' for all types of learners. Its tools distill abstract ideas into a shared vocabulary that lets students express themselves with empowering precision. That lucidity also helps them connect classroom learning to the outside world and tackle challenges in their daily lives."[19]

Applying Systems Thinking Beyond School

"Systems" as the context for learning is also appearing in a variety of out-of-school learning settings, including museums, after-school clubs, high school internships, and farms. In Lincoln, Massachusetts, educators at the Drumlin Farm Wildlife Sanctuary use a systems-thinking play kit with visiting children ages six to sixteen. The kit encourages the children to think deliberately about living systems in a farm setting and gives them a mental framework to apply in other contexts. Through games, discussions, and system-mapping activities using bendable wax-coated yarn, students explore the interconnections and dynamics surrounding a mobile chicken coop known as the Egg Mobile. Concepts such as feedback loops, time horizons, and stocks and flows are illustrated through a study of the relationships among elements of a farm pasture, including chickens, cows, soil, plants, and manure.[20]

At the university level, the Social System Design Lab at Washington University in St. Louis, Missouri, applies these same systems concepts and tools to social issues. Students become experts in developing simulation models of problems facing complex social systems. A resource for students, professionals, and researchers, the Design Lab builds the capacity of those who want to learn and apply system dynamics in order to understand and address specific problems within an organization and community. At the 2016 Changing Systems Student Summit—designed, led, and facilitated by youth leaders—students tackled the issue of gun violence in St. Louis. They used system dynamics concepts and tools to uncover the structures that underlie neighborhood gun violence, identifying and prioritizing leverage points for change and engaging and involving the stakeholders who have the power to influence broad institutional change.[21]

Finally, systems thinking is emerging as a partner to critical thinking, problem solving, and social innovation frameworks, such as in the area of design thinking. A promising trend is the spread of "makerspaces" (also known as Fab labs and hacker spaces) in libraries, museums, community centers, private organizations, and schools. Packed with craft and hardware supplies, electronics, and a variety of tools, as well as three-dimensional (3-D) printers, these do-it-yourself stations invite peer learning, knowledge sharing, and social innovation. The integration of a systems-thinking framework within makerspaces, in or out of schools, helps to address the limits of purely technical solutions and raises the question of context: in what context are the "inventions" being designed?

CSM Library

Demonstration of a 3-D printer at the College of San Mateo makerspace, California.

In the case of oysters in Baltimore's city harbor, the context is a living system. Oyster reefs play a valuable role in the harbor's overall health (health being an emergent property, shaped by the interaction among diverse elements within the system). Oysters help to filter waste and provide habitat for a variety of species, as well as offering other benefits. Yet pollution, dredging, overharvesting, and other human impacts have damaged the Chesapeake Bay's oyster population. In a 2016 FabSLAM event, organized by Baltimore's Digital Harbor Foundation, a team of middle schoolers tackled the challenge of a dwindling oyster population by creating 3-D-printed "reef balls"—perforated hollow spheres to which organisms can attach—to serve as oyster habitats.[22]

Closing the Circle

Mythologist Joseph Campbell once said, "People who don't have a concept of the whole can do very unfortunate things." But the corollary is rarely considered: *People who understand the whole can do very fortunate things.* If we raise young people who have a concept of the whole—of how systems work—they will be geared toward seeing the systems around them and will not, by nature and training, see things in isolation. They will not stand for silos but will reach out over silos because they know better. They will be indignant when conversations become narrowly linear and will look for a wider variety of causal connections.[23]

So much in our culture forces us into compartments. But just as we teach kids not to be victims of advertising (see Chapter 13), we can teach them to see beyond the obvious, to see the systems all around us. We can awaken the innate systems intelligence in young people and encourage them to recognize what the Reverend Dr. Martin Luther King, Jr. described in his 1967 *Christmas Sermon on Peace*: "It really boils down to this: that all life is interrelated. We are all caught in an inescapable network of mutuality, tied into a single garment of destiny. Whatever affects one directly, affects all indirectly. . . . We aren't going to have peace on Earth until we recognize this basic fact of the interrelated structure of all reality."[24]

As systems become the context for learning, students will move beyond discrete lists to seeing patterns of interaction that more closely match the interdependent, complex world in which they live. This understanding, coupled with curiosity and stamina, has the power to unleash the vast human potential to navigate and ultimately transform the interlinked social, environmental, and economic issues generated by the Anthropocene.

CHAPTER 13

Reining in the Commercialization of Childhood

Josh Golin and Melissa Campbell

In 2012, *The Lorax*, an animated big-budget version of Dr. Seuss's classic cautionary tale about overconsumption and greed, arrived in theaters. Accompanying its release was a slew of brand-licensed cross-promotions targeted directly to kids. Children were encouraged to visit stores like Pottery Barn Kids, Target, and Whole Foods for Lorax-themed promotions, to eat Truffula Chip Pancakes at the breakfast chain IHOP, to pack their lunches with Lorax YoKids Yogurt, and to urge their parents to buy Hewlett-Packard printers.[1]

The most troubling brand partnership was with car manufacturer Mazda. Commercials on children's television called the Mazda CX-5 the only sport utility vehicle with the "Truffula Seal of Approval." At sponsored assemblies at schools across the United States, a Lorax handed out hugs while Mazda sales representatives told children that the CX-5 was "the kind of car the Lorax would drive"—and urged them to go home and ask their parents to test drive one. For each test drive, Mazda donated a meager $25 to the school's library.[2]

Years later, *The Lorax*'s cross-promotions remain notable because the disconnect is so striking: A beloved Dr. Seuss character designed to warn children against overconsumption and consumerism was being used to get kids to buy, buy, buy. But *The Lorax* is a drop in a vast sea of media properties designed to sell kids not just on products and brands, but on values. And Mazda is just one of countless companies selling products to children in schools, encouraging them to nag their parents, and teaching them to associate school spirit with brand loyalty.

Over the past thirty years, childhood has been transformed by the media

Josh Golin is executive director, and **Melissa Campbell** is program manager, at the Campaign for a Commercial-Free Childhood.

and marketing industries. Marketers spend billions targeting children each year, creating a commercialized childhood that is unhealthy, unsustainable, and leaves kids woefully unprepared for a future that will require new kinds of behaviors, skills, and values. For these reasons, ending the commercialization of childhood should be a high priority not only for parents and educators, but for anyone concerned about the future of our planet. To do so will require strong regulation of the media and marketing industries, establishing schools as commercial-free zones, and helping children spend less time with screens and more time in creative play.

Commercial Culture as Education

Education doesn't just happen in schools. All systems—familial, economic, social—have their own knowledges, which they both produce and teach to their participants. Capitalism and consumer culture are no exception; as children and as adults, we are routinely taught that success can be measured through consumption, and that the brands and products we buy reveal some truth about who we are as people. This education begins at birth: branded products for newborns, from diapers to blankets, are designed to encourage babies to form attachments to corporate mascots and logos, ensuring that they will be requesting brands by the time they can speak.[3]

Capitalism requires constant growth, which requires finding and developing new markets—and it turns out that children are an excellent market. They have unprecedented amounts of spending money, influence their parents to make purchases big and small, and, most importantly, are uniquely vulnerable to advertising messages. Very young children cannot distinguish between commercials and programming, and until the age of about eight, children do not understand advertising's persuasive intent. Even older children sometimes fail to recognize product placement and online advertising: in a 2015 study, only one-third of British twelve- to fifteen-year-olds successfully identified which search results on Google were paid ads.[4]

Because kids are such a lucrative market, there has been a concerted effort over the past three decades to target them with advertising. In 2008, companies spent $17 billion advertising to children, a startling increase from the $216 million (adjusted for inflation) that they spent in 1983. For companies, if not for children, families, and the Earth, these efforts have paid off. It is estimated that children under twelve influence about $1.2 trillion worth of purchases in the United States each year.[5]

The dramatic increase in marketing aimed at children has been facilitated by a huge increase in media made for children. (Or perhaps more accurately, the realization that marketing to children is profitable has led companies to create more kids' media.) And kids are tuning in: screen time often begins in infancy, despite public health warnings to avoid screen media until age two. On any given day, 64 percent of babies and toddlers spend time with screens, averaging just over two hours per day. Conservative estimates find that preschoolers spend an average of 2 hours a day with screens, and some research puts that number as high as 4.6 hours. Kids ages eight to eighteen spend more time with screen media than they do in classrooms or with parents.[6]

Screens are far and away the number-one way that marketers reach children. On television, kids see about sixty-eight commercials per day, not counting product placements. On YouTube, one of the three most popular websites for kids, brands create long-form ads disguised as programming, and children are paid to sell products to other children in the form of "unboxing" videos

Screen power: three youngsters in front of a television display in a shop in Löwental, Germany.

and other influencer advertising. More than half of apps used by children contain advertising, and almost 100 percent of mobile games played by kids serve them ads.[7]

As digital video recorders (DVRs), streaming services, and adblockers allow some children to avoid ads, marketers have developed new ways to sell things to kids. Product placement is increasingly prevalent in kids' video games and online programming. Brand licensing has become an essential part of the children's media economy, and nearly every popular show has toy lines and other licensing agreements, where characters appear on food packaging, apparel, toothbrushes, bedding, and more. And digital platforms let marketers quickly launch toy lines by creating short, inexpensive videos instead of costly, time-intensive television shows or movies.[8]

Online and mobile media are designed to track children and serve them personalized advertising based on their browsing and viewing habits. While laws in the United States and the European Union prohibit the collection of personally identifiable information from children under thirteen, in 2016, the websites of leading marketers Mattel, Hasbro, and Nickelodeon were discovered to be tracking kids. Other apps and websites popular with children simply ignore existing prohibitions on data collection and targeted marketing by pretending that their users are thirteen and older. And there are no laws protecting the information of teenagers. On social networks, friendships are mined for data, sold to marketers, and then used to design ads capitalizing on kids' relationships.[9]

Even traditional educational spaces are commercialized. Thousands of U.S. schools show Channel One News, a current-events program with specially designed student-targeted commercials. Schools that use Channel One lose a full day of school each year to ads. Ads appear in school gyms, on walls, and even across students' lockers. And under the guise of promoting everything from reading to healthy lifestyles, Ronald McDonald regularly visits preschools and elementary schools in Australia, Brazil, China, The Netherlands, and the United States.[10]

In recent years, a wealth of research has confirmed that a commercialized childhood undermines children's social, emotional, and physical well-being. Exposure to marketing is a factor in childhood obesity, eating disorders, unhealthy body image, youth violence and aggression, sexualization, and family conflict. Less understood is how marketing indoctrinates children into consumer capitalism, developing habits and values that are unsustainable not only for kids' physical and emotional health, but also for our planet.[11]

The Unsustainability of Commercial Culture

As an outgrowth of consumer capitalism, the commercialization of childhood has contributed to the rapid changing of our planet. It is unhealthy and unsustainable, and it imparts values and behaviors to children that are directly at odds with the values and skills they will need in a changing world.

First, and most simply, the products marketed to children are wasteful. Take the Happy Meal, the centerpiece of McDonald's comprehensive strategy to hook kids on a lifelong habit of resource-intensive fast food. In 2011, McDonald's sold 1.2 billion Happy Meals, making the company the largest toy distributor in the world. Happy Meal toys are generally small, plastic, and disposable. Most are tie-ins for kids' movies, television shows, and other media properties. Because each of these properties also has countless other toy lines, the appeal of Happy Meal toys is transient at best, sometimes lasting only as long as the meal itself. As one giveaway replaces another, toys inevitably end up in landfills.[12]

It's not just McDonald's that sells kids on this cycle of never-ending consumption. The proliferation of brand licensing means that children's favorite media characters beckon from grocery store shelves, toy store displays, and in nearly every department of big-box stores like Walmart. Licensed characters star in kids' television shows, which, in turn, feature commercials for toys and media starring other licensed characters. These programs and commercials reinforce one another and encourage children to express their love for a character by owning it in as many forms as possible. This is particularly dangerous at a time when the need to consume fewer nonessentials has never been more urgent, but it also signals a larger, deeper problem with the commercialization of childhood: marketers are selling not just products and brands, but also the notion that consumption is a path to happiness. Yet evidence suggests that the opposite is true: the pressure to spend and consume actually makes people *less* happy.[13]

A materialistic orientation, or the idea that buying things will make us fulfilled and successful, has been repeatedly linked to advertising. One study found that materialism among high school seniors was linked over time to national advertising expenditures, while another found that exposure to advertising in schools is associated with increased materialism in students. Studies from around the world consistently demonstrate a relationship between children's media use and materialism, and a recent study found that materialism increases among American and Arab youth as their social media use increases.[14]

Materialism, in turn, is linked to ecologically destructive behavior in children and adults alike. Materialistic people place less value on protecting the environment and are less concerned about the effects of environmental degradation on people living today, as well as the impact it may have on future generations. Consumer capitalism, which drives materialism, prioritizes values such as hierarchy, authority, status, and wealth over unity with nature, social justice, curiosity, and creativity. It is unsurprising, then, that materialistic values are negatively correlated with the prosocial values that are crucial to functioning communities.[15]

These are particularly sobering thoughts as our climate becomes increasingly unpredictable and as our ability to rely on one another, and to respect and understand the natural world, becomes more important. Through a nearly constant barrage of marketing, children are internalizing the dangerous values that led us here in the first place: that we need to buy things to be happy and that our own happiness comes first, even if it is harmful to others or to the planet.

It is now more important than ever to create advertising-free spaces where children can experience nature, build community, and develop a sense of curiosity. Historically, and with varying levels of success, schools have played a role in this kind of value-building, which makes the influx of in-school marketing particularly troubling. As school budgets shrink and as funding gaps expand, brands and industries are able to insinuate them-

A Happy Meal in Madrid, Spain.

selves and their agendas into the lives of children. Sometimes these insinuations are blatant—such as fundraisers where teachers "work" at a McDonald's for a night and encourage their students to buy fast food—but often, they are more difficult to discern.[16]

In particular, sponsored curricula and classroom resources pose a threat to children's education. A review of seventy-seven corporate-sponsored classroom kits found that nearly 80 percent were biased or incomplete, "promoting a viewpoint that favors consumption of the sponsor's product or service or a position that favors the company or its economic agenda." For example, the International Food Information Council (an industry group funded by Monsanto, DuPont, Nestlé, and McDonald's, among others) and the American Farm Bureau Federation (an agribusiness lobbying group) developed lesson plans about the benefits of biotechnology and genetically modified organisms (GMOs) that gave no space for students to debate or discuss the environmental implications of those technologies. This use of classrooms by corporations to promote industry-friendly viewpoints and brand loyalty is not limited to the United States: in Australia, Apple and Mazda are among the companies that actively promote their brands in schools under the guise of education.[17]

Claudio Lobos

Reducing and Reversing the Impact of Commercialism in Children's Lives

Given the impact that the commercialization of childhood has on the health of children and our planet, ending it should be a high priority for educators, advocates, and parents concerned about the future of the world. A three-pronged process for reducing and eventually eliminating child-targeted marketing includes: strong regulation, the designation of schools as commercial-free zones, and a reduction of screen time in favor of physical, creative play.

Regulating Kid-targeted Marketing

The commercialization of childhood is a fairly recent phenomenon and, in the United States, is a direct result of deregulation. In 1980, the Federal Trade Commission was stripped of much of its authority to regulate marketing to children. A few years later, the Federal Communications Commission gutted the rules governing advertising on children's television. Today, the minimal ad regulations on children's television do not apply to programming that kids consume on the Internet or mobile devices. As a result, YouTube and other online platforms are awash with content that would be illegal on children's television, including "host selling," where the host of a program endorses a toy or product.[18]

Other countries offer, or are beginning to develop, better protections for children. Most notably, Sweden and Finland prohibit any advertising to kids. In 2014, Brazil declared marketing to children under thirteen abusive and therefore illegal, although it remains to be seen whether this policy will be enforced. And in 2016, a bill was introduced in Canada that would ban *all* food and beverage marketing targeted at children, an acknowledgement that commercialism—and not just advertising for junk food—is harmful to them.[19]

For countries that are not ready for all-out bans, policies can be adopted that would greatly limit children's exposure to marketing and lessen its negative impacts. In the United States, the Center for Digital Democracy and the Campaign for a Commercial-Free Childhood (CCFC) are leading efforts to advocate for cross-platform rules that limit and govern marketing to children. Such policies would include time limits on the amount of advertising that can be shown to children, clear separation of programming and commercial content, and significantly more-robust prohibitions on collecting data from children on the Internet.[20]

Commercial-Free Schools

In order for children to become critical thinkers and active problem solvers—two skills that will be invaluable in the coming years—it is crucial that they have time and space to learn and develop without being subjected to commercial pressures or corporate propaganda. Children need to learn how to objectively evaluate the companies and industries whose actions are creating or affecting the social and material conditions of the future. In addition, allowing advertising and sponsored educational materials in schools teaches children and adults to rely on the largesse of corporations rather than to view education as a public good. For these reasons, it is critical to enact policies that prohibit all forms of marketing in schools.

There is considerable public and global support for such prohibitions. Eight countries, including Finland, France, and Vietnam, already ban in-school advertising. In 2014, the United Nations Special Rapporteur in the Field of Cultural Rights issued a report calling for a complete ban on advertising and marketing in both public and private schools. In the United States, chronic underfunding of public education has led many school districts to experiment with advertising as a source of revenue. Yet public opinion remains firmly opposed to marketing in schools, even in difficult economic times: in a 2014 U.S. survey, 66 percent of respondents supported bans on advertising in schools, textbooks, and school buses.[21]

Grassroots efforts can play an important part in ridding schools of harmful commercialism. Consider, for example, the campaign against a 2011 curriculum paid for by the American Coal Foundation and distributed by the publishing company Scholastic to seventy thousand fourth-grade classrooms. Scholastic, a highly trusted brand among educators, claimed that "The United States of Energy" would teach kids about the advantages and disadvantages of different types of energy. Yet the curriculum failed to mention a single negative aspect of coal—not a word about greenhouse gases or the massive pollution caused by mining.[22]

CCFC partnered with Rethinking Schools and leading environmental groups to mobilize parents and educators against the biased materials. Faced with the possibility of long-term damage to its brand, Scholastic pulled the curriculum. The coalition then turned its attention to Scholastic's InSchool Marketing division, which produced "The United States of Energy" and other sponsored materials on behalf of clients like Nestlé, Shell, and the U.S. Chamber of Commerce. Once again, Scholastic relented and agreed to significantly limit its practice of partnering with corporations to produce sponsored teaching materials.[23]

Replacing Screen Time with Unstructured, Physical, Creative Play

Screens are by far the number-one way that marketers reach children. Children and teens spend more time with screens than on any other activity except for sleeping, and most media are ad-supported—if not through visible advertising, then through data collection that is used to serve advertising elsewhere, or through brand licensing that makes entire media programs essentially one long commercial. The simplest and quickest way to limit advertisers' access to children is to reduce the amount of time that children use screens.[24]

In place of screen time, educators, parents, and caregivers should focus on increasing the amount of time that children spend in unstructured, physical, creative play. (See Chapter 7.) Research links decreased screen time and increased play to numerous emotional and physical health benefits, including better sleep, better connection within families, and healthier Body Mass Index. In one study, just one week at a screen-free (read: ad-free), nature-based camp increased preteens' abilities to empathize with one another and to read emotion in other people's faces. (Increasingly, screen-based games, such as Pokémon Go, sell themselves as a kind of nature-based play, but there are reasons to be skeptical of such claims; see Box 13–1.)[25]

Children learn best through unstructured, creative play. Play is where kids learn to problem solve, to figure out how their bodies move through space, and to understand the world around them. Unguided play lets children work through their anxieties, explore their surroundings, and build relationships with their peers. In contrast, play that is shaped by branded and licensed characters—or by toys that have computer-driven "personalities"—can limit children's creativity, reducing a universe of potential imaginative scenarios to a small set of narratives shaped by corporations.

Moving Toward an Ad-Free Future

A social, political, and economic system built on a foundation of ever-expanding consumption has created a world where our shared and individual futures are precarious. Children occupy a unique position in this system. They are trusting, eager, and developmentally vulnerable to persuasion. When children hear more from marketers than they do from parents, caregivers, educators, and friends, it stunts their radical potential, limiting their creativity and

Box 13–1. Pokémon No Go

For many years, and particularly since the 2005 publication of Richard Louv's *Last Child in the Woods*, there has been increasing concern that children spend too much time indoors and, as a result, are disconnected from the natural world. Thus, many welcomed the phenomenal success of Pokémon Go, a location-based, augmented reality game where players visit real-world places to capture, battle, and train virtual creatures. Reports of players of all ages spending more time outside and joining with strangers on cooperative Pokémon hunts filled the news, and many hoped that this would be the start of a new age—one where technology would be used to get kids outside, into nature and interacting with one another, rather than immersed in solitary, indoor screen pursuits.

Within weeks, however, the game's producer, Niantic, announced that it would start selecting specific "PokeStops" and "Pokémon Gyms"—the real-world locations that players must visit to succeed in the game—based on paid sponsorships. The first sponsor was McDonald's, and in Japan, every McDonald's became a Pokémon Go hotspot. When children playing the game arrived at the restaurant, they were enticed to buy Happy Meals with Pokémon Go toys. The promotion was wildly successful and a boon to McDonald's sales.

Perhaps there are ways to use new digital technologies to connect children to nature and each other. But as the Pokémon Go experiment demonstrates, under our current economic model, apps are more likely to drive children to familiar commercial establishments than to unexplored places.

Source: See endnote 25.

their ability to imagine a different world. Marketers teach children to continue the values and behaviors—the overconsumption, the allegiance to destructive corporations, the focus on individual attainment, the dedication to *objects* and *things* over people and planet—that have brought us to this brink.

As we move into an unpredictable and almost certainly turbulent future, seeking out individual success in the form of consumption will not get us very far. Neither will trusting industry to know what is best, preserving hierarchies at the expense of creative problem solving, or looking to shopping as a means to fulfill our emotional needs and desires. The values imparted to children by marketers are necessarily at odds with our rapidly approaching future, and require us to take action now.

CHAPTER 14

Home Economics Education: Preparation for a Sustainable and Healthy Future

Helen Maguire and Amanda McCloat

As complex societal and ecological challenges increasingly jeopardize the future of the planet, it is critical that humans, and especially younger generations, develop new ways of being in the world. All global citizens urgently require new modes of thinking and doing. As we settle into the realities of the Anthropocene—an epoch in which human beings are changing the Earth in profound and potentially irreversible ways—fundamental transformations in learning are required to enable all citizens to adapt. People everywhere will need to develop applicable life skills, appropriate competencies in specific domains, and improved critical and reflective capabilities.[1]

The discipline of Home Economics has distinct qualities and progressive potential in enabling such a future-oriented education and practice toward global sustainable well-being. It enables students to develop essential skills related to living a better quality of life sustainably and to acting globally across a range of daily activities. Home Economics professionals—including researchers, teachers, and curriculum designers—are in a unique position to evaluate alternative futures, to develop innovative skills leading to effective citizenship, and to negotiate alternative visions of development. The central tenet of Home Economics is that the world is our home, and the discipline's body of knowledge serves as a fundamental tool for influencing and enabling sustainable behavior patterns in everyday life.[2]

Helen Maguire is a lecturer of Home Economics, and **Amanda McCloat** is head of Home Economics, at St. Angela's College in Sligo, Ireland.

What Is Home Economics?

Home Economics, as a discipline, aims to achieve healthy and sustainable living for individuals, families, and societies. To support the achievement of this fundamental aim, Home Economics integrates knowledge, problem solving, and practical skills for everyday life with an emphasis on taking decisive action to enhance the overall health and well-being of learners. The discipline of Home Economics incorporates diverse content from the social, physical, and human sciences, contextualizing, applying, and consolidating knowledge related to everyday healthy, resourceful, and sustainable living for individuals. Ultimately, the field of Home Economics is transdisciplinary, employing a wide range of approaches to address topics related, in one way or another, to the phenomena and challenges of the everyday life of individuals and families, both within the home and in the wider environment.[3]

Home Economics encapsulates three broad areas of study: Family Resource Management, Food Studies, and Textiles, Fashion, and Design. (See Figure 14–1.) Within these three areas, it explores diverse elements, including family and childcare studies; family resource management; consumer studies; shelter and housing; sustainable design; food, nutrition, and health promotion; food science and hospitality; textiles, craft, fashion, and design; education; and technology. In Home Economics, the teaching of each element is approached with an appreciation of the centrality of the family, whatever its formation, as a basic unit in the world, with a corresponding mindfulness to the progress and prosperity of the entire human family globally. Core to the discipline is an effort to develop an awareness of the interdependence of the individual or family and the environment and to promote a sense of responsibility to global issues. The systems approach employed in

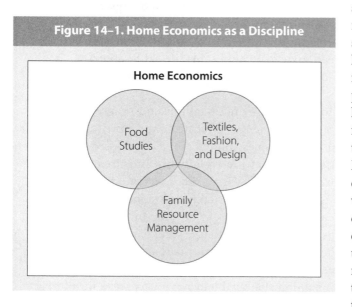

Figure 14–1. Home Economics as a Discipline

Home Economics

Food Studies

Textiles, Fashion, and Design

Family Resource Management

Home Economics encourages students to make sustainable change in addressing everyday practical concerns.[4]

Home Economics as a Curriculum Area

The holistic and integrative nature of Home Economics education across the core areas of Family Resource Management; Food Studies; and Textiles, Fashion, and Design ensures that the subject is ideally placed to integrate a wide range of sustainable development principles and appropriate practical life skills. (See Table 14–1.) Home Economics education empowers students to take action in order to enhance the well-being of individuals, families, and communities at all levels and sectors of society. Ultimately, Home Economics education seeks to prepare students for life by enabling them to develop their resources, capabilities, and decision-making skills. Home Economics teachers also have an influence and can play a key role in shaping the attitudes and values of their students.[5]

Internationally, many Home Economics curricula have identified sustainability-related themes as a guiding principle. In Malta, sustainability is one of four main tenets underpinning the study of Home Economics in schools, where students are encouraged to develop an understanding of their role as advocates for a sustainable future. In Japan, where Home Economics is compulsory for all students, sustainability is a core theme, and "Daily Consumption and the Environment" is one of four areas of learning in the Home Economics curriculum. Home Economics students in Finland study "Home and the Environment," and those in Sweden study "Environment and Lifestyle," as one of three areas of learning. In Ireland, sustainability underpins the Home Economics curriculum across the three areas of food, family resource management, and textiles. Around the world, teachers of Home Economics use a variety of active learning strategies as well as practical activities to engage students and prepare them for a sustainable and healthy future.[6]

Preparing for a Sustainable, Healthy Future Through Family Resource Management

Through Family Resource Management, Home Economics students develop an appreciation of the complexities associated with managing home and family resources in light of a multitude of influences from contemporary society. Family Resource Management affords students the opportunity to appreciate

	Table 14–1. The Home Economics Curriculum: Integrating Sustainable Development Principles and Developing Key Life Skills	
Area of Home Economics	**Elements of the Curriculum That Integrate Sustainability Themes and Values**	**Key Life Skills Developed**
Family Resource Management	The family, family functions, and roles and responsibilities	Understanding roles and responsibilities within the family and society
	Social and cultural issues	Developing effective social, communication, and decision-making skills
	Communication, decision making, and resource management in the home	
	Financial literacy	Managing personal and household resources effectively and sustainably
	Consumer discernment	Developing financial literacy skills
	Household technology	Appreciating and managing the impacts of house design, construction, and day-to-day use on the environment
	Sustainable home and interior design	
Food Studies	Nutrition, diet, and health	Eating a healthy and sustainable diet for a healthy lifestyle
	Cooking	
	Menu planning	Being conscious of food issues from an ecological and ethical perspective
	Budgeting and shopping for food	Developing into a discerning and responsible consumer when making food and health choices
	Food waste	
	Food systems and sustainable food issues	
	Ethical food issues	Saving money and making better use of resources
Textiles, Fashion, and Design	Hand and machine sewing, craftwork, and garment construction	Gaining consumer competencies with regard to the purchasing, wearing, care, repair, and disposal of textiles and clothing
	Textile resource management	
	Ecological and ethical impact of textiles industry	Reducing the ecological and ethical impact of textile production and care
	Sustainable care of textiles and clothing	Developing creativity and skills in extending the lifecycle of clothing and textile products through use and reuse
	Consumer textile purchasing behaviors	

the relationship between the family system, resources, and the environment and is based on systems thinking. It encompasses aspects of the Home Economics curriculum such as consumer competence; financial literacy; family, child, and social studies; and home and resource management. A key role of Family Resource Management is to facilitate and empower students to make wise and sustainable choices and decisions regarding the use of financial, material, and other resources, while being aware of the complex challenges that families encounter in modern society.

The development of Family Resource Management literacies such as consumer responsibility, discernment, and empowerment is particularly important today amid the myriad challenges that individuals and families face in a fast-paced, consumer-oriented society. (See Chapter 13.) In applying sustainable and responsible personal or family purchasing behavior, students are often asked to keep a consumption diary for a selected time period and then to collectively review their consumption patterns with an eye toward considering "needs" versus "wants." Another typical lesson is to select one recently purchased item and to review the sustainability impact of that item against alternative products.[7]

In the classroom, Family Resource Management activities are centered on the real and ongoing concerns of families and communities, and they include concepts for resolving these concerns through problem solving, core reasoning, and responsible and ethical action. Such authentic learning activities are designed to teach transferable processes and to prepare students for their roles and responsibilities when living on their own as adults. Common components that are explored include "the family as a caring unit" and "family functions and structures across societies." For example, an instructor might use a collection of images depicting families from around the world to stimulate discussion about the challenges and opportunities for sustainable and responsible living in a variety of family contexts.

Frequently, Home Economists work with families in the local community to promote and develop everyday literacy skills in family resource management. In the United States, this is conducted through established extension services. For example, the Family and Consumer Sciences Extension of the University of Kentucky conducts seven broad initiatives in the community in an effort to build stronger families, including initiatives such as Making Healthy Lifestyle Choices, Securing Financial Stability, and Accessing Nutritious Food. In Thailand, the Mechai Pattana secondary school actively includes students' families in the education process, providing trainings on resource management and involving them in community service projects and local conservation efforts. (See Box 14–1.)[8]

Classroom activities in Family Resource Management also engage students in practical reasoning that is relevant to everyday life and incorporate problems from students' actual experiences. Working in groups, students may, for example, assess the energy efficiency of a variety of small household or food

Box 14–1. Providing Environmental Consciousness Through Life Skills Training

The education needed to build a sustainable future need not be explicitly environmental. In some cases, a more life skills-based education can develop in students the same academic and personal qualities that a more orthodox and explicitly environmental education would seek to cultivate.

Supporting skills-based environmental education is one way of lowering the barriers to entry for students seeking environmental fluency and hoping to engage in environmental work. By focusing on the means (the personal qualities and vocational skills that support environmental fluency) rather than on the ends (the explicit connections of these qualities and skills to larger environmental narratives), schools can expand the opportunity for people from different classes, cultures, political affiliations, and disciplines to engage with environmental challenges.

In terms of skills-based environmental education, there is perhaps no better example than the Mechai Pattana "Bamboo" secondary school in Buriram, Thailand. Mechai Pattana was established in 2009 by Mechai Viravaidya, who believed in the need for radical rural education reform in Thailand, with a focus on changing: 1) what we teach, prioritizing material that is directly applicable in and relevant to the future lives of students; 2) how we teach, emphasizing student-centered and student-directed learning; and 3) the role of the school, reimagined as a "lifelong learning center" for students and families and as a hub for improving life in surrounding communities.

From this vision, Mechai Pattana was born, with the objective of fostering "good citizens"— learners who are not only academically capable, but also honest, generous, competent in vocational and life skills, confident as leaders in community development, and avid promoters of gender equality. Mechai Pattana currently has one hundred and sixty students from all seventy-seven provinces in Thailand, including students who represent ethnic minorities and some stateless students from the Thai border regions.

On school days, students at Mechai Pattana attend academic classes from 8 a.m. to 2 p.m., and from 2 p.m. onward they engage in a range of student-run businesses and activities. Some students tend to the chickens or goats on campus, others grow melons or off-season limes to sell for profit, others make ice cream or print t-shirts, and others greet guests and plan events. Through these activities, students learn the skills of growing food, working with others, starting and running businesses, managing finances, and more. Students pay their school fees not in money but by completing four hundred hours of community service and by planting four hundred trees per year. The parents of the students must fulfill the same requirements in their own communities.

While there is no environmental curriculum at Mechai Pattana *per se*, all students graduate with certain skills and knowledge that are central to environmental fluency. They are all proficient at growing food and planting trees, and they understand—on an intuitive

Box 14–1. continued

level—the limits of uninhibited population growth, both from regular discussions with the school's founder (the former health minister and an active promoter of safe sex and family planning) and from the lessons of "Rabbit Island," a now-barren island in the middle of a campus pond that has fallen victim to a surging population of rabbits who, as one student explained, "don't practice effective family planning."

The school also has partnered with environmental organizations to organize tree-planting initiatives in communities, conduct biodiversity surveys and water-quality testing on campus, and research the potential for bamboo as an alternative material for construction. In terms of "soft skills," Mechai Pattana inculcates in students generosity, curiosity, respect for one another, and a strong sense of responsibility for the well-being of their community (human and otherwise).

These qualities, while not explicitly linked to larger environmental narratives, foster in students a deep environmental consciousness and fluency that will stay with them throughout their lives. Students grasp the direct personal relevance of environmental-ism to their well-being and that of their families and rural communities. Mechai Pattana stands as one model of a less-explicit form of environmental education—one that teaches through the cultivation of personal qualities and vocational skills—and as a reminder to broaden our expectations of what exemplary environmental education might look like.

—*Kei Franklin, Yale-NUS College, Singapore*

Source: See endnote 8.

preparation appliances. At Presentation Secondary School in Wexford, Ireland, Home Economics students have reviewed appliance energy labels, evaluated the efficiency of the appliances both when in use and on standby, and identified ways to improve the energy performance of the units.[9]

Home design aspects of Family Resource Management can include a review of different housing styles and options for individuals and families, local amenities for sustainable living, passive home design and green building, sustainable materials selection, and energy efficiency in the home (exploring, for example, housing systems and services, insulation and ventilation, energy-efficient spaces, and water heating and cooking practices). Students could, for example, apply this information to real-life analysis of the impact of different housing styles, sizes, and construction types on the energy ratings of buildings. They also could be challenged to design an "ideal" room featuring greater energy efficiency, more-sustainable materials selection, and the use of natural or environmentally preferable alternatives. Students at Glenamaddy

Community School in Galway, Ireland, redesigned their bedrooms to be more sustainable, drawing a scaled floorplan for the room and presenting details of their preferences for the materials, furniture, and fittings selected.[10]

Many Home Economics programs incorporate a focus on care and cleaning in the home, including an awareness of natural versus chemical cleaning agents. Students can apply such eco-efficient practices regularly during the cleanup activities that follow hands-on sessions on food skills. Being an environmentally conscious consumer—through activities such as efficient waste management, materials reuse and recycling, waste reduction, selection of renewable resources, and composting—also can be incorporated in an applied way in both food and textile practicals. Such applied learning enables students to develop self-efficacy—a belief in their own ability to accomplish a task or to succeed in specific situations—together with the skills of consumer discernment, problem solving, reflection, advocacy, and empowerment. Importantly, such instruction can foster in students the ability to synthesize relevant information, exercise critical judgement, and make more-sustainable choices and decisions that can lead to an improvement in the quality of everyday life.

Preparing for a Sustainable, Healthy Future Through Food Studies

The development of food literacy skills is a key component of Home Economics. Students study the theoretical aspects of nutrition, diet, health, and ethical and ecological food systems and then apply this knowledge in daily meal planning, food choices, shopping for food, and cooking. In schools, the emphasis is on the acquisition and development of food literacy skills and promoting a positive attitude toward food so that students are equipped with essential practical life skills. Students are taught how to make healthy, nutritious meals in a hands-on environment, with the lesson underpinned by nutritional theory.

A key aim of the Food Studies component of Home Economics is to ensure that students are food and health literate. This involves a range of life skills, including developing the ability to plan a healthy and nutritious meal, understanding where food comes from (including issues such as food processing and genetically engineered foods), making discerning choices around food selection and purchasing, and possessing the culinary skills to cook meals in a safe and hygienic manner for oneself and one's family. Often, Home Economics teachers work in the community setting to develop food literacy life skills. Through the Recipes for Success project in Sligo, Ireland, conducted jointly by St. Angela's College and the Gaelic Athletic Association (GAA)—Ireland's largest sporting organization—Home Economists work with young athletes in the community to teach the practical life skills of menu planning and cooking healthy, nutritious meals.[11]

In the Home Economics classroom, the advancement of culinary skills is achieved through experiential, student-centered learning activities. Students apply their nutritional knowledge and meal planning skills to plan, prepare, and cook healthy and nutritious meals and snacks in the Home Economics kitchen. This is supplemented by group work to discuss personal or family food choices and the factors that may influence those choices, such as budgetary limitations, food availability, food seasonality, cultural or personal dietary needs, environmental or ethical considerations, religious and media influences, and health and nutritional considerations. The Utah Education Network, a consortium of public education partners in the U.S. state of Utah, has developed lesson guides on planning and cooking meals and emphasizes that skills learned in the Home Economics kitchen should be transferable to the students' experiences in the settings of work, home, and family.[12]

Another important aspect of the Food Studies curriculum is industrial food processing—including food processing techniques, food adulteration, food additives, biotechnology and genetically modified organisms (GMOs), and animal welfare. For example, teachers might present food labels from a range of commercially prepared foods and ask students to identify the ingredients added to food through the use of industrial processing techniques, as well as any ingredients that might be genetically modified. Students could then engage in a debate arguing the case for and against the use of GMOs in food production, referring to resources from the United Nations Food and Agriculture Organization, the World Health Organization, and the European Food Safety Authority, among others. Home Economics students also can explore food adulteration and the ethical standards of the food industry, using case studies of recent food scandals and animal welfare issues.

When discussing the food supply and food systems, students can work in groups and use images to explore the diverse factors that affect the availability of food—including weather, fuel costs, organic production, and the growing season. The group also can examine the impacts of over- and underproduction and the global challenge of food waste. They then can cook a variety of healthy dishes in the kitchen using in-season food. Home Economics courses also commonly investigate social issues related to food, using a case-study approach to encourage students to reflect on food poverty, food insecurity, and malnutrition, as well as the role of the food system in contributing to these issues.

In Ireland and the United Kingdom, Home Economics curricula apply a strong focus on minimizing food waste in the home. This skill is developed in the classroom through effective resource management: students are encouraged to be conscious of planning meals to prevent waste, including by cooking in bulk and preparing food for storage (freezing, preserving, etc.). They learn how to prepare simple, healthy dishes using leftovers. Many teachers use the *Education for Sustainable Development "Images and Objects" Toolkit*, developed by the Consumer Citizenship Network, to teach students about the sustainability of their food choices—including issues such as food miles, organic production, fair trade, local food, climate change, pest control, fertilizers, and packaging. For example, during "fair trade week," students at Little Flower Girls' School in Belfast, Northern Ireland, display and arrange a variety of fair-trade food products for their fellow students to sample and learn about.[13]

Preparing for a Sustainable, Healthy Future Through Textiles, Fashion, and Design

Textiles and clothing are an essential practical and renewable resource for daily life in both the home and the built environment. In Home Economics curricula, students develop practical skills such as hand and machine sewing and embellishment techniques and apply them in the design and execution of a variety of textile craft and fashion projects. Project-based learning and the use of "design briefs"—a written outline of the deliverables and scope of the task including any specific requirements, timing, and budget allocation—is valued as a core approach in Home Economics classes, with the aim of developing the identified textiles skills while also nurturing key competencies such as a sense of inquiry, research and analysis skills, problem solving, judicious decision making, consumer discernment, responsible materials use, and reflective evaluation. Students may be required to undertake the project task in consultation with or to meet the needs of a specific real-world "client."

Sustainable textile resource management is a central component of Textiles, Fashion, and Design in Home Economics. Common topics include product/garment repair; the reuse, repurposing, and recycling of textiles; sustainable material selection; and design innovation. Students might apply hand and machine sewing skills to carry out a range of simple repairs to their favorite garments, or they can bring from home an old garment or product that is suitable for upcycling and use it to design and produce a new item. At St. Aidan's Comprehensive School in eastern Ireland, Home Economics students upcycled old shirts into artful cushions, using hand appliqué and decorative embroidery.[14]

While engaged in such practical activities, students also can be encouraged to discuss acceptable garment lifecycles, "fast fashion" trends (as a counterpoint to longer-term emotional attachments to certain garments), globalization in the fashion industry, and ethical issues associated with the production of textiles and clothing. Maintaining a textile bin in the Home Economics classroom and encouraging students to bring in and reuse textiles-based products from home can enhance this approach.[15]

The care of textiles and clothing has strong potential for focusing discussion on the sustainability of laundering practices. Textile care labeling, sustainable laundry practices, the ecological impact of the selection and use of detergents, efficient water use, water pollution, and washing and drying efficiencies are all elements of this aspect of Textiles, Fashion, and Design. In applying such theoretical knowledge, students might examine care labels on a variety of household products and garments and identify their fabric types and appropriate care practices. Other useful classroom applications may include conducting a group discussion on household laundry practices and routines or exploring the potential for improved sustainability and resourcefulness in their identified everyday practices.

Textile Science is an underpinning theoretical component of the Textiles, Fashion, and Design core area, and students can be empowered to make more sustainable and efficient choices in daily life by understanding related content such as fabric types and properties, production methods, treatments and finishes, natural and organic fabrics and dyes, and the ecological impacts of textiles. In applying such theoretical knowledge, students might undertake simple experiments to review the composition, properties, suitability, and durability of a range of fabrics for use in their own individual design briefs. They also can classify samples of products and clothing by fabric composition and discuss the properties and suitable use for a range of different fabric types.

Textiles, Fashion, and Design also encompasses the development of consumer competency and discernment, integrating sustainable and responsible themes such as purchasing behaviors, consumer decision making, garment and product lifecycles, cultural diversity, and the influences of interculturalism on clothing styles.

Home Economics as Transformative Learning

Like the planet, the subject of Home Economics is now at a convergent moment, as a variety of interconnected societal factors—such as sustainability, global consumption, health, and social justice—are positioning around the profession. Home Economics education is significant in the provision of life skills for sustainable and healthy everyday living. Managing everyday life is not inevitably easy: it requires effecting a variety of cognitive, social, emotional, and practical skills. Home Economics education enables students to participate actively in everyday life, now and in the future, in a sustainable and healthy manner.[16]

Finnish Home Economics professor Terttu Tuomi-Grohn has observed that it is the regular, everyday actions that we take today that create and shape the future. Conversely, because our everyday actions are future-oriented, our expectations and visions for the future also will affect the choices and decisions that we make today. Ultimately, individual choices and behaviors do affect the wider planet. The application of learning with a local and global focus in Home Economics empowers learners to respond to such diverse, complex issues and to consider and practice alternative outcomes.[17]

Helping students recognize the interaction of family systems, the relationship between individuals and their environment, and the global influence on well-being is core to the philosophy of Home Economics. This approach is now regarded more widely as transformative learning and is exemplified in the real-world activities now taking place in Home Economics classrooms, which challenge students to evaluate alternative healthy and sustainable futures, as well as in less-formal mentoring experiences. (See Box 14–2.) Such activities and engagement ultimately lead to effective citizenship, with students experiencing, developing, and negotiating alternative visions of healthy and sustainable development for themselves, their families, their communities, and society.[18]

There is growing recognition of the pivotal role that Home Economics can play in the development of an informed global citizenry that is prepared to adapt and thrive amid the more-challenging Anthropocene era. Home Economics should be available for all students and should be regarded as a central component of a future-oriented curriculum. By uniting the discourses of well-being and sustainability, Home Economics is a critical component of the curriculum that can serve as a key enabler and influencer in this new epoch.

Box 14–2. The Value of Intergenerational Mentoring

Older people in the city of Volograd in Russia are teaching sewing, cooking, and house-keeping to orphaned children. They socialize, play, share knowledge of the history of the country, and encourage the orphans to reach their potential by introducing them to different jobs. The intergenerational learning program—a collaboration between the city and the Elderly People Club—strengthens the connections among generations through mentoring and the teaching of skills and crafts. The children gain love and knowledge, and the elders feel personal fulfillment and contribute to the community.

Intergenerational learning, writes academic Tine Buffel, "brings people of different generations together and builds on the positive resources they can offer each other via purposeful, mutually beneficial activities, which can also benefit participants' communities." It includes the exchange of knowledge, skills, attitudes, and values.

This practice of intergenerational learning is as old as human civilization, with elders in Indigenous communities passing on critical wisdom to the young members of their community. These practices continue today. Young aboriginal girls come together with their elder women to learn Indigenous cultural practices in remote Australia, organized by Kapululangu Aboriginal Women's Law and Culture Centre. In Canada, elders and youth from British Columbia's Seabird Island Band take weekly walks sharing knowledge of plants, cultural landmarks, and traditional stories while building identity and a sense of belonging among youth. There is a growing recognition that traditional knowledge provides essential insights about the ecosystems that are home to Indigenous peoples, and there is equal interest in the value of multigenerational learning as a critical part of sustainability.

Historically, intergenerational learning occurred informally in families. In places such as India, multigenerational families remain the norm. Elsewhere, multigenerational families are on the rise due to economic pressures or health crises. In South Africa, grandparents now raise children whose parents succumbed to the HIV/AIDS epidemic. Increasingly, intergenerational practice is becoming more relevant in extra-familial contexts including schools and educational institutions, cities, communities, and companies.

City infrastructure can support intergenerational exchange—as encouraged by the World Health Organization's Age-Friendly Cities network—through public space design, multigenerational housing options, and institutions such as science museums. Repair cafés, popping up in cities around the world, engage seniors in sharing their expertise in repairing goods, which is building a new movement among youth to demand repairable products.

Professional organizations are recognizing the benefit of mixed-age teams, informal apprenticeships, and mentorship programs. Companies such as GE and Target are engaging in "reciprocal mentoring," where older participants share their rich experience and insights on company strategy while youth share fresh approaches and technological savvy. Technology is supporting mutual learning across continents: the Sole project, also

continued on next page

Box 14–2. continued

known as the "granny cloud," links adults of all ages to classrooms in Cambodia, Greenland, India, Mexico, and elsewhere to read stories, share cross-cultural exchange, and provide a "virtual granny" to the students. In intergenerational meetings in Portugal (known as Foz Côa, or "more social"), elders teach children how to make bread, raise farm animals, and play traditional games while the youth teach the seniors how to use computers, paint, and apply techniques for monitoring their health.

Intergenerational exchanges not only build skills and capacities in the face of sustainability challenges, but also support rich and creative approaches that are essential for creating a just and sustainable world.

—*Vanessa Timmer, Executive Director, One Earth*

Source: See endnote 18.

CHAPTER 15

Our Bodies, Our Future: Expanding Comprehensive Sexuality Education

Mona Kaidbey and Robert Engelman

It is morning at the St. Jan de Groper elementary school, in the Dutch city of Utrecht. Students in all grades are participating in "Spring Fever" week, dedicated to a focused education on sexuality. Lesson plans differ by grade. Eleven-year-olds address sexual orientation and contraception. Eight-year-olds talk about self-image and gender stereotypes. Kindergarteners, some as young as four years old, discuss crushes. Small giggling children raise their hands and talk about how it feels when they hug someone they like. Some talk about embracing their parents or a sibling, while a few mention other students in the class. One boy says a good hug is like "feeling butterflies in my stomach."[1]

The weeklong initiative is part of "Long Live Love," the Netherlands' all-ages curriculum in comprehensive sexuality education (CSE). CSE, as defined jointly by several United Nations agencies, is an "age-appropriate, culturally relevant approach to teaching about sexuality and relationships by providing scientifically accurate, realistic, non-judgmental information." The UN Population Fund (UNFPA) has broadened the definition to cover elements of human rights, gender, and overall life skills, rather than focusing solely on the avoidance of pregnancy and sexually transmitted infections. (See Box 15–1.)[2]

As in the Netherlands, course names for CSE in several other European countries indicate a positive, respectful, and nonjudgmental view toward what children will be learning about sexuality and attitudes toward it: "Good Lovers"

Mona Kaidbey recently retired from the United Nations Population Fund, where she specialized in comprehensive sexuality education, among other topics. She continues to speak and write on CSE.
Robert Engelman is a senior fellow and former president of the Worldwatch Institute. He directs the Institute's Family Planning and Environmental Sustainability Assessment project.

Box 15–1. Comprehensive Sexuality Education: The United Nations Definition

The United Nations Population Fund defines comprehensive sexuality education as a rights-based and gender-focused approach to sexuality education, whether in school or out of school. CSE is curriculum-based education that aims to equip children and young people with the knowledge, skills, attitudes, and values that will enable them to develop a positive view of their sexuality, in the context of their emotional and social development. By embracing a holistic vision of sexuality and sexual behavior, which goes beyond a focus on prevention of pregnancy and sexually transmitted infections, CSE enables children and young people to:

- Acquire accurate information about human sexuality, sexual and reproductive health, and human rights, including about: sexual anatomy and physiology; reproduction, contraception, pregnancy, and childbirth; sexually transmitted infections and HIV/AIDS; family life and interpersonal relationships; culture and sexuality; human rights empowerment, nondiscrimination, equality, and gender roles; sexual behavior and sexual diversity; and sexual abuse, gender-based violence, and harmful practices.
- Explore and nurture positive values and attitudes toward their sexual and reproductive health, and develop self-esteem, respect for human rights, and gender equality. CSE empowers young people to take control of their own behavior and, in turn, treat others with respect, acceptance, tolerance, and empathy, regardless of their gender, ethnicity, race, or sexual orientation.
- Develop life skills that encourage critical thinking, communication and negotiation, decision making, and assertiveness. These skills can contribute to better and more productive relationships with family members, peers, friends, and romantic or sexual partners.

Source: See endnote 2.

in Belgium; "Love Talks" in Austria. There is no furtiveness, embarrassment, or disapproval in these curricula: just basic education about human bodies, sexual feelings, and sexual behavior—and how to communicate clearly, making sure that such behavior is intentional, safe, pleasurable, and respectful of others, and results in pregnancy only when pregnancy is wanted. This education also takes on issues of gender, rights, and how to relate confidently to romantic and sexual partners—and everyone else.

Welcome to "sex ed," European-style. Yet in many places worldwide, sexual education still carries the euphemistic label "family life education," as though sexuality were solely about family formation rather than about individual feelings and expression, emotional and physical health, and social relations. At its worst, CSE is absent, hushed by negative attitudes, taboos, and stigma. In El Salvador, despite passage of a 2012 law guaranteeing the right to sexuality education, "[b]oys and girls come to have their first sexual relationship without having had any professional information," Deputy Health Minister Eduardo Espinoza told a reporter for Reuters in mid-2016. "Generally, the information they have comes from other children who are just as misinformed as they are." More than one-third of all pregnancies in the country occur among girls nineteen or younger, and 41 percent of HIV-positive individuals are from fifteen to nineteen years old.[3]

At its best, as in parts of Europe and increasingly elsewhere, CSE embraces sexual health as essential to physical, emotional, mental, and social well-being. As often expressed by young people, "sexuality is who we are, not what we do." Sexuality is a dimension of who we are as human beings, experienced throughout life with diversity of expression, encompassing sexual orientation, pleasure, intimacy, and reproduction. Yet no safe, intentional, and harm-free variation of sexuality or its expression is inherently bad or abnormal. It is simply an integral, if sometimes challenging and troublesome, part of each person's being. To recognize this is an important step toward learning to respect the right of all individuals to express their sexual and gender orientations, free from violence or loss of privilege, and without interference or coercion from others. Learning this is also about learning fundamental values regarding tolerance, compassion, and the dignity of all human beings—values that may serve humanity well as environmental and other stresses accumulate in the rest of this century.[4]

Not surprisingly, such a perspective is controversial and even abhorrent in many cultures and communities. In Colombia, thousands of parents marched in the streets in August 2016 to oppose new national guidelines designed to enforce a court decision that CSE must address sexual diversity and eliminate gender-based harassment in schools. The protestors blamed the Minister of Education for working with nongovernmental organizations (NGOs) and the UN and others to promote homosexuality in Colombia's schools. Faced with the protests and pressure from leaders of the Catholic Church, the government backed down and disowned the guidelines. Even within the demographic research community, which widely credits expanding education as among the most important contributors to fertility decline, the contribution of CSE itself rarely rates any mention. In the United States, many school curricula on sex, fueled by federal government dollars, limit their teaching to the value of abstinence, despite abundant evidence that teaching abstinence fails to delay the initiation of sexual intercourse or to reduce rates of teen pregnancy, whereas CSE tends to do both.[5]

Given the rancor of opponents and the apparent indifference of demographic experts, international consensus on the value and importance of CSE is hard to assemble. The UN Sustainable Development Goals (SDGs), promulgated in 2015, include among their one hundred and sixty-nine targets for 2030 "to ensure that all learners acquire the knowledge and skills needed to promote sustainable development." This target mentions the importance of education on "gender equality, . . . a culture of peace and nonviolence, . . . and appreciation of cultural diversity"—values that inform many CSE programs. But the SDGs and their targets are largely silent on sexuality education itself and the role it could play in sustainable development. This has inspired more than forty youth organizations to launch a global campaign, "Have You Seen My Rights?" to advocate for young people's right to sexual and reproductive health—including CSE—during the development of the SDGs.[6]

Why Teach Sex?

From the vantage point of educating young people for life in the twenty-first century, it is not hard to see why CSE deserves a secure niche among mandatory courses in an "EarthEd" sustainability curriculum. Environmental and social change show few signs of slowing. In an increasingly crowded and resource-stressed world, young people are more mobile than ever, more socially and economically vulnerable, and more likely to live in insecure urban environments. They are sexually active at early ages, and girls are often forced into early marriage and violent sexual relationships. They are facing the need to adapt to rapid change—in the climate, in technology, and in their prospects for employment and secure shelter—the likes of which no earlier generation has experienced. In such a world, the personal and community consequences of sexually transmitted disease and unintended pregnancy may be more costly than ever.

Human numbers, too, are unprecedented: 7.5 billion people live on Earth in 2017, and more than 1.8 billion of them are from ten to twenty-four years old. More than 2.5 billion people, just over one-third of the total, are children or teenagers, and this proportion is much higher in most less-developed countries. Four-fifths of the world's teenagers and children live in these countries, where both poverty and fertility tend to be higher and where young people are more likely to encounter obstacles to their rights to education, health, and freedom from violence. Absent catastrophic increases in death rates, the reproductive decisions that these young people make—and are able to put into effect—will determine the future of world population. And since no human can avoid interacting with the environment—some more intensively than others—the path of population will play a major role in determining the path of ongoing environmental change.[7]

Today, two out of five pregnancies worldwide are unintended. Roughly half of these unintended pregnancies result in live births. Although intentional reproduction is a globally recognized right and is not under siege in any country outside China, some researchers have argued that childbearing does confer a legacy of long-term environmental impacts that parents should ponder when thinking about family size and their own ecological footprint. A world in which every pregnancy is wanted—a key objective of sexuality education—would undoubtedly support a more environmentally and socially sustainable population than today's.[8]

United Nations Photo

A young mother with her child at the Nutrition Rehabilitation Unit for Children at Schiphra Hospital in Ouagadougou, Burkina Faso.

The need for CSE is equally obvious on an individual level. About 16 million adolescent girls between the ages of fifteen and nineteen give birth each year, most of them in low- and middle-income countries. Complications from pregnancy and birth are among the leading causes of death among adolescent girls, with unsafe abortion contributing significantly to their mortality. Sexual and gender-based violence is rampant across the world. Half of all sexual assaults take place against girls below the age of sixteen. One out of every three girls and women at some point is a victim of violence induced by an intimate partner.[9]

Dealing with such issues requires recognition of the diversity of the world's young people. Geographic and cultural conditions affect their well-being, the protection of their rights, and their opportunities. Young people may be married girls, migrants, refugees, internally displaced, disabled, ethnic minorities, indigenous, lesbian, gay, transgender, or some combination of these. This diversity requires the closest possible understanding of the needs of each group and tailoring teaching to their circumstances and the barriers they face. The need for teacher training and sensitivity to individual student needs is acute.

Can and should sexuality education really begin in kindergarten? A key term in the definition of CSE is "age-appropriate." Attitudes about gender and sexuality are formed early in childhood and consolidate around puberty. Sexual activity of some kind generally begins in the teen years. At the very least, children need to learn at the earliest possible age that no one but family members, close friends, and trusted caregivers should touch them, and that at this age some kinds of touches—those that cause harm or that sexually gratify the toucher—are wrong from anyone. Young children need to learn to respect the bodies of other children and not to bully or otherwise harm others.

One replicable example of working with very young children is the Indonesian program "You and Me," which was inspired by a similar Dutch program. "You and Me" was developed by the CSE-advocacy group Rutgers WPF and the Indonesia national affiliate of the International Planned Parenthood Federation (IPPF). The program is aimed at four-to-six-year-olds and their parents. Topics include "Me and My Body," "Boys and Girls," and "My Feelings and Your Feelings," and the lessons employ games, dolls, and comic books—not only in classes but in schoolyards and parks.[10]

As children grow older, the realm of the age-appropriate expands to the maturing bodies of boys and girls and to sex itself, along with abstinence, contraception, sexual safety, and the critical importance of distinguishing consent from refusal in social, emotional, and physical interactions. No means no, which is something all human beings must learn how to decide, to say clearly, to hear accurately, and to respect and obey when heard. These are the ABCs of CSE. When they are not offered in schools, these lessons have no secure place outside the home itself, where such instruction varies dramatically in both quantity and quality.

Even where there is governmental and popular support for CSE, obstacles are formidable. After a three-year legislative debate, Argentina passed a law requiring CSE courses in public schools, only to discover that 60 percent of teachers said they had no idea how to teach it. Beginning in 2008, the government focused on teacher training, prioritizing parts of the country with the highest adolescent pregnancy rates. In the United States, a study of teen pregnancy prevention programs, a core component of CSE, found that parental indifference or opposition and lack of school-district financial support often made the programs difficult to sustain once limited grant funding expired.[11]

Gaining Support

Defying such barriers, governments and intergovernmental organizations around the world are increasingly expressing support for at least some form of sexuality education. Most recently, in the context of the framing of the SDGs, governments, intergovernmental organizations (including the UN and the World Bank), and civil society groups worked together over two years to develop an international declaration on education, called *Education 2030*. Many among these groups saw sexuality education as a critical dimension of an education that prepares the individual to manage actively and responsibly his or her present and future life trajectory and to navigate safely in a complex physical, social, and economic environment.[12]

Equally important, the UN and other intergovernmental bodies and NGOs have cooperated in offering guidance on CSE—often promoting more holistic approaches that combine traditional sex education with a greater focus on how respect for gender equality and human rights can inform broader tolerance for social diversity and interest in civil participation. A 2010 set of standards for CSE in Europe stressed the importance of sexual well-being beyond health, shifting the mentality of behavior change in favor of facilitating growth and development, a "sex-positive" educational orientation, and respect for rights and diversity.[13]

A 2014 UNFPA document assembled evidence that positive attitudes on gender equality are not just a matter of feel-good political correctness, but actually encourage healthy sexual behavior. One researcher concluded that classes emphasizing critical thinking on gender and power relationships were more than five times as effective in reducing unintended pregnancy and sexually transmitted infections as classes that failed to address gender or power. Similarly, while some governments feel political pressure to focus almost entirely on the prevention of sexually transmitted infection and unintended pregnancy, evidence suggests that more holistic and positive approaches that stress healthy sexuality achieve the same prevention results. Moreover, they may achieve them more effectively, along with a wider range of benefits related to respect for rights and gender and other personal relationships.[14]

This evidence adds to a large body of data—often based on self-reporting by young people—demonstrating that exposure to sexuality education decreases high-risk sexual behavior, delays sexual initiation, reduces the numbers of sexual partners, and increases the use of condoms and other forms of contraception. Studies indicate that young people who adopt more egalitarian attitudes about gender roles or who form more equal intimate heterosexual relationships are more likely than their peers to delay sexual debut, use condoms, and practice contraception. They also have lower rates of sexually transmitted infections, HIV, and unintended pregnancy and are less likely to be in relationships characterized by violence. While there appears to be no peer-reviewed scientific literature exploring the implications of such links to the future of the world's population, let alone its environment, it seems self-evident that the behavior that CSE encourages contributes to a population that is changing more sustainably than currently and that is characterized by healthier human relationships.[15]

Sexuality Education in the World

The advance of this research, along with the growth of technical guidance, appears to be having an impact in school classrooms around the world. A recent report by the United Nations Educational, Social and Cultural Organization (UNESCO) surveyed forty-eight countries and concluded that "[t]he vast majority of countries [surveyed] are now actively embracing the concept [of CSE] and engaging in the process of supporting—or strengthening—its implementation at a national level. This has resulted specifically in ongoing attention to curricula revision in many countries, integration of CSE into the national curriculum, and the development and roll-out of effective teacher training."[16]

The commitment and quality of these efforts vary by region and among and within nations. In much of Latin America and the Caribbean, adolescent pregnancy rates are among the highest in the world. In 2008, health and education ministers in the region agreed on a declaration that combined commitments both to CSE and to the provision of contraceptive services. The region's ministers thus modeled a key linkage in sexuality education between what students learn in class and youth-friendly sexual and reproductive health services. Both are needed for healthy sexuality. A recent review by the Western Hemisphere affiliate of IPPF found the highest levels of implementation of "Education for Prevention" in Costa Rica, Argentina, and Uruguay, followed by midlevel implementation in Colombia, Honduras, El Salvador, Peru, the Dominican Republic, and Mexico. The lowest implementation was seen in Chile. Latin American civil society groups have played an important role in ensuring the continuity of efforts at times of political transition and have influenced a new wave of reform to improve the quality and reach of sexuality education across the region.[17]

In 2015, a regional declaration related to population and development in Latin America and the Caribbean went so far as to offer a definition of sexual rights, a highly controversial concept internationally. These were defined as "the right to a safe and full sex life, as well as the right to take free, informed, voluntary and responsible decisions on [individuals'] sexuality, sexual orientation, and gender identity." The declaration stated that CSE should "ensure the effective implementation from early childhood of comprehensive sexuality education programs . . . with respect for the evolving capacity of the child and the informed decisions of adolescents and young people regarding their sexuality."[18]

This ministerial-level leadership on the issue cascaded across the southern Atlantic Ocean. Health and education ministers from twenty governments in East and Southern Africa, where decades of devastation by the HIV pandemic had already led to some advances in sex education, explicitly cited their Western Hemisphere colleagues when they endorsed, in 2013, a declaration whose sole purpose

A volunteer demonstrates correct condom use to secondary school students in Chandigarh, India.

© 2007 Pradeep Tewari, Courtesy of Photoshare

was to bolster the case that their region's young people, too, need access to CSE and to sexual and reproductive health services. The impact was much less influential in Central and West Africa, however, where opposition to sexuality education runs high. In that region, girls are far less likely than boys to go to school at all, and child marriage rates are among the highest in the world.[19]

Currently, UNFPA and UNESCO are undertaking efforts to engage governments elsewhere in expanding CSE, but no ministerial gatherings similar to those in Latin America and the Caribbean and in Eastern and Southern Africa have yet occurred. In the Middle East and North Africa, only a few countries—notably Algeria, Lebanon, and Tunisia—have in-school sexuality programs. In this region, youth-led networks such as Y-PEER have played the primary role in reaching out-of-school children and young people.[20]

In Asia and the Pacific, approaches to CSE appear to be as diverse as the region itself, where 57 percent of the world's under-twenty population lives. A

UNESCO review of the experience of twenty-eight countries in the region found that fewer than half had any integration of sexuality education at the primary school level, while slightly more countries included it in secondary schools. Content varies by country, with conservative views about sexuality often governing the educational approach. In some countries, however, gender-related violence has created openings for strengthening the programs, often due to civil society pressure on governments and schools. The movement toward greater emphasis on gender equality appears to be gaining ground in existing programs. Where no programs exist, volunteers often go into schools on their own to engage students on sexuality—as has happened in Kyrgyzstan, with assistance from IPPF. "The response from the children is great because no one else talks to them about these topics," reports one twenty-three-year-old volunteer.[21]

The leading region for CSE, however, remains Europe—including Eastern Europe. Estonia has pioneered efforts to link sexuality education and reproductive health services, demonstrating the cost-effectiveness of this approach. The national program began in 1997, and two revisions have occurred since then. By 2009, CSE had reached twenty-eight thousand students in three hundred and twenty-nine schools, and research documented significant reductions in pregnancy, abortion, and sexually transmitted infections.[22]

Western European countries such as Belgium, Denmark, Finland, the Netherlands, Norway, and Sweden are demonstrating the value of holistic sexuality education that stresses not so much behavior change as personal growth and sexual development. While not ignoring the risks that sexual behavior can pose, these risks are subordinated to the development of personal skills such as assertiveness, empathy, and clear communication. The programs teach respect for diversity and a sense of responsibility for one's own sexual health and well-being as well as that of others. And they encourage young people to grow freely in their understanding and enjoyment of sexuality. For many in the public health community, the European approach offers a gold standard for CSE that the rest of the world's countries can model for successful outcomes.[23]

CSE and the Future

What are successful outcomes toward which sexuality education should grow? Happiness and health for the young as they turn into adults, for starters. But the benefits are also catalytically global and long-term—and, as it happens, especially germane to an EarthEd curriculum. The future of world population and its social and environmental impacts depends largely on the behavior and decisions that today's 2.5 billion children and teenagers will make. The families they form, the friendships they forge, the ways that they deal with cultural and gender differences, their health, and their overall well-being are critical for humanity and the Earth, now and for decades to come. The critical thinking skills that young people develop through sexuality education will enable them to make informed decisions about some of the most intimate aspects of their being: their bodies, their relationships, and their sexuality. And those skills will inform the decisions they make about everything else in their lives, including their environmental behavior, their political engagement, and their decisions about what they buy and consume and how they live.

How to teach about sexuality is a decision best made with the fullest participation of all stakeholders: educators, parents, and students themselves. Transparency in the development of CSE is essential, with availability of the details of the curricula available to anyone wishing to review them. Conversations with all parties, including opponents, can often dispel myths and misconceptions. It is unlikely that opposition will soon disappear in spite of such openness, however. That is best addressed through continued communication of the well-documented benefits—and, conversely, the lack of any evidence of harm to young people—of CSE and through skilled mobilization of public support for its presence at all levels of education, everywhere.

The best of the European programs point the way to success. In the Netherlands, rates of teen pregnancy, birth, abortion, and sexually transmitted disease are far lower than those in the United States and even most other European countries. It can hardly be coincidence that the country offers few barriers to access to contraceptive services and reproductive health care generally, and that it accepts the principle that encouraging healthy sexual development is an essential component of education. The country's course name for its CSE program in schools could well be a motto for the essence of sustainable human life in a challenged century, where conflict and intolerance are on the rise and environmental change threatens the long-term survival of civilization. Learners who internalize—along with their reading, writing, and arithmetic—*long live love* are likely to become the citizens that the rest of the twenty-first century desperately needs.[24]

PART TWO

Higher Education Reimagined

CHAPTER 16

Suddenly More Than Academic: Higher Education for a Post-Growth World

Michael Maniates

Over the past twenty years, higher education has undergone an environmental revolution. Campus sustainability offices that track resource use and promote eco-efficiency are becoming the norm. The number of academic programs in sustainability science and environmental studies has increased, as have student enrollments and passion. New academic journals have flourished, and with them venues for publication by young academics aspiring to become tenured professors in the field. It is true that many colleges and universities worldwide have not fully embraced this momentum: they have yet to incorporate tenets of sustainability into their hiring, curriculum, infrastructure planning, or investment strategies. But these institutions are viewed increasingly as outliers that poorly serve their students and the social good. Anthony Cortese, an early advocate of environmental stewardship within higher education, gets it right when he recently observed that "higher education's rapidly expanding response to this [environmental] challenge over the last two decades is a beacon of hope in a sea of turbulence."[1]

Cortese's beacon of hope is powered by an array of college and university sustainability programs, some of which are described here. It is a bright light that is not easily dismissed, but it must now evolve if it is to guide us though the coming turbulence of environmental change and social turmoil. The reason is both straightforward and stark: we are getting our first real taste of a post-growth world where rapid, sustained economic growth is a thing of the past, and it looks to be a bumpy ride.

Higher education is uniquely positioned to nurture and disseminate the social innovations that we must embrace to make our way to a world free from

Michael Maniates is professor of environmental studies at Yale-NUS College in Singapore.

the environmentally destructive imperative of rapid and sustained economic growth. After all, colleges and universities are globally distributed, loosely networked around an expanding agenda of sustainability, and open to new ideas. They command respect. But they also are creatures of the high-growth world from which we must exit: they depend upon economic growth and often promote it, and, as a result, the sustainability efforts that flow from them are often tailored to it. If how we school our children is an important part of the puzzle of human prosperity in a turbulent twenty-first century, few tasks are more important than reorienting higher education toward a post-growth future.

Thriving in a Post-Growth World?

For almost forty-five years, since the 1972 publication of *The Limits to Growth*, a sliver of the environmental movement has struggled to frame unending economic growth as a core driver of environmental harm. They argue that an ever-expanding economy generates exponentially increasing pressures on environmental systems that will inevitably carry us beyond the "safe operating space" of the planet. From the depletion of forests and fisheries, to ocean plastic pollution and climate change, it is the staggering growth in the volume of materials extracted, products consumed, and waste produced that brings us to the doorstep of an environmental unraveling. Emergent "green" technologies—renewable energy, polyculture agriculture, decarbonized transportation systems, reduced product packaging, and the like—can blunt these impacts, but not for long, since their ameliorative effects are quickly swamped by the environmental penalty of additional growth. To reap lasting benefits from these technological innovations, we need a suite of parallel social innovations to release us from political and economic systems addicted to growth.[2]

These arguments have largely fallen on deaf ears. Rather than viewing economic growth as a source of environmental degradation, most see it as essential to sustainability. To them, an ever-expanding economy drives technological innovation while lifting billions out of environmentally destructive poverty. Growth also means more government revenue, which supports ambitious environmental initiatives at no cost to other programs, thus avoiding nasty political conflict. From this vantage point, growth is the solution, not the problem, so long as it is environmentally sound growth. The job of higher education is to train students and to conduct research that produces the technologies and practices that fuel this growth and turns it from "brown" to "green."

For the past few years, this pro-growth logic has been under especially fierce assault by scholars and activists. (See Chapter 20.) One is Boston College sociologist Julie Schor, who calls for a "slow consumption" movement inspired by the push toward "slow food" and writes persuasively about a "plentitude economy" divorced from the growth imperative. Another is economist Richard Norgaard, professor emeritus of energy and resources at the University of California at Berkeley, who likens today's obsession with growth to a modern-day religion built around a disastrous faith. "The economy," Norgaard suggests, "really is the world's greatest faith-based organization." And Gus Speth, former dean of the prestigious Yale School of Forestry and Environmental Studies, recently wrote that "it is time for Americans to move to a post-growth society, where our communities and families are no longer sacrificed for the sake of mere GDP growth."[3]

The rough outline of social innovations for this post-growth society is clear. As Speth explains: "[W]e already know the types of policies that move us toward a post-growth economy that sustains both human and natural communities . . . . [There is] a long list of public policies that would slow GDP growth, thus sparing the environment, while simultaneously improving social and individual well-being." Speth's list, which draws on work by numerous scholars, includes shorter workweeks, longer vacations, and more investment in local, small-scale economic enterprises that prosper by staying small. A shift to worker cooperatives and community banking with a strong commitment to social equity and environmental limits also makes the list. So too do progressive taxation policies, seed grants to promote community entrepreneurship, and guarantees for part-time workers.[4]

Speth's recommendations could easily be dismissed by those unpersuaded by the post-growth argument, except for one glaring reality: for more than a decade, we have been living in the very low-growth world that many dismiss as impossible, hopelessly dismal, or a retreat to some dark age. "Economic growth," explains Neil Irwin of the *New York Times*, "has been weaker for longer than it has been in the lifetime of most people on Earth." Since 2001, U.S. economic growth per capita rose 0.9 percent a year, almost a 60 percent decline from the 2.2 percent annual increase between 1947 and 2000. Economic growth in Western Europe and Japan has been even lower. Because of a number of still-unclear factors—aging populations, slowing population growth, and the intermittency of economically transformative technologies, among others—there is good reason to expect this tepid growth to continue, with some ups and downs, for the foreseeable future. These will be turbulent times that call for a particular kind of education across colleges and universities. (See Box 16–1.)[5]

It is tempting to double-down on the economic growth machine by mobilizing multiple forces in society, including higher education, to get us back to the time of 2 percent-plus growth per year. In the short run, this might work, but it ultimately will heighten the conflict between exponential economic growth and the integrity of environmental systems upon which human prosperity rests. Much of this growth, after all, delivers ambiguous benefits, and some of it actively undermines human prosperity. In the end, we will still need to deal with the implications of persistent low- or no-growth—the material base of the economy cannot continue to grow exponentially—with an even more despoiled environment on our hands.[6]

More important, pining for the "good old days" of robust growth diverts us from the critical task of adjusting, now, to a low-growth world in ways that are just, equitable, democratic, and environmentally restorative. Even if we believed that a return to muscular growth was just a few years away, wouldn't we want to explore how to gracefully adapt to our current conditions, if only as an insurance policy against the possibility that the days of high growth are behind us? Few, if any, of the social innovations described by Speth are inherently anti-growth, so there is little to lose by assessing and spreading them as we are able.

For reasons still opaque to economists, slow growth is no longer a fuzzy wish tossed about by environmental scholars. It is here, among us, in our communities, on the ground, affecting our pocketbooks and driving our politics. It is no longer just academic. Rather than treating tepid growth as a problem to be solved ("how do we get the economy growing again?"), higher education can reclaim its beacon of sustainability by attacking an altogether different but immediately relevant question: How do complex human societies thrive—environmentally, equitably, and justly—in a post-growth world?

Higher Education and the Growth Machine

Under normal circumstances, it would be foolhardy to expect colleges and universities to tackle this question. After all, higher education has long been understood as an engine of economic growth. When educational researchers Anna Valero and John Van Reenan show that universities around the world drive economic growth, their results are publicized by higher education leaders as evidence that the university is doing its job. No one raises an eyebrow when

Box 16–1. Running the Rapids

The past fifteen years of low economic growth helps us envision a post-growth society. But a sluggish economy is no cause for celebration. Slow growth constrains government spending and drives politically toxic zero-sum thinking, and it can exacerbate social inequality, leading to deep feelings of marginalization. That makes it difficult to marshal the shared commitment critical to a sustainable future. These forces also can engender a yearning for strong leadership, and even authoritarianism, that offers comforting answers to complex questions. Spin grows more important than fact, and elite decision making informed by careful analysis falls into disrepute. All this arises at a moment of environmental instability and human hardship spawned by climate instability, water scarcity, collapsing fisheries, and stark economic inequality.

It is going to be a bumpy ride to a post-growth future of prosperity and justice. Navigating this turbulence requires college and university students that are imbued with a special set of skills and temperaments: a steely equanimity, adept at conflict management, familiar with notions of social change, well versed in the science of sustainability with a rootedness in values of justice and community, and more at home in the metaphorical turbulence of whitewater rafting than the placid predictability of canoeing on a gentle summer's day.

Alas, these are not the sort of people that higher education typically graduates. Most institutions focus on producing experts who will command the respect of policy makers and citizens by virtue of their training. They are canoeists, poorly acclimated to a world of surprise, unpredictability, and opportunity.

Teaching for this coming turbulence does not mean skimping on analytic rigor. But it does require more than getting the facts right in the classroom. Students must become practiced in coping with ambiguity and diffusing conflict around contentious environmental issues, drawing with ease on a healthy mix of qualitative and quantitative insight. They will be well served by their instructors if they come to understand themselves not as "I have the right answer" elites, ready to assume their place in the halls (or cubicles) of power, but as "knowledge brokers" tasked with creating and disseminating knowledge in ways that privilege values of precaution, systems thinking, and advocacy for the defenseless—typically the poor, the environment, and future generations.

Cultivating these orientations calls for a special breed of professor, one that is open to curating student experiences that foster boldness and humility. Fortunately, professors like these exist, and they are no longer restricted to institutions such as College of the Atlantic or Prescott College, both exemplars of sustainability. Higher education is changing, and for the better, but it must quicken its pace if students are to confidently run the rapids to come.

Source: See endnote 5.

administrators such as Elisa Stephens, president of the Arts Academy in San Francisco, assert that "higher education, job creation, and economic growth" are inextricably linked. History explains much. In the United States, public universities were launched in the nineteenth century to innovate agricultural and engineering practices in service of economic growth. (See Chapters 21 and 22.) The contemporary incarnation of European universities follows a similar path. It is by design, not by happenstance, that higher education in its modern form is a core component of "the great acceleration"—the exponential increase in production, consumption, and environmental assault since 1950.[7]

Students at the University of Michigan take part in a "Waste Audit & Education Day" sponsored by the Ross School of Business.

Dave Brenner, SNFE

Higher education's reading of sustainability reflects this marriage to growth. Despite lofty and often genuine commitments to planetary health by many colleges and universities, the bulk of their sustainability initiatives center on four practical goals: increasing efficiency, reducing waste, decarbonizing energy use where affordable, and improving the institution's environmental image. Programs that trim energy use, water consumption, and waste production typically take center stage, and for good reason: they generate positive publicity and cultivate student goodwill while producing financial savings. All three then can be redirected to support the overall growth of the institution.[8]

This pattern surfaces in UI Green Metric's "World University Ranking," which assesses campuses on waste generation, water use, carbon footprint, transportation choices, infrastructure innovation, and a catch-all "education" category. The top institutions are technologically innovative, are sensitive to the sourcing of food and energy, and demonstrate how to do more with less, so that they can then grow in other ways. The same largely holds true for the Association for the Advancement of Sustainability in Higher Education (AASHE), which by July 2016 had analyzed information from almost four hundred colleges and universities in nine countries. For AASHE, the best colleges and universities are becoming smarter in their resource use and are rewarding faculty for helping their students and the larger world do the same. They are good green consumers at an all-campus level, much like individual households that try to "save the planet" by buying recycled products or using renewable energy, and sharing their experiences with their neighbors.[9]

These accomplishments are not trivial. Twenty years ago, most observers would have dismissed them as impossible. They reveal remarkable flexibility and innovation, proving that higher education can indeed become a "beacon of hope in a turbulent world."

But with few exceptions, these initiatives accept and often facilitate a social logic of unrestrained economic growth. Recycling initiatives on campus marginalize questions about the growth of disposables in industrial society; instead, recycling is often experienced as a reward for consumption. Composting of food waste is admirable, but it may sideline questions about the drivers of waste or the ecological affordability of meat. Energy- and water-efficiency savings are redirected to facilitate growth in other areas of campus operations. A much-needed shift to decarbonized energy sources skirts more fundamental questions of how much energy is enough. And divestment from fossil fuel providers, a new and important feature of campus sustainability (see Box 16–2), nevertheless normalizes a broader logic of growth-driven investments in private firms that themselves are wellsprings of growth.[10]

By accepting growth as given, higher education undercuts its considerable power to drive lasting sustainability. Consider, for example, the common scenario where funds generated by energy-efficiency improvements in academic buildings are redirected to faculty research. Professors are delighted, as is the admissions office, which can trumpet the greening of the university. But the overall carbon footprint of the campus grows as happy faculty fly to more international conferences to share their research with colleagues. Expand this example, and it becomes apparent that, when growth is king, the environmental benefits of sustainability initiatives in one sector of the economy can be swamped by growth in another. This is not an argument against energy efficiency or other smart technological innovations; it is a plea to combine the familiar focus on eco-efficiency and decarbonization with searching initiatives for a post-growth world.

Now is the moment for higher education to mobilize around this plea, but it will not be easy. Colleges and universities are not just agents of economic growth; they also depend upon it, which makes it doubly hard for them to envision a post-growth world. Bigger budgets, new buildings, better-paid faculty, an expanding student body—all are markers of institutional success,

Box 16–2. Student Activism and Training Within the Fossil Fuel Divestment Movement

Student organizers on college and university campuses have begun pushing—hard—for institutional endowments and other funds to be stripped clean of investments that support the fossil fuel industry.

Today, a movement that began in the United States is finding global reach. The effort also has moved beyond higher education to challenge other institutions to consider their role in supporting fossil fuel extraction. By the end of 2016, more than six hundred institutions, including seventy-five colleges and universities, dozens of religious institutions, over a hundred foundations, and well over fifty municipalities, together representing more than $3.4 trillion in assets, had committed to some level of divestment from fossil fuels.

Although unlikely to make a big dent in company bottom lines, the divestment push ratchets up the moral stakes. The campaign raises hard questions for fossil fuel companies and for the investments and investors that support their operations.

More than this, the fossil fuel divestment effort has proved an extraordinary training ground for a new breed of student climate activists. The campaign has discovered a way to take the intractable challenge of climate change and to direct the energies of students toward direct, creative forms of action against an identifiable target. Students are learning what it takes to move stubborn institutions in positive directions, by employing insider/outsider campaigning strategies, by embracing a climate justice framing that broadens the set of constituencies interested in working for divestment, and by radicalizing the very notion of campus sustainability.

The student leaders of the divestment push are suggesting that for a campus to carry the label "sustainable," it must do more than commit to green buildings and on-campus composting. Instead, these students are saying, a truly sustainable campus is one that contributes to tackling rather than perpetuating the world's most critical problems.

Colleges and universities remain our principal institutions for post-secondary education and for the creation of new knowledge. Traditional forms of teaching and learning matter for the transition to sustainability. The fossil fuel divestment campaign shows, however, that some of the most important learning is happening outside the walls of classrooms, as students define for themselves the opportunities that exist in taking up the climate challenge.

—*Eve Bratman, Franklin & Marshall College*

—*Kate Brunette, Raise Up WA*

—*Simon Nicholson, American University*

—*Deirdre Shelly, 350.org*

Source: See endnote 10.

and all become difficult to achieve amid slow economic growth and pinched public funding.

It is never clear, moreover, how much of each is enough given the spiraling "arms race" in higher education, where new facilities and programs at one university must be matched by other institutions to avoid losing ground in the battle for good ratings, strong students, and top faculty. Cornell University economist Robert Frank captures it perfectly when he notes that "universities face increased pressure to pay higher salaries to star faculty; to spend more on marketing, student services, and amenities; and to offer ever-more generous financial aid to top-ranked students from high-income families. It is little wonder, then, that their financial situations have grown more precarious."[11]

Happily, we do not live under normal circumstances. Our current bout of low growth creates opportunities to reorient the ivory tower toward a post-growth world. Smart institutions will not bet the farm on a low-growth future—that would run counter to the DNA of higher education. But with prodding, colleges and universities will become increasingly receptive to initiatives that offer a Plan B for their own financial struggles, and for society as a whole, should low growth become the norm. If pursued successfully, these initiatives will make human prosperity in a post-growth world more realistic, more tangible, and—one hopes—more desirable. And this can drive momentum for change.

Building Momentum

Seeds of this momentum are now sprouting as colleges and universities become more comfortable with resilience as a guiding strategic concept. Notions of resilience—the capacity to absorb shock and bounce back, perhaps better than before—have been prominent in higher education since the attacks of September 11, 2001. Originally preoccupied with how colleges and universities might recover from terrorist attacks, disruptions to information technology services, or natural disasters, resilience thinking among campus administrators has gradually expanded to include disruptions from climate change. In May 2014, Boston-based Second Nature—a nonprofit organization collaborating with colleges and universities on climate issues—sharpened this focus on climate by launching the Alliance for Resilient Campuses (ARC). Still in its infancy, ARC helps colleges and universities formulate programs that respond proactively to the effects of climate change on their own operations and the surrounding community.[12]

The growing prominence of resilience within higher education, reflected in projects like ARC, offers a striking opportunity to nudge colleges and universities toward a post-growth mindset. That is because campus and community resilience, especially in response to climate change, is not primarily about economic growth or enhancing an institution's reputation. Focusing on resilience means zeroing in on noneconomic foundations of human prosperity: social capital, mutual trust, strong community, loving and respectful relationships, local knowledge, community self-reliance, and limited inequality. As colleges and universities cultivate these elements in their own operations and within their neighboring communities, they are laying the groundwork for human flourishing in a post-growth world.

Those associated with higher education—students, staff, faculty, alumni, administrators, and funders—would thus do well to promote climate resilience in campus communities known to them, and then to encourage the school to infuse resilience thinking into existing environmental initiatives. Together, they could steer the sustainability conversation away from asking "how can we be more efficient (so that, perhaps, we can keep growing)?" and toward "how can we enrich human connections and a strong sense of collective self-reliance to reduce our impact on the planet (in ways that make us more resilient and, coincidentally, help us thrive in a post-growth world)?"

As interest in resilience begins to supplant a campus focus on eco-efficiency, it will become easier to draw colleges and universities into developing, testing, and disseminating the policies and norms we need for a post-growth world. One place to begin is with campus-sponsored experiments in economic reorganization, especially around locally based worker and community cooperatives. In the future, these enterprises will need to be the norm, not the exception. They deliver human prosperity and environmental sustainability without an intrinsic need to grow, and they enjoy citizen support across the political spectrum—an important quality in these politically fractured times. Higher education commands the expertise, capital, and experience to assess and disseminate several variations of these business models. Lasting sustainability demands nothing less.[13]

Inspiration abounds. Take, for instance, the Evergreen Cooperative Initiative in Cleveland, Ohio, which enjoys support from Case Western Reserve University. Modeled after the eighty-five thousand person Mondragon Cooperative in Spain, Evergreen includes a greenhouse, a large-scale environmentally advanced laundry, and a solar panel and weatherization company. It is cooperatively governed, hews to core notions of sustainability, and provides good jobs for some of the city's most challenged neighborhoods. Community interventions such as these appear to be on the uptick: the Democracy Collaborative and the Responsible Endowments Coalition report that 16 percent of U.S. colleges and universities invest locally, although few with the ambition and effect demonstrated by Case Western Reserve. However, far more must be done in the domain of community investment.[14]

Horticulture Group

The Evergreen Cooperative Institute in Cleveland, Ohio, includes Green City Growers, a 1.3 hectare hydroponic greenhouse filled with leafy greens.

Retooling older forms of economic organization for the twenty-first century is one way for higher education to respond effectively to today's slow growth while sliding us toward a future post-growth world. Another is to aggressively model necessary changes in work life. Scholars agree that workweek reductions are central to any transition to a post-growth society. The standard forty-hour workweek adopted in the United States in 1940 is not etched in stone, just like the hundred-hour workweek in 1890 was not sacred either. At low or no economic growth, we will all need to work less—to spread the work around to ensure an acceptable degree of income distribution—while finding satisfying ways of swapping leisure for consumption.[15]

Corporations already are experimenting with workweek modifications. Amazon.com, for instance, is piloting a thirty-hour workweek for select employees, who will receive full benefits and a 75 percent salary. Given the positive influence of shorter workweeks on employee productivity, retention, and absenteeism, this makes good business sense. But we cannot rely solely on organizations guided by profit to create and disseminate workplace arrangements for the future. Higher education must pitch in, too, with innovative workweek programs of its own that look beyond next quarter's balance sheet. These programs would push the frontiers of campus sustainability far more than another campus community garden or array of solar panels.[16]

Universities and colleges also can continue to pioneer and spread consumption-reducing "choice edits" that are critical to a post-growth future. Chatham University in Pittsburgh, Pennsylvania, has banned the sale of bottled water on its sprawling campus, installed filtered water dispensers, and distributed reusable water bottles, helping students save money and modeling a needed shift from recycling to reuse. Like many schools, Yale-NUS College in Singapore has done away with trays in its campus dining rooms and is preparing a shift to smaller plate sizes, moves that can cut food waste up to 30 percent. Additionally, an electricity meter in every dorm room means that students receive individual bills for their air conditioning use, which prompts them to cool their rooms only during the hottest hours of the day, or not at all. Suddenly, it is not so cool to use air conditioning. At the University of Wisconsin-Madison, bicycles are the primary mode of transportation for almost one-quarter of students, in part because of the university's support of the city's bike sharing program. This helps make biking feel like the natural thing to do in Madison, which can lead to more bikers, stronger support for biking, and falling demand for other economically and environmentally costly transportation options.[17]

Overcoming Obstacles

If these initiatives—cooperative businesses, worktime reductions, and choice editing, all supported by resilience thinking—are to thrive within the halls of academia, they must resonate with the core research and teaching missions of the university. That will be difficult if these measures are understood as fundamentally oppositional to growth, corporate capitalism, or material acquisition. Higher education, after all, identifies as an engine of growth, and the deep insinuation of corporate interests in the modern university is well documented.[18]

Fortunately for academic researchers, today's lukewarm economic growth offers safe haven. Inquiry into post-growth alternative work and business models, and other post-growth policies, can be framed as applied research motivated by our current economic doldrums: it is all about searching for solutions to today's problems rather than arguing for low growth later. Characterizing research as solution-driven also could free up needed funding, which is key, since less research money exists for post-growth research than for studying how to address pollutants from an expanding economy. One notion that is attractive for its symmetry is to divert all savings accrued from campus eco-efficiency projects to faculty research on the transition to a post-growth world.

It is more difficult to address conflicts with the teaching mission of environmental studies and sustainability science. Do these programs recruit students, generate research money, confer prestige on the institution (by impressing business and government elites), and situate students for good, well-paying corporate jobs or entry into prestigious technocratic graduate programs? If so, then all is well. But programs that question the underlying rationale of growth-centered economies often are mismatched to these imperatives of higher education. These programs, vital to our transition to a post-growth world, are especially vulnerable to being labeled as "anti-corporate," "insufficiently scientific," or "too ideological," and then marginalized, as if faith in the perpetual growth of industrial economies is somehow objective or reasonable.

One piece of a solution is to connect these programs to external networks of credibility and prestige. A project like the Ecosphere Studies initiative (see Box 16–3) thus becomes important not just for the radical curriculum that it is producing, but also for the impressive credentials of its participants and the wide dissemination of results. When highly respected academics develop hard-hitting curricula and build networks around their delivery, it is easier to view like-minded curricula as "state of the art" and "cutting edge" rather than "overly normative" or "unscientific." That creates the space that innovative programs need to work and thrive.[19]

And let us not forget AASHE and UI Green Metric, as well as groups like The Sierra Club that also publish environmental rankings. They can help by highlighting colleges and universities that rigorously explore post-growth options for the future. Schools no longer should be lauded for making cost-effective investments in water, energy, and waste efficiencies. Nor should they receive high praise for moving to renewable energy when it is increasingly affordable to do so. To paraphrase Robert Reich, professor at the University of California at Berkeley and former U.S. Secretary of Labor, that is not socially responsible behavior—these days, it is just good business. The highest marks and the best publicity must be reserved for those colleges and universities at the edge of social innovation for the planet.[20]

The next frontier of sustainability in higher education, now at our doorstep, revolves around charting new paths to a post-growth future in which we all would want to live. With fifteen years of low growth behind us, higher education is finally starting to move in this direction, despite its affections for a high-growth world. Now all it needs is a good strong push.

Box 16–3. Reframing Higher Education Around Ecosphere Studies

The seeds of an initiative to develop a transformative curriculum for higher education have been sown by The Land Institute in Kansas and its cofounder, Wes Jackson, a plant geneticist and international leader in sustainable agriculture. The Institute is best known for developing perennial grains to grow in polycultures, to help shift food production away from industrialized agriculture's unsustainable reliance on monoculture annual grains, such as wheat.

Now Jackson and the Institute have assembled an inaugural core faculty of twelve educators and scientists from eleven colleges and universities and The Nature Institute, a nonprofit organization, to design a curriculum that transcends disciplinary boundaries while drawing from the different sciences, philosophy, history, and the arts. The goal: to propose a comprehensively holistic approach to radically realign teaching, learning, and research around the theme of "ecosphere studies." Reorganizing around a vision of the living world itself as the ultimate educational authority, collaborators believe, is a remedy for higher education's current core contributions to propping up extractive economies. Such economies threaten ecosystems around the world and even the super-ecosystem—what Jackson calls our planet's "ecosphere"—the entire global web of relationships that comprise life's only home.

In Jackson's view, modern agriculture lies at the heart of the problem, and a radical new way of perceiving and orchestrating our relations within the ecosphere is essential. The Ecosphere Studies initiative will explore how to make "nature as measure" the new paradigm for human action and specifically in connection with feeding the growing world population. That means consciously striving to fit agriculture and all other activities into the realities of natural systems, rather than assuming that nature's limits can be conquered or exploited.

After an exploratory conference with forty scholars in 2015, the core faculty and invited guests met in 2016 to plan further for such a new curriculum. In the coming years, the focus will be on developing teaching materials that are both radical and relevant. Some core faculty members are already testing new courses and methods on their own campuses or revising existing courses. Their practical experiences will inform the initiative as it moves forward.

—Craig Holdrege, director of The Nature Institute and core faculty member of the Ecosphere Studies initiative

Source: See endnote 19.

CHAPTER 17

Bringing the Classroom Back to Life

Jonathan Dawson and Hugo Oliveira

As part of their first-year curriculum, students at Swaraj University in western India head off on a ten-day "cycle yatra," a low-technology bicycle journey through nearby villages that provides them with an immersion opportunity to see and experience the world differently. The invitation to the students is to "leave behind your money, credit cards, cell phone, iPod, snack food, and all things plastic" and to "secure your food and shelter with the gifts of your labor, your creativity, and your capacity to build relationships with strangers. You will practice surrender."[1]

At Schumacher College in Devon, England, master's students complement theoretical studies in holistic science with a range of more experiential activities, including a Deep Time Walk, a guided 4.6 kilometer walk along a stretch of coastal path that traces the 4.6 billion years of the history of the Earth, with each meter representing 1 million years.[2]

In southern Italy, participants in the educational and artistic experiment known as the Free Home University seek to "generate new ways of sharing and creating knowledge by experiencing life in common." Among the activities that they have designed as part of their collaborative curriculum is a Walk With the Dead. Drawing on wisdom traditions from around the world associated with the Day of the Dead, they each invoke the spirit of a recently deceased loved one and walk "in conversation" with them from sun-up to sundown, sharing insights at the end of the day.[3]

Jonathan Dawson is head of economics at Schumacher College in England, where he coordinates and teaches/facilitates on the innovative Economics for Transition postgraduate program. **Hugo Oliveira** is a landscape ecologist and permaculture specialist at OrlaDesign and a researcher at the Center for Ecology, Evolution, and Environmental Change at the University of Lisbon in Portugal.

These examples from around the world illustrate a growing trend in education toward a more innovative, whole-person, and experiential approach that, explicitly or implicitly, is challenging the fundamental assumptions that have underlain the entire history of mass education.

To better understand the significance of this new wave of educational experimentation, and the philosophy underlying it, one must dig into what educational scholar Stephen Sterling calls "the subterranean geology of education"—the hidden assumptions that underpin today's dominant educational paradigm. These include the assumption that knowledge is more or less fixed and can be divided into relatively autonomous subject silos; that the most effective form of knowledge acquisition is transmission from expert teacher to student; that the intellect is the only legitimate and scientifically verifiable mode of learning; and that education is primarily a private rather than a community-based activity.[4]

Michael Carcus

Attendees at an Association for Experiential Education conference experience tree climbing.

In mainstream academia and beyond, each of these underlying assumptions is coming under increasing scrutiny as a growing number of educationalists suggest that it is time for a revolution in our understanding of how learning takes place, informed by insights derived from systems thinking and complexity theory. (See Chapter 12.) These perspectives introduce a much more dynamic understanding of how the world works by shifting the focus from things to relationships and patterns, helping us understand how living systems self-organize and self-sustain, are rich in feedback, and manifest emergent qualities that cannot be predicted from a study of the component parts. In consequence of this emerging shift in perspective, the dominant metaphor is beginning to transition from machine to living system, from dead matter to animate Earth.

This is especially true at this juncture, since, as sustainability researcher Judi Marshall and her coauthors note, "in modern western societies, we are caught in a conceptual trap that renders our accepted ways of understanding our world unequal to the task that the sustainability challenge offers." There are solid grounds for believing that most of the critical crises currently converging on our civilization are "wicked," or complex in nature, lending themselves well to apprehension and engagement through a systems/complexity lens. In this light, it is useful to look at some of the key shifts in educational theory and practice that such a reorientation will entail, as well as examples from pioneering educational institutions of how this is already manifesting on the ground.[5]

From Objective Truths to Emergent Meaning-making

New approaches to teaching and learning reject the conventional neat separation between the observer and the observed that is central to traditional, mechanistic, educational philosophy. Rather, it is becoming increasingly recognized that the researcher and the object of study affect each other mutually and continually in the research process. There is, in short, no "objective truth" waiting to be cognitively uncovered and communicated to the student as a fixed body of knowledge. Rather, meaning necessarily emerges out of an ongoing, iterative process of experimentation, questioning, and reflection.[6]

Allan Kaplan, cofounder of The Proteus Initiative, captures the essence of this way of understanding the dance of inquiry between students and the object of their study: "[T]here is a delicate relationship between the world 'out there' . . . and the sense-making that we bring to that world; that the phenomenal world we live in arises from the conversation between sense and sense-making."[7]

In this context, rather than being rigidly determined in advance, the curriculum takes on a provisional character, demanding the space and flexibility to evolve in directions required by the flow of the inquiries. In the words of educational theorist John Dewey, "the learner and curriculum are each transformed as they interact with each other." Thereby, the center of gravity of the classroom shifts from the authority of the teacher to the distributed intelligence of the learning community, with students taking ever greater responsibility for the framing and management of their own learning journeys, a process known as heutagogy. (See Box 17–1.)[8]

A growing number of educational centers have explicitly adopted this approach, encouraging their students to tailor and evolve their own learning journeys in collaboration with coaches and peer learners. In addition to the cycle yatra, students at India's Swaraj University undertake a series of visits to "new economy" pioneers and social entrepreneurs and then are encouraged to

Box 17–1. Global Sustainability Through Local Heutagogy

Research indicates that learning outcomes improve when students can relate the material that they encounter in the classroom to their own life world. This suggests that attention to how we teach is as important as what we teach.

An educational philosophy that expressly highlights the contributions of students as context experts and co-teachers is "heutagogy." A heutagogical strategy views the instructor not simply as the person who transfers information to the student, but as the person who develops learning skills within learners to prepare them to take responsibility—as co-instructors—for the overall learning experience.

Heutagogical strategies lend themselves uniquely to bringing the multiple contexts, experiences, and perspectives of diverse learners into the learning process. The learners become co-creators of knowledge within an enhanced overall learning experience, since the contexts and perspectives of the participating learners invariably expand and diversify the insights that even the most informed instructor can offer.

In a course that is designed according to heutagogical principles, the role of the instructor is to facilitate learning and to design the course in such a way that the selected learning activities make optimal use of the diverse perspectives that the course participant can contribute. As researchers Jane Eberle and Marcus Childress explain, "rigidly structured environments are not conducive to heutagogy." Heutagogy requires creativity as it shifts responsibility in the learning process to a mutual commitment to achieving learning goals. A heutagogical learning environment can be created in any setting, including face-to-face, online, and blended modes of instruction. Given the level of responsibility that it places on the learner, it also is uniquely suited for graduate and professional studies that can benefit greatly from the experiences and expertise of the participants.

A curriculum project that employed heutagogical strategies to meaningfully bring into focus the larger economic, social, and environmental context of the learning experience was designed and first implemented in 2013 by then-Global Ecology principal Sabine O'Hara, in collaboration with the University of South Africa (UNISA), a mega-university serving 320,000 students via distance learning. A key objective of the project was to design six signature courses that use heutagogical strategies for the six colleges of UNISA. The signature courses are designed to engage students as co-instructors and "context" experts. Every student is encouraged to bring his or her own life world context into the shared learning experience and to serve as the expert for that particular context and its relevance to the course content.

For example, students were given assignments to collect specific data in their home communities; to identify a pollution problem in their region; to use clear instructions to write a case study about a development issue in their community; to post information gathered in their community; and to comment on the information collected by at least three fellow course participants. These kinds of localized and personalized assignments engage students with each other and with their own context and its demographic,

Box 17–1. continued

social, cultural, historical, economic, environmental, and spatial characteristics. As a result, students are more socially and environmentally aware, self-motivated, and committed to civic engagement, sustainability, and social justice. Preliminary assessment results from the UNISA signature courses indicate that a significantly higher percentage of students successfully complete the signature courses compared to other UNISA courses, despite the challenges associated with the online delivery format.

The College of Agriculture, Urban Sustainability and Environmental Sciences (CAUSES) of the University of the District of Columbia, too, has implemented heutagogical strategies in several of its courses. These strategies are especially successful in some of the required science and general education courses that tend to be challenging for the university's large number of first-generation college students. A Sustainable Urban Agriculture course, for example, requires students to take soil samples from two different locations and to research the land-use history in those locations before they analyse the soil samples and assess the consistency of the test results with the land-use histories. A General Education capstone course organizes students into interdisciplinary teams and assigns them research projects that improve the quality of life of specific Washington, D.C. neighborhoods.

In addition to its students who are seeking academic degrees, CAUSES serves a large number of non-degree-seeking students through its land-grant programs that offer workforce development, certificate, and continuing education courses. (See Chapter 18.) These tend to be short courses that emphasize practical learning and skills development. Yet often, learners are eager to also gain a better understanding of their own place and contribution to the larger community. Several of the land-grant programs have developed a heutagogical approach to address these learning interests of engaged citizens who are committed to sustainability and to improvements in quality of life for their communities.

—*Sabine O'Hara, Dean of CAUSES, University of the District of Columbia in Washington, D.C.*

Source: See endnote 8.

develop their own ideas for social enterprises with the support of a personal feedback council comprising at least five peer learners; a team of learning facilitators; a mentor; and the leaders of the projects where the students go on two-month placements.[9]

The online Gaia University follows a similar model, with students supported in their evolving learning journeys as collaborative, self-directed learners by peer guilds, mentors, and supervisors. Student projects range from the design and creation of community currencies to smallholder agriculture projects and the design of ecovillages. The university has no fixed campus, with short courses offered at a range of partner institutions close to the students' homes.[10]

Central to this research and learning ethic is a suspicion of claims that language can, in any meaningful sense, provide an objective description of a pre-existing reality. Language is understood not as describing some autonomous, value-free progression of pure, abstract knowledge, but rather as being embedded in structural power relations, which, when unquestioned and unchallenged, tend to be insidiously reinforced.

The Ecoversities Network, an international grouping of activist educational initiatives that are reimagining higher education, facilitates student-centered, self-tailored, intercultural exchange programs, enabling young adults to encounter new cultural norms, practices, and teaching methods as a way of widening their worldviews and skill sets. One of the network members, Red Crow Community College, a Blackfoot Nation community-learning initiative in Alberta, Canada, approaches environmental studies and the other-than-human world from a worldview of respect and interdependence that is firmly rooted in their Indigenous traditions and cosmologies. This includes spending time with tribal elders learning about traditional approaches to medicine and healing.[11]

Universidad de la Tierra, or Unitierra, in Oaxaca, Mexico, was created in response to a belief that "[t]he school has been the main tool of the State to destroy the Indigenous people." Unitierra has created a learning ethic that is more closely based on Indigenous educational practice and that emphasizes informal, peer-supported project-based education over the traditional model of the teacher-student relationship. Local people, with support from the school's staff, work together on projects such as building clay ovens and other artisanal appropriate technology, exchanging knowledge on design ideas, and gaining practical skills while developing ideas for further common projects.[12]

All of the above serves to revolutionize the role of the teacher. The primary role of the teachers shifts from being a transmitter of a fixed body of knowledge to being an "educator" (etymology: "to draw out from"), helping students engage creatively and intelligently in their sense-making enquiries. Teachers play the roles of catalyst, mentor, provocateur, and, to some degree, also that of peer within the learning community of which they form a part.[13]

From Cognitive Rationality Alone to Multiple Ways of Knowing and Learning

The western scientific tradition leans heavily on rationality and validation based on empirical evidence. In the words of John Dewey, "The old center of the universe was the mind knowing by means of an equipment of powers complete within itself and merely exercised upon an antecedent external material equally complete within itself."[14]

The new center relocates the learner as being embodied and deeply implicated in multiple relationships within the human and other-than-human world, exploring it with the full range of human faculties: rational and cognitive, experiential, intuitive, relational, and embodied. Linguist George Lakoff goes so far as to argue that *all* cognition is based on knowledge that comes initially from the body, and that other domains are mapped onto our embodied knowledge using primarily conceptual metaphors.[15]

In a review of the United Nations Decade of Education for Sustainable Development (2005–14), sustainability researcher Daniella Tilbury cites more than twenty studies from scholars worldwide that highlight the alignment of sustainability education with active and participatory approaches to conveying knowledge. Active learning processes of conceptualizing, planning, acting, and reflecting were found to better enable students to engage critically and creatively with the values, skills, and knowledge requirements of sustainable development.[16]

Meanwhile, a recent study into education for sustainable development found that in the absence of emotional engagement, "cognitive understanding is not enough to foster behavioral changes. . . . Emotions concern what gives meaning to life; they frame, transform and make sense of our perceptions, thoughts and activities." Student engagement on an emotional level, the study found, was essential for behavioral change.[17]

The validation of the "subjective" that this entails brings the classroom back to life. Students are no longer asked to park their emotions, their intuition, and their bodies at the classroom door. Rather, they are invited into a space that welcomes their creativity and playfulness, their passions and their tears. The student's role shifts from that of object to be operated on to a subject within relationships. British journalist George Monbiot captures this beautifully: "Acknowledging our love for the living world does something that a library full of papers on sustainable development and ecosystem services cannot: it engages the imagination as well as the intellect."[18]

Schumacher College in England has experimented with introducing applied, hands-on learning curricula before engaging on a more theoretical level. The aim is to enable the world to reveal itself to the students afresh, rather than being prepackaged and defined in advance by abstract theoretical formulations. For example, an economics class taught by Jonathan Dawson includes theatrical elements, such as having students "embody" the ecological footprints of different lifestyles associated with different clusters of nations: huge, fingertip to fingertip circles representing North American lifestyles, and tight knots of bodies representing African lifestyles. One student-led session on the British enclosure movement saw the class being locked out of their teaching space, giving students a deeper visceral experience that had a much greater impact than a purely conceptual treatment would have achieved.[19]

From Learning as an Individual to a Community-based Pursuit

The endpoint of conventional approaches to education where knowledge is transmitted from teacher to student is the acquisition of knowledge by the individual student, who has succeeded in absorbing what has been presented. In so doing, the student becomes what educational theorist Ken Gergen calls "a simulacra [superficial likeness] of the authority." He describes this process of imprinting knowledge on students as "an obliteration of identity and an invitation to lethargy."[20]

An alternative interpretation of the learning process that sees knowledge as being socially constructed—shaped through cultural or social practice—is gaining ground. This view challenges the ideology of the self-contained and essentially independent individual learner and asserts, in Gergen's words, that there can be "no individuality without collectivity, no independence without interdependence." The student is embedded within multiple relationships with the human and other-than-human world, and it is these relationships that enable and catalyze the emergence of knowledge.[21]

A learning community promotes thinking across disciplines, enables the diversity of relationships necessary for the generation of quality knowledge, and meets our innate need for sociability. (It is precisely for these reasons that we are seeing the proliferation of innovation hubs and other shared work spaces, which are proving especially popular with young people. This trend mirrors and helps to drive the transition from centralized to distributed organizational forms that we are witnessing across society, not least in education.)[22]

In this context, it is unsurprising that a common feature of many of today's pioneering educational initiatives is their rootedness in community. A growing number of educational programs, most notably those offered by Gaia Education, are designed explicitly to be embedded within "living and learning" communities. In these programs, students and staff are often found working side by side managing the education center: growing food, cooking, washing dishes, cleaning, and maintaining the buildings. This expands the "living classroom" to include all dimensions of the life of the college, enabling a breaking down of the artificial boundaries that conventionally exist between the theory and practice of sustainability. Students learn to grapple with issues relating to decision making and conflict resolution, sourcing and cooking food, and relating to others in respectful and regenerative ways.[23]

Such practices are not limited to small-scale, non-formal, residential settings. Under the aegis of a recently ended European Union-funded project, CEAL (Community-based Entrepreneurship Action Learning), university students in six European countries, as part of their accredited programs, worked with residents of nearby deprived neighborhoods in the development of projects and facilities for community benefit. Students at Ideen3 in Berlin, Germany, one of the participating educational institutions, worked closely with local communities in a series of innovation labs to design and build several creative sculpture pieces that today populate the neighborhoods in which they were conceived.[24]

End-of-program evaluations frequently indicate that students that take part in such initiatives prize the immersion experience in the practice of sustainable living above all else. In Japan, assessments have found that students studying in rural villages and participating in village life through experiential learning programs have positive, life-changing experiences and leave more committed to sustainable practices. (See Box 17–2.)[25]

The embodied and community-based nature of learning described here has implications for the types of assessment associated with accredited education. Schumacher College, among other places, has experimented extensively with allowing students to submit accredited assignments both for collaborative work and for creative, artistic projects—including musical compositions, art exhibitions, video work, and audio podcasts. Growing numbers of schools also are experimenting with self- and peer-assessment as a complement to assessment by the teaching staff, helping students hone their skills in providing insightful, compassionate feedback to their peers.

Revolution in Teaching Practice

The celebrated English anthropologist Gregory Bateson suggested that the root of the various crises converging on our societies derives from a worldview that is founded upon an "epistemological error"—an error in how we conceptualize knowledge—that has led to a perception of, and belief in, separateness that in turn manifests separateness and fragmentation. Education has had a substantial role to play in perpetuating this error.[26]

Box 17–2. Experiential Learning Helps Change Behaviors

"The paddy was not only for making rice but also a home for a bunch of creatures. I feel that we live together on the same Earth," wrote a twenty-year-old Japanese university student in a report describing her experience weeding a rice paddy in the farming village of Tochikubo. Because the paddy was kept free of agricultural chemicals to protect consumer health and the ecosystem, the student was able to observe a variety of living creatures—including frogs, leeches, and water striders—when she stepped into the marshy water with her bare feet.

Carefully designed experiential learning programs, which provide opportunities to become immersed in or directly engaged with the natural world and with people's lives supported by nature, have the potential to raise awareness of sustainability and to increase sustainable behaviors. Research shows that knowledge alone rarely leads people to act, but when they gain direct experiences, they tend to form an emotional connection to other people and to other species and to convert this knowledge into action.

In a 2010 study of a series of experiential learning programs held in Tochikubo, 80 percent of participants indicated an intention to alter their lifestyles in the direction of sustainability based on what they had learned during the programs. The experiences also helped the participants connect more closely with global issues, such as understanding what it means to farmers and consumers when they hear a statement such as, "A one-degree Celsius rise in air temperature can reduce the grain harvest by 10 percent."

Experiences have the power to connect learning and actions, but intentional reflection—as embedded in the Tochikubo programs—is a critical element of the process. Reflection on and processing of experiences can take many forms, such as guided journal writing, questionnaires, and daily reflection sessions with peers and instructors.

For example, the student who marveled at the diversity of the rice paddy had believed initially that nature did not exist in the city of Tokyo, where she grew up. During a reflection session at the end of the program, she lamented that she would not be able to apply anything she had learned because her home had nothing to do with the natural environment. Some of the participants questioned her comments, and the instructor shared her own experiences with noticing nature in cities. A few days later, in personal communications with the instructor, the student wrote that, having returned home, she continued to find nature everywhere in Tokyo: "This is the first time I have felt this in my life; now, my view of this place is changed."

—*Takako Takano, Professor at Waseda University, Tokyo, and Chair of ECOPLUS*

Source: See endnote 25.

Our educational systems are in crisis, beset by multiple sources of disruption. These include the growing gulf between the mindsets, competencies, and skills required to address our converging "wicked" crises and those provided by a conventional university education, as well as growing student dissatisfaction with what is being offered. Other challenges include increasing commercialization and fees associated with education in many parts of the world (as government funding drops or fails to keep pace with the rise in student numbers), a flourishing non-formal education sector that is to some extent experimenting with new-paradigm teaching practices, and the availability of no- or low-cost, high-quality courses on the Internet.[27]

This crisis also offers opportunities. The core challenge today facing education systems around the world is that of addressing Bateson's epistemological error and its faulty perceptions of separateness and fragmentation. This is something that, by definition, cannot be tackled using the old methods alone. Replacing one set of textbooks with another is not going to do the trick. The field of engagement needs to include that subterranean territory that houses our understanding of how knowledge is generated and how students learn. The revolution that is required in our educational practice needs to be felt on an embodied level as much as understood cognitively. (See Box 17–3.) It can be described as nothing short of bringing the classroom back to life.[28]

Box 17–3. Education for Ecosocial Change

The conception and predominant practices of education are far removed from the principal problem faced by the world today: environmental destruction and its collateral effects in terms of land and social expulsions. And yet this destruction, which affects all life on Earth, mandates an ecological conversion of education and the use of Earth-centric teaching practices that have the capacity to convert all students into agents of ecosocial change.

Education for ecosocial change advocates a holistic education that reformulates the mission of educational action. Education is not preparation for the work of the market society, and learning is not the accumulation of academic knowledge in obsolete curricula—even if this is done through new, innovative teaching methods. Instead, education should imbue all educational activity with ecology and should consist of the comprehensive training of students with multiple intelligences, including ecological intelligence, social and emotional intelligence (see Chapter 8), and moral intelligence.

In its 1996 report, *Learning: The Treasure Within*, the United Nations Educational, Scientific and Cultural Organization (UNESCO) proposed four educational pillars that need to be recovered: learning to be, learning to coexist, learning to know, and learning to do. To meet the challenges of the twenty-first century, these pillars have to be applied to and

continued on next page

Box 17–3. continued

reworked from an Earth-centric perspective. This involves:

- Learning to be part of the Earth by discovering our deep links with nature in order to construct ourselves with an ecological identity.
- Learning to live in harmony with human beings—especially those who suffer the most from environmental destruction—and with the plants, animal, forests, oceans, etc.
- Learning to know the sciences, arts, and humanities, and knowing how to relate them to the ecological dimension and to environmental problems, so that knowledge comprises this reality.
- Learning to "do" using ecosocial practices, and encouraging an orientation toward professions that facilitate the ecological reconstruction of societies.

It is important to establish interconnections among all of the agents of education, including schools and universities, families, youth organizations, civic associations, religious institutions, social movements, and the media. For the transition to a new ecosocial education that links ecology and social problems, it also is vital to establish the nature of its aims. Ecosocial teachings should be common to school-based education as well as to the education undertaken by other agents, such as youth organizations, religious institutions, and families.

In the school setting, the creation of a three-part ecosocial curriculum is a priority:

First, the curriculum needs to be holistic and comprehensive, teaching about ecology, the root and immediate causes of ecological problems, and their consequences and solutions. A holistic look at these will include history, international relations, economics, science, philosophy, and more. Ecosocial education must be incorporated into the very heart of school teaching, forming a new curriculum in which the teaching of mathematics, history, language, physics, and other core subjects is shaped by taking into account the world's social problems.

Education also cannot focus exclusively on providing information about ecological problems and describing international proposals to reduce them, little by little; instead, it must be transformative. Ecosocial education needs to show the connections between the most important social problems and the political, social, and economic structures that generate them. For example, it is not enough to reduce climate change; society must transform the economic system that changes the climate.

Second, this ecosocial curriculum should cultivate the character of students—teaching morality, ecological connections, the study of consumerism and the manipulation of individuals and societies by advertising, and other influences that shape societal and individual values.

Box 17–3. continued

Third, and finally, the ecosocial curriculum should actively encourage ecological activism and offer ample opportunities to engage in activist efforts. This involves ecosocial political mobilization of children, adolescents, and young adults—encouraging them to discover their position as citizens of the Earth, as responsible for its destiny, and as caretakers of the planet and its inhabitants. Ecosocial mobilization is also about educating people to be aware of the structural causes of the Earth's destruction, to build ecological societies, to take up professions that ecologically rebuild the lives of cities and towns, and to adopt ecological lifestyles in their own everyday lives.

Every family, every school, and every city can play a part in organizing annual projects for ecosocial practices that feed off each other. Ecologist educators and movements should build networks, create stores of good ecological practices, and generate effective systems for communicating these practices. This will facilitate mass ecoliteracy and help environmental movements grow. Partnership between families, schools, and ecological movements is necessary. Schools can be the hub of this partnership.

Educators are at a major crossroads: either they help to reproduce the system that created the ecological crisis, or they intervene actively to change it. They are not—nor can they be—neutral.

—*Rafael Díaz-Salazar, Professor of Sociology and International Relations in the Faculty of Political Sciences of the Complutense University of Madrid, Spain.*

Source: See endnote 28.

CHAPTER 18

Preparing Vocational Training for the Eco-Technical Transition

Nancy Lee Wood

Over the coming decades, in the face of climate change, resource depletion, and economic contraction, the kinds of things that people will need to know and the standard approaches to accessing that knowledge will change. As people seek new and innovative ways of building sustainability and resilience into their lives and within their communities, they will encounter a widening variety of educational choices.

In the foreseeable future, the role of conventional university and graduate programs as credentialing agencies will undoubtedly continue, but this will become less standard as learners tap into alternative educational venues that offer streamlined instruction within shorter time frames and at relatively lower cost. Some of those alternative learning experiences will take place through new or reworked programs in academia, while others will occur in nontraditional settings such as Internet-based instruction, institutes, workshops, and community-based opportunities. Some will lead to degrees and certificates, while others will not. Regardless of the venue, the focus most likely will be on eco-technical education: applying science to meet human needs while minimizing ecological disruption.

In short, education is in the process of transformation, not only in what is learned, but in how and where learning takes place. The critical questions are: "What kinds of knowledge and skills do people need for the Anthropocene, an epoch in which humans are dramatically altering the climate and the environment? Are there educational models currently in existence that could be ramped up quickly to meet the challenges before us? And

Nancy Lee Wood is a professor of sociology and director of the Institute for Sustainability and Post-carbon Education at Bristol Community College in Fall River, Massachusetts.

how might we hasten their implementation in the midst of conventional cultural lag?

Tackling the numerous societal and ecological challenges that lie ahead requires a fundamental shift in educational priorities. New teaching practices and training programs are needed to enable growth in four key areas of vocational education: workforce development; carbon mitigation through energy efficiency and green buildings; regenerative land management and sustainable agriculture; and peace and conflict management. By providing students with the practical skills and knowledge that they need in these critical areas, we can better prepare them for the potentially very different world of tomorrow.

Education for Workforce Development

As we move further into the Anthropocene, work is likely to become more localized as fossil fuel energy constraints alter labor activity and mobility. Some work, such as agriculture, will become more labor intensive, and the types of preparation for work will be more streamlined, with training and education concentrated on information that is relevant to a specific job or set of skills. The awarding of four-year and advanced college and university degrees is likely to decline as large numbers of young people seek this training and knowledge through other avenues.

Already in countries such as the United States, where a college education has been touted as the sure way to a financially secure future, young people and their families are increasingly questioning the value of an expensive four-year degree, as graduates face strangling student loan debt, underemployment or unemployment, and stagnating wages. Worldwide, youth labor markets have suffered since the 1970s, but they were particularly hard hit with the Great Recession of 2007. In the European Union, just under one in five young workers (18.8 percent) are unemployed on average, with the highest rates of joblessness in Greece (50.3 percent), Spain (43.9 percent), and Italy (39.2 percent). Youth unemployment averages are even higher in poorer regions of the world, particularly in the Middle East (28.2 percent) and North Africa (30.5 percent).[1]

Undoubtedly, many types of labor need to be performed, and many workers need to be trained, to ensure a sustainable future. Ultimately, all sectors of labor—from bread baking to carpentry to banking—will have to go through some degree of transformation to coalesce with ecological demands. In the meantime, numerous lucrative midlevel jobs that need to be done and that require specific vocational skills remain unfilled and could be "greened." This gap can and must be closed.

The Organisation for Economic Co-operation and Development (OECD), in its *Skills Beyond School Synthesis Report,* emphasizes the critical role that post-secondary vocational education will play in the coming years. The OECD reports that nearly two-thirds of employment growth in the European Union will be within the category of technicians and associate professionals. The projection for the United States forecasts that by 2018, one-third of job vacancies will require some type of post-secondary qualification, but not a four-year college degree. Student populations are likely to decline at four-year institutions as learners seek more affordable and more targeted options.[2]

As high school vocational training programs have waned over the past several decades in the United States, community colleges have increasingly become training centers of choice in traditional areas such as criminal justice, culinary arts, dental hygiene, and nursing. Many community colleges now offer courses or program concentrations in environmental studies, green engineering, water quality training, and sustainable agriculture. While most community colleges also offer degree programs in liberal arts and general studies with the intention of students going on to four-year institutions, the vocational degrees and certificates are designed to move students into the labor force soon after graduation. The structure of community colleges is ideal for relatively short-term, low-cost vocational training; however, many of these programs fail to provide substantial hands-on experience in the field.[3]

Students in the Austin Community College Veterinary Tech program learn skills while helping out at the Bastrop County Animal Shelter, Texas.

One way to provide such experience is through apprenticeships. The Technical and Vocational Education Training (TVET) strategy developed by the United Nations Educational, Scientific and Cultural Organization (UNESCO) addresses the need for young people throughout the world, particularly in developing countries, to become skilled workers through quality apprenticeships. TVET focuses on the crisis of widespread youth unemployment in much of the world, along with the problem of skill mismatches. It aims to promote work-learning schemes as a necessary way to reduce poverty, stimulate economic recovery, and foster sustainable development. Recognizing that "one size does not fit all," TVET encourages a wide range of skills development from basic to advanced levels that are congruent with regional, national, and local socioeconomic, geopolitical, and cultural realities. Moreover, TVET, connecting with UNESCO's Global Action Programme on Education for Sustainable Development (ESD), promotes skills and knowledge for sustainable development.[4]

One of the most successful efforts to provide hands-on apprenticeships as a viable path from school to career comes from Germany. Through the country's highly respected dual-education model, learners can take on apprenticeships in numerous fields where they combine one or two days a week of academic classroom instruction at a vocational school with three or four days of practical experience within a company or organization that offers vocational training. More than three hundred and fifty such programs are available, and most apprenticeships require a two- to three-and-a-half-year commitment, although a few require longer periods of training. While learning their trade and becoming socialized into the expectations of their field, apprentices receive an allowance from their training company that averages about 650 euros per month, depending on the field. Frequently, apprentices who successfully complete their training are offered permanent positions within the company, and, provided that they are successful workers, they can later train to become masters in their fields, qualifying them for managerial positions. These hands-on apprenticeship experiences are important in developing a well-trained workforce while providing meaningful and lucrative employment for youth.[5]

In concert with TVET goals, Barefoot College, a unique and highly successful model of education for self-sufficiency in rural communities, has been developed in India and is spreading to other countries. It empowers the poor—especially women—by demystifying and making technical knowledge accessible, helping to improve the lives of thousands of villagers in some of the poorest, most remote parts of the world. (See Box 18–1.)[6]

Worker cooperatives—enterprises that are owned and run jointly by their members, who also share the profits and benefits—span many economic sectors and can play an important role in vocational training that is geared toward the coming eco-technical transition. New York City's Green Worker Cooperatives, a nonprofit organization that helps worker-owned green businesses get

Box 18–1. The Barefoot Model

Barefoot College is a not-for-profit, grassroots social enterprise that has been providing basic services and solutions to the challenges facing rural poor communities for more than forty years, with the objective of making them self-sufficient and sustainable, valuing and respecting the knowledge and wisdom that they already possess. The College was founded in 1972, and today it has campuses in four countries, including its main campus in Rajasthan, India.

Barefoot College believes in keeping alive the lifestyle and work ethic of Mahatma Gandhi by offering down-to-earth, collective and applied, practical learning experiences. The College demonstrates—from the knowledge, skills, and wisdom of village elders and young participants alike—that simple, inexpensive solutions are more sustainable. The College often is shown as an example of what is possible if very poor people, who have never been to school, are allowed to develop themselves. It is a concept that has stood the test of time.

What the College has effectively demonstrated is how sustainable the combination of traditional knowledge ("barefoot") and demystified modern skills can be when the tools are in the hands of those who are considered "very ordinary" and are written off by urban society. Just because someone is illiterate does not mean she cannot be a solar engineer, architect, designer, communicator, computer technician, or builder of rainwater harvesting tanks. There are many more powerful ways of learning than the written word.

The formal system of education demeans and devalues traditional knowledge, village skills, and practical wisdom that the poor value, respect, and apply for their own development. Just because it is not "certified" does not mean it is inferior. Village knowledge and skills have been respected and applied for hundreds of years, well before the certified urban doctor, teacher, and engineer turned up in the villages.

Barefoot professionals include educators, doctors, night school teachers, solar engineers, water drillers, architects, designers, midwives, masons, communicators, hand pump mechanics, computer programmers, and accountants. Tens of thousands of students have passed through the College and are productive, responsible members of rural society.

In rural India, 60–70 percent of children do not attend school during the day. Instead, they help their families with essential activities, such as collecting firewood or drinking water. Formal school traditionally has not been perceived as a valuable use of the children's time. But they have time to attend school at night.

Since 1979, the Barefoot Night Schools have educated more than seventy-five thousand children, three-quarters of them girls, making learning accessible to all and relevant to young people in poor rural areas. The Night Schools' unique approach exposes the children to how village institutions work. Through an elected children's parliament with a twelve-year-old prime minister, they learn about democracy. Currently, some eight

continued on next page

Box 18–1. continued

hundred and fifty villages have solar-lit night schools, and more than three thousand Barefoot teachers are reaching twenty-five thousand children who are going to school for the first time. In addition, the Night Schools employ more than nine hundred traditional communicators, with evening puppet shows helping to educate more than one hundred thousand people in remote villages where there is no television and no newspapers.

Barefoot College also has been in the process of expanding the role of education to help empower women not only to survive, but to thrive. It is the only college of its kind in the world. There is no other place of learning where illiterate rural women from some of the most inaccessible villages in the world can learn how to be solar engineers in six months using only sign language. The focus is on giving opportunities to rural women and providing them with skills usually identified with men. Sophisticated technologies such as fabricating solar lanterns, cookers, and heaters are no longer complicated for the illiterate rural women to handle, repair, and maintain.

Barefoot College trains women only between ages forty and fifty, because more mature village women tend to stay with their families in the village. Men, on the other hand, tend to use skills to find work outside the village and to send money home. Women often have more patience, are more skilled with their hands, and, the older they are, are more respected and listened to more seriously.

Barefoot College's solar engineering students come from poor families with little or no educational qualifications. No translation, no interpreters, and no written language is required. They learn with their hands. They are practical and intelligent and willing to learn, showing enormous patience. Once they are trained, they are less likely to leave their communities, because they have been given a chance to prove their worth to their own communities. Many graduate knowing more about fabricating charge controllers, inverters, and solar lanterns than graduates of five-year university programs.

Using the Barefoot College approach, nine hundred and twenty-four Solar Mamas have been trained from seventy-two countries. They have solar-electrified more than sixteen hundred remote villages all over the world, and, in the process, they have become powerful change agents of the future.

—*Bunker Roy, founder of Barefoot College*

Source: See endnote 6.

started in underresourced communities, serves as a boot camp for budding entrepreneurs. Trainees get five months of expert training, which includes one-on-one mentoring with successful entrepreneurs, legal assistance with structuring and business incorporation, support in fundraising and branding, access to a peer-support network, and name recognition and visibility for their cooperative. Worker cooperatives likely will become increasingly popular as traditional privately owned businesses face economic contraction and continue to downsize their operations. Green worker cooperatives could become a critical solution both to youth unemployment and to meeting sustainability needs.[7]

Creating a Generation of Carbon Mitigation Specialists

Reducing carbon emissions through the use of renewable energy sources, the redesign and retrofitting of buildings, the adoption of green transportation systems, and the development of sustainable technological innovations must be at the core of the transformation to a viable future. Fortunately, in many of these areas, work is already under way. According to a report by the International Renewable Energy Agency, there were 8.1 million jobs worldwide in the field of renewable energy in 2015, and this figure is likely to double by 2030.[8]

Numerous degree and nondegree programs in North America and Europe are geared toward providing renewable energy training and education. Examples include one- to nine-day workshops on renewable energy, green buildings, sustainable ecological design, and alternative construction methods at the Solar Living Institute in Hopland, California; a bachelor of science degree in Energy and Environmental Management at the University of Central Lancashire in the United Kingdom; and a master's degree specializing in wind energy at the National Technical University in Athens, Greece.[9]

In the United States, several community colleges have developed training programs awarding certificates and two-year associate of science degrees in "green" engineering. (See Chapter 22.) In one of the most impressive efforts, Bristol Community College (BCC) in Fall River, Massachusetts, has greened its engineering curricula thanks to a four-year, $900,000 National Science Foundation grant awarded in 2010. The degrees are in mechanical engineering focused on wind technology, electrical engineering focused on solar energy, and environmental engineering focused on energy conservation. BCC also grants certificates in solar and wind energy.[10]

All BCC engineering students are required to take two basic courses. The first, "Introduction to Sustainable and Green Energy Technologies," includes the construction/engineering design and implementation process, green building practices focused on energy efficiency, environmental conservation and resource management, and renewable energy. In the second course, "Green Building Practices," students study the methods, materials, and equipment used in the construction of residential and commercial buildings, roads, and highways, approaching these from a green lens. They learn about the proper use, selection, specifications, strength and limitations, fire resistance, and code conformity of basic construction materials and fabrication processes. A lab requirement includes fieldwork and basic laboratory testing procedures.[11]

Some students graduating from these programs go on to four-year institutions to continue their studies in engineering, having obtained a firm grounding in green technologies. Others go directly into the labor force, finding employment in areas such as wastewater operations, hazardous waste management, underwater remotely operated vehicle design (such as was used in the cleanup operations following the Deep Horizon oil spill off the U.S. Gulf Coast in 2010), computer-assisted drafting, geological surveying, and technical consulting in manufacturing. Such training and work is critical in creating a viable ecological future, and the time investment and cost for preparation at a community college is relatively minimal. If face-to-face tutelage is not possible, many options are available for online and distance learning training in the fields of renewable energy and building retrofitting. The American Society of Civil Engineers and Green Training both offer short-term online trainings.[12]

In the area of sustainable transportation, the Transportation@MIT initiative at the Massachusetts Institute of Technology offers undergraduate and graduate students exciting and innovative opportunities to study dynamic systems of mobility for the twenty-first century. Drawing from fifteen departments within the School of Engineering, the Sloan School of Management, and the School of Architecture and Planning, the initiative is multidisciplinary and offers coursework in areas such as sustainable energy, the environment, safety, green technology and infrastructure, information systems, economics, settlement patterns, urbanization, social structure, globalization, and political and personal behaviors. The emphasis is on moving people efficiently and promoting access to goods and services while simultaneously preserving and restoring the environment. Graduates typically go into government, business, and academia with careers in research, design, development, planning, and consulting.[13]

The Institute for Transportation Studies at the University of California-Davis also is paving the way in education toward sustainable transportation. Although its main focus is on research, it offers an interdisciplinary graduate program in Transportation Technology and Policy that draws from thirty-four disciplines. Emphasis is on improvements in network efficiency, use of high-efficiency vehicles, substitution of alternative fuels, use of alternative modes and multimodal transport, incorporation of sustainable designs, and efficiency in integrating transportation systems with land use. Graduates go on to work in diverse sectors including public agencies specializing in ecological research and natural resources, city planning and environmental analysis, conservation organizations, environmental consulting firms, and legal practices specializing in environmental issues.[14]

Regenerative Land-Use Management and Sustainable Agriculture

Reducing carbon emissions through the use of renewable energy sources, green buildings and transportation systems, and sustainable technologies have become familiar soundbites for responding to climate change. In contrast, little-to-no public discourse has focused on regenerative land management, a strategy that goes beyond adapting to the "new normals" of worsening heat waves, droughts, intense storms, and floods and offers a means to "re-green" the planet through holistic land-use practices—such as regenerative agriculture, holistic grazing, restoration of prairies and wetlands, and reforestation—that capture and retain carbon in the soil. Carbon sequestration holds the promise of restoring ecosystems that can actually reverse global warming while also developing healthy soils through the reintegration of biodiversity, the reestablishment of nutrient cycles, and increased water retention. This process also can enhance food production, helping to improve food security and to diminish hunger and famine.[15]

Although regenerative land management is not mainstream, it is starting to develop. A growing number of college and university courses and programs are focused on sustainable agriculture, and some also specialize in agroecology, or the growing of harvestable trees and shrubs among or around crops or on pastureland as a way to preserve or enhance the productivity of the land. Beyond the university setting, scores of organizations are working to train communities in regenerating their lands. Some institutions offer study-abroad programs that bring students into rural communities to learn sustainable management practices directly from village farmers and fishermen. (See Box 18–2.)[16]

Perhaps the name most closely associated with regenerative land management is Allan Savory, a Zimbabwean wildlife biologist-farmer who, in the 1960s, identified the causes of degraded and desertified grassland ecosystems throughout the world. By managing livestock to mimic behaviors practiced in the past by vast herds of grazing animals, Savory developed a method to restore

Box 18–2. Community-based Education in Thailand

Understanding the link between culture (how we live) and ecology (where we live) is a critical part of education for sustainability. The best people to teach about this are the community members who are on the front lines of sustainability. In Southeast Asia, this includes people such as village organic farmers, coastal dwellers who are working to restore mangroves, and tribal people who are conserving watersheds.

Study-abroad programs that focus on experiential learning in specific ecosystems, watersheds, mountain ranges, and island archipelagos have shown that students learn best about ecology and sustainability when exposed to the complexities of the real world beyond the classroom. Community-based education moves beyond this and has the community members themselves play a key role in the education process, from course design to teaching in the field.

By engaging local residents directly and using the tools of participatory research, educators can work as facilitators to help communities discover what they need to teach outsiders about their lives, their struggles, and the ecosystems of which they are a part. Letting "the farmer be the teacher" is tremendously empowering for communities, as they are given a chance to tell their stories and to pass on their knowledge and wisdom. Students benefit from learning directly from practitioners, and community members benefit by learning to better tell their stories to others, from development workers to government officials.

In Thailand, the International Sustainable Development Studies Institute (ISDSI) worked with Bak'er'ya tribal community leaders in Mae Hong Son province to create a course examining upland rotational farming, a practice that is sustainable if managed correctly, but that government officials have vilified as "slash and burn." These communities, as well as others that ISDSI works with, have found it helpful to participate actively in the teaching of visiting U.S. college students, as this helps them better articulate their issues to skeptical government officials and park administrators. Two local community leaders, Padti Saju and Loong Prapat, noted that this engagement encourages them to keep fighting for community rights to sustainably manage the watershed forests, knowing that outsiders are aware of and support their struggle.

The key to effective community-based education is handing over real control to shape course development to the villagers, empowering them to teach and giving them an opportunity to pass on their local knowledge and their passion for sustainability to visitors from around the world.

—*Mark A. Ritchie, Executive Director, International Sustainable Development Studies Institute*

Source: See endnote 16.

devastated land back to health. The Savory Institute, which he founded in 2009 with a small group of associates, is working to develop one hundred self-sustaining Savory Hubs—focused on holistic management—around the world by 2025. By partnering with local individuals and organizations, these Hubs are composed of farmers, ranchers, and land managers who receive, as part of the curriculum, hands-on training and implementation support for both commercial and communal land-based operations. Nongovernmental organizations and government agencies that work with pastoralists also receive training.[17]

The Savory Institute estimates that providing training and support for thousands of practitioners will improve the management of 1 billion hectares of land—approximately one-fifth of all grasslands worldwide. As of 2015, thirty Savory Hubs had been established in fourteen countries (including Argentina, Malawi, Mexico, Turkey, and the United States), with 6.5 million hectares of grasslands improved and more than three thousand land managers trained. The first of these Hubs, the Africa Centre for Holistic Management in Zimbabwe, was cofounded by Savory and provides seminars, workshops, and training programs to help

Joy of Urban Farming is an urban agriculture project managed by the office of the vice mayor of Quezon City, Philippines.

local farmers improve their food and water security and livelihoods by using livestock to regenerate damaged watersheds, wildlife habitats, and croplands.[18]

In urban settings, the U.S. nonprofit Growing Power offers a grassroots, community-oriented model for educating about sustainable agriculture through urban farming. Established in 1993 and based in the Midwestern cities of Milwaukee and Chicago, Growing Power was started by the father-daughter team of Will and Erika Allen and helps urban residents of all income levels grow affordable, high-quality protein and vegetables using sustainable methods. The organization provides job training and life skills to underserved youth via its Youth Corps; it also offers challenging farm-practice internships and develops workshops centered on building community through agriculture.[19]

Reforestation is another critical element in the carbon sequestration equation. The nonprofit group Trees for the Future, based in the U.S. state of Maryland, provides technical knowledge in agroforestry, reforestation, and sustainable development and has trained over three hundred thousand families in two thousand communities, mostly in Africa. More than 60 million trees have been planted by these families, and some eighty-seven thousand acres of degraded land have been restored to productive health.[20]

Through concerted efforts to re-green the planet, communities could go a long way toward mitigating the impacts of climate change, promoting biodiversity, and reestablishing the human connection with nature. This work is so critical to the future of humankind that, in the United States, it would be well worth implementing a National Community Resilience Corps, ignited with the vigor and determination of the Civilian Conservation Corps that was mobilized to expand infrastructure and create jobs in the aftermath of the Great Depression. (See Box 18–3.)[21]

Education for Peace and Conflict Management

In the Anthropocene, we must make peace with the planet as well as find ways to make peace with each other. Regenerative soil management goes a long way toward making peace with the planet, and providing education models that prepare people for meaningful work with viable incomes can contribute to interpersonal and community well-being and reduction in tensions.

Nonetheless, conflicts between and among peoples, groups, and nations are inevitable and can be constructively transformative, taking people to higher levels of understanding and cooperation when the conflicts are well managed. Scholars in peace studies most often define peace not simply as an absence of conflict and violence, but as cooperation that engenders justice and freedom rooted in human rights, equal access to education, and sociopolitical structures that are fair and just. Peace studies as a discipline typically combines anthropology, history, political science, psychology, sociology, and theology as a means to find the roots of conflict, identify underlying causes, create preventive strategies, and teach conflict management skills. More recently, environmental studies has been part of the peace studies academic mix as well.

Many universities around the world offer undergraduate and graduate programs in some aspect of peace and conflict management. These include programs focused on the theory and practice of conflict management, dispute resolution, mediation, and negotiation, as well as programs that offer internships and other practical field experience in areas of conflict so that students can learn these skills firsthand. The United World College of the American West (UWC-USA), a two-year pre-university boarding school in the state of

Box 18–3. Building a National Community Resilience Corps

In the dark days of May 1934, U.S. First Lady Eleanor Roosevelt expressed deep concern about the lasting impact of the Great Depression on young people, who had no previous experience of rewarding work or prosperous times to recall. "I have moments of real terror when I think we might be losing this generation," she told the *New York Times*. "We have got to bring these young people into the active life of the community and make them feel that they are necessary."

Mrs. Roosevelt championed the formation of the National Youth Administration (NYA), which joined the better-known Civilian Conservation Corps (CCS) in providing incomes, skill building, and meaningful work for millions of young people. CCS members planted billions of trees, built or improved facilities at more than eight hundred parks across the United States, and profoundly changed Americans' appreciation of nature and the value of environmental conservation.

Young people today face their own share of economic and financial challenges, along with other, very real, fears about their future—most notably, the climate crisis. Much has been made of the recent economic recovery, but sixteen- to twenty-four-year-olds have found it more difficult than most other age groups to enter the workforce, with employment in this group falling 9.7 percent from 2000 to 2016. They also are hamstrung by catastrophic levels of student debt, which has risen 423 percent since 2003. At the same time, multiple polls have shown that millennials are more worried than older generations about climate change and strongly favor reducing the use of fossil fuels.

Considering the economic insecurity that even a smooth transition to a post-carbon economy will bring, might it be time to bring back the NYA or the CCS, or to establish something like these for the twenty-first century?

The idea of government-supported service never really went away in the United States. The CCS helped inspire the Peace Corps and continues its legacy in the form of state and local conservation corps across the country. The federal government operates a number of national service programs—most notably AmeriCorps—through the Corporation for National and Community Service. Many of these, along with nonprofit programs such as Green Corps, offer young people opportunities to engage in conservation, environmental education and activism, and even climate adaptation and disaster response efforts.

Although these valuable programs engage thousands of young people every year, we desperately need much more, lest we lose not only this generation, but countless generations to come. The climate crisis presents a challenge that justifies a vast mobilization of the public as much as the Great Depression did. One means of doing this would be to form a National Community Resilience Corps (NCRC), which would harness the untapped passion, creativity, and labor of millions of young people to implement projects to grow

continued on next page

Box 18–3. continued

resilience and build sustainability in tens of thousands of communities across the country. The NCRC could expand from the new Resilience AmeriCorps program and provide full-time employment for eighteen- to twenty-five-year-olds for a period of two years.

NCRC members would support the development of a range of community projects, such as:

- Community Resilience Assessments, designed to engage community members in identifying shared values, vulnerabilities, and opportunities for intervention;
- community education, including in the areas of systems thinking, ecoliteracy, and resilience science;
- hands-on soil and energy conservation programs;
- the development of distributed, community-owned renewable energy and food enterprises, along with other critical local and regional services and goods; and
- growing community connections and engagement.

Like other service programs, the NCRC would deliver basic training in community organizing, communications, and specific project-related tasks. Critically, it also would invest in providing foundational education for its members, so that they can: 1) understand the true nature of interrelated environmental, energy, economy, and equity crises of the twenty-first century, and 2) build their capacity to think systemically and apply resilience science and practice to their lives and work. After all, thriving on a changing planet will require a whole new way of thinking.

—*Asher Miller, Executive Director, Post Carbon Institute*

Source: See endnote 21.

New Mexico, offers an innovative international baccalaureate diploma with a special focus on peace studies and conflict resolution. Each year, UWC-USA admits some two hundred students, ages sixteen to nineteen, from approximately eighty countries. Respect for diverse cultural, social, and religious backgrounds is central to the school's educational mission.[22]

The Martin Luther King, Jr. Center for Nonviolent Social Change in Atlanta, Georgia, founded in 1968, provides education and training based on Dr. King's nonviolence philosophy and methodology. The Center's Nonviolence 365 Program focuses on youth and helps students realize nonviolence as a lifestyle by facing problems, tensions, conflicts, and violence in nonviolent ways. The program empowers learners to think, speak, and act grounded in nonviolence while fostering leadership skills. The Center has developed training work-shops for youth as well as educational programs for schools: for example, the King Scholars Curriculum aims to help classroom teachers from kindergarten through grade twelve incorporate freedom, justice, and peace education into their lesson plans. Auxiliary tools such as online e-learning modules and speaker series make it possible for people around the world to tap into learning about Dr. King's work and how individual, group, and institutional conflicts can be addressed through nonviolent means.[23]

No doubt, education as we have known it will change as the ecological challenges that we face demand creative thinking and innovative adaptations. The realities of climate change, resource depletion, and economic contraction are upon us—and the clock is ticking. We do not have the luxury of time to postpone doing the work that is needed to stabilize our future. Many short-term educational practices and vocational contributions can lead humankind to long-term gains. We are capable of mustering the courage and stamina that are needed to see us through some destabilizing times ahead, but we must act now.

CHAPTER 19

Sustainability Education in Prisons: Transforming Lives, Transforming the World

Joslyn Rose Trivett, Raquel Pinderhughes, Kelli Bush, Liliana Caughman, and Carri J. LeRoy

Fifty kilometers northeast of Seattle, Washington, staff and inmates at the Monroe Correctional Complex made plans for a "worm farm," a composting program that would provide education and training while mitigating the institution's $65,000 annual expense for food waste disposal. This simple idea ballooned in ambition and scale due to the energy and charisma of an inmate in the program. He engaged with leading experts in vermiculture (breeding worms) and vermicomposting, asked for investment in scientific resources and equipment, and fostered a program culture of education and outreach.[1]

With the support of staff from the Department of Corrections, the program's inmate technicians built breeding and composting bins from reclaimed materials and learned how to brew "worm tea." The team grew the worm population from two hundred starter worms in 2010 to 7 million by 2016, and they now process nine tons of food waste from the facility each month. Twenty-two men have served as worm farm technicians so far. They commonly describe the job as their first experience with science and the first place that they "experienced peace" in prison.[2]

After six-and-a-half years, the program continues to innovate. Trials using anaerobic fermentation to preprocess fats and oils have led to the program accepting meat and dairy wastes. The most recent pilot project involves feeding postconsumer food waste to black soldier fly larvae and then donating

Joslyn Rose Trivett is education and outreach manager at the Sustainability in Prisons Project (SPP). **Raquel Pinderhughes** is a professor in urban studies and planning at San Francisco State University and executive director of Roots of Success. **Kelli Bush** is SPP program manager, **Liliana Caughman** is SPP lecture series coordinator, and **Carri J. LeRoy** is SPP codirector and a faculty member at The Evergreen State College.

Sadie Gilliom

A worm farm technician at the Monroe Correctional Complex shows the prison's vermicomposting program to visitors from the region's leading organic gardening nonprofit organization.

the excess larvae to nearby zoos as high-quality animal feed. The program now serves as an international model for worm composting, and hundreds of visitors have been impressed by the environmental expertise of prisoners who work in the program.[3]

Prisons are designed to contain and control the people they house, and corrections facilities may be the last place where one would expect to find innovative environmental programs. Communities outside the prison fence typically ignore and discount the people within. In stark contrast to these everyday realities, however, beautiful and productive environmental education initiatives are blooming in such facilities across the United States and around the world. Extraordinary partnerships have emerged among incarcerated people, corrections staff, nonprofit organizations, visiting students, faculty, scientists, and community groups. Together, they are creating programs that provide environmental education, gardening, recycling, composting, materials repurposing, habitat restoration, endangered species recovery, and job training.

The Sustainability in Prisons Project (SPP), the U.S. umbrella organization under which many of these programs exist, was founded in Washington State by the Department of Corrections and The Evergreen State College in 2003. In fourteen years, the effort has grown dramatically, supporting more than one hundred and seventy sustainability programs in Washington State prisons and gaining national and international recognition for leadership in the prison-based environmental movement. With support from the National Science Foundation, the SPP Network was formed in 2012 to support expansion of these programs. SPP-modeled programs have been established in state prisons in Maryland, Ohio, Oregon, and Utah and in county jails in Santa Clara, Los Angeles, Multnomah, and Salt Lake counties. Twenty-five more U.S. states have similar programs or aspirations. SPP's work demonstrates the vast potential to expand environmental education initiatives that engage underserved populations in corrections institutions throughout the world.[4]

SPP programs provide multiple examples of innovative initiatives and partnerships that bring environmental education into prisons in the United States. New ideas become successful programming when the interests of all stakeholders overlap, when there are benefits for all involved, and when adequate resources are available. The original spark of an idea can come from an inmate, sustainability expert, nonprofit, graduate student, academic, prison staff member, or volunteer citizen. The programs that develop are often informal and low-cost, and thus are more accessible to a broad group of contributors and more affordable to correctional facilities.

In the United States, a country that has more corrections facilities than colleges, providing environmental education to prisoners has at least two powerful effects. First, it creates opportunities for prisoners to heal and redeem themselves, increase their knowledge and skills, and reduce their chances of returning to prison after release. Second, expanding environmental education to currently excluded populations is a necessary part of developing and nurturing diverse and talented environmental stewards and environmental justice activists. Doing so will increase the number of people who are committed to the struggle for a more just and sustainable world and will lead to transformational social and environmental partnerships.[5]

Environmental Education in Prisons

Environmental education must be relevant, holistic, socially responsible, considerate of all life forms, issue-based, action-oriented, reflective, and democratic. At the 2002 World Summit on Sustainable Development in Johannesburg, South Africa, participants issued a multinational plea "for partnership that would allow communities, professionals, and governments to jointly take action," recognizing that we are stronger if we are inclusive and united. This approach to environmental education, often referred to as "social learning," begins by recognizing every person's ecological knowledge and by building shared beliefs through storytelling, dialogue, debate, and problem solving. It requires that mainstream environmentalists move outside of familiar circles, visit new communities, and be maximally prosocial—taking action to benefit others. Marginalized groups are present in every community, and so long as one is willing to step into a new neighborhood, onto a reservation, or through a prison gate, there is access to original ideas and wisdom.[6]

When it comes to the challenge of extending environmental education to previously excluded groups (see Box 19–1), there are few groups more

People from low-income communities and communities of color bear the brunt of environmental problems and injustices. Environmental justice expands environmentalism so that all groups of people have opportunities for benefit and influence; in this context, increased access to environmental education is a means to empowerment. The environmental movement also has much to gain from inclusiveness. There is a far greater hope of meeting global interdisciplinary and intersectionality issues and challenges if all stakeholders are on board.

An environmental movement that represents all stakeholders will be stronger and more resilient. When people who traditionally have been excluded from the mainstream environmental movement are included, they contribute a much-needed diversity of experiences and new ideas. Ensuring diversity in environmental education, sustainability initiatives, and environmental movements will open up myriad ways of tackling environmental problems and creating solutions.

Source: See endnote 7.

marginalized than prisoners. The U.S. crisis of mass incarceration currently imprisons 2.4 million adults, and 92 million Americans have criminal records. African Americans and Latinos are disproportionately affected by environmental problems and injustices, are underrepresented in the mainstream environmental movement, and are disproportionately represented in the incarcerated population. One in twelve African-American men and one in thirty-six Hispanic men serve time in prison. Incarcerated adults have less education than the general public, and, without education, their prospects of post-incarceration employment are discouragingly low.[7]

In-prison education is the only proven method of reducing recidivism (return to incarceration), yet access to educational programming in prison is a rare privilege. In the United States, only 6 percent of incarcerated adults receive in-prison postsecondary education, nearly all of which is vocational. A focus on environmental issues or science in prison is even more uncommon.[8]

Degree-awarding programs in prison provide an invaluable opportunity to their students, but they are offered in very few facilities and are available to very few incarcerated individuals. A meta-analysis of the effects of correctional education on post-release employment and recidivism shows that informal or vocational programs convey as many advantages as formal academic programs in prison. Although increasing access to traditional academic programs is important, nontraditional, lower-cost education programs, such as the ones facilitated under SPP's umbrella, must be made available on a broader scale to serve a larger number of incarcerated individuals.[9]

SPP's programs incorporate three educational approaches, as described below:

Experiential learning opportunities. Hands-on practice and experimentation is hugely motivating and engaging for most students. Environmental literacy students at Washington State Penitentiary follow mornings in the classroom with afternoons in the adjoining Sustainable Practice Lab (SPL). They receive hands-on training in areas such as bicycle and wheelchair restoration, plant propagation, aquaponics, composting, and making crafts from reclaimed materials. Many graduates continue in the SPL's job programs, and all are encouraged to contribute to improving and adding to the Lab's eighteen program areas. A similar, less formal, initiative exists in Santiago, Chile, where the nonprofit organization Casa de la Paz empowers inmates to refashion discarded materials into valuable tools and crafts.[10]

Prosocial motivation. Students are motivated by topics that are practical and valuable in the real world and to their communities, and most participants in SPP programs want to be helpful to others. In the U.S. state of Ohio, incarcerated women enrolled in an environmental literacy and justice class called Roots of Success use recycled bags to make crocheted mats that are donated to people sleeping on the streets. The mats provide a barrier between the street and a sleeper's body and deter bed bugs and other insects. At the same time, the women's efforts reduce waste at the prison.[11]

Community-based approach. Emphases on community building and mentoring create a learning community that supports and reinforces education. Incarcerated women at Mission Creek Corrections Center for Women outside of Belfair, Washington, collaborate with SPP graduate students and staff, corrections staff, biologists, zookeepers, and others to rear federally endangered Taylor's checkerspot butterflies. Since 2012, the program has reared and released more than ten thousand caterpillars and adult butterflies. Incarcerated technicians receive extensive education and training on the technical work, contribute to program trials and protocols, and increase the capacity of an endangered species recovery effort. They also develop connections with partners outside the prison that expand their education and employment support network.[12]

These three educational approaches are important because they inform the learner's identity and help integrate new ideas and skills into a sense of self.

Research has shown that educational programs that foster shared learning significantly reduce achievement gaps, and they motivate individuals to work as a team for the good of the group. SPP partnerships are founded on shared beliefs and intentions that acknowledge academic potential and that support students' sense of belonging.[13]

Butterfly technicians at Mission Creek Corrections Center for Women in Belfair, Washington, pose in front of a poster set up for visiting daughters of inmates through the Girl Scouts Beyond Bars program.

It is difficult to find environmental curricula that satisfy all three of the components above. Few curricula contain content that is directly relevant to the everyday experiences of people who come from marginalized or low-income communities, or from communities of color. Environmental curricula that are designed for students who have weaker academic skills are even more difficult to come by. It also is challenging to find curricula that give sufficient attention to achievable and local actions and solutions. Often, environmental education focuses on global-scale "doom and gloom" that the average student does not find compelling or motivating.[14]

In 2012, Ohio's Department of Corrections began offering Roots of Success, a ten-module environmental literacy and work readiness curriculum that is designed to meet the needs of students who have not been well-served by the education system. Students graduate from the course understanding a wide range of environmental issues, with the job and reentry skills needed to work in the green economy, and with a certificate to show potential employers. Each module focuses on a different environmental sector, such as water, waste, transportation, energy, building, and food. Videos and case studies feature individuals from marginalized communities—including incarcerated youth and adults—in leadership positions. The teaching approach encourages students to extend previous knowledge and experiences and to connect what they are learning in the classroom to real-world issues and employment opportunities. Intensive student engagement, relevant content, multimedia materials, and group activities stimulate students' intellectual curiosity and ignite their interest in the subject matter. The course challenges incarcerated individuals to think critically about environmental issues and problems and is solutions oriented.[15]

Roots of Success has been adopted widely and is used in prisons throughout the United States. It has been customized for other types of corrections institutions (for example, jails and juvenile facilities) as well as for reentry programs, high schools, job training programs, and other settings. In prisons, one of the most powerful and unique aspects of the course is that it is taught by incarcerated individuals who are trained and certified to teach the curriculum. Roots of Success instructors and students report that this approach to teaching leads to classroom interactions unlike any others in prison. For example, experience has shown that, within a very short time, incarcerated individuals who have been hostile to one another outside the class are willing to sit and learn alongside each other, setting aside their gang, religious, and other affiliations.[16]

Environmental education is an ideal subject for education in prisons. The environment is a universal topic—everyone lives in a place and interacts with the built environment, natural resources, and the weather on a daily basis—but one that may have been invisible to most students before the class. Roots of Success instructors and students take what they learn in the classroom and apply it in their everyday lives, proudly sharing, for example, how they have helped family members reduce household energy costs by conveying new knowledge about how to conserve resources. As Roots of Success instructor Cyril Waldron explains:

> [W]e are not only bringing a new world to our students but are introducing them to the world, a world they never knew existed, by exposing them to concepts that were previously foreign to the vast majority of them. It is not that they do not have the aptitude or attitude to learn, but have been denied the opportunities. These previously unreached students can no longer use that as an excuse because they have been touched by the gospel of sustainability.[17]

Transformation and Opportunity in Prisons

The challenge to support environmental education in prison is also a call for mutual transformation. For incarcerated individuals, environmental education and practice can contribute to both redemption and healing. Imprisonment often results from two violations of the social contract: 1) in which a citizen has violated people or property, and 2) in which society failed to ensure a citizen's basic human rights. While incarcerated, many individuals seek opportunities to redeem and heal themselves, and environmentally focused programs can provide a venue for both. As one inmate at Santa Clara County Jail in California observed, "Being in jail is sorta like being on one of my combat tours. Working in the landscape program definitely helps me escape the environment, and puts me at ease and relaxes my mind, and takes me away for the few hours that we're working, lets me feel like I'm on the streets [out of prison] again. . . . I am happy to give back."[18]

To do this work successfully, prisoners must be afforded trust and opportunity so that they can build a sense of safety—what the Universal Declaration of Human Rights calls "security of person." Building a sense of well-being requires basic physical safety as well. Prison administrators affirm that SPP programs and Roots of Success classes reduce violence and increase safety within the prisons, and preliminary measures of inmate behavior have reinforced these observations. Collaborative, meaningful work itself can be healing, offering all of the positive benefits that humans require to flourish: positive emotions, engagement, relationships, meaning, and accomplishment.[19]

Researchers have tied positive environmental attitudes to greater prosocial behavior, and prosocial behavior is key to success in society. There is growing evidence that environmental programs for inmates are therapeutic and reduce recidivism. The Insight Garden Program, now offered in three California state prisons, uses a restorative justice model to talk about gardening principles and design, while at the same time cultivating personal integrity, accountability, emotional intelligence, relationships, and other valuable life skills. The flagship garden program—at San Quentin State Prison north of San Francisco—boasts a low rate of recidivism among participants following their release.[20]

Since 1989, the Horticultural Society of New York has provided the Horticultural Therapy Partnership to prisoners at Riker's Island, one of the largest jail complexes in the world, housing an average of thirteen thousand inmates daily. Through the program, incarcerated students design, install, and maintain a one-hectare area of landscaped and productive gardens, and, upon their release from prison, they have the option to join the GreenTeam, a vocational internship program. An analysis of the program showed that one- and three-year reconviction rates among participants were significantly lower than among all comparison groups. The restorative benefits of nature in other institutional environments, such as workplaces and hospitals, has been well established, and connection to nature also has been shown to reduce violence in a maximum-security prison environment.[21]

Education, in and of itself, has been shown to reduce recidivism more than any other kind of intervention. Those who are afforded education while in prison have substantially better chances for employment post-release, and also for avoiding criminal activity and a return to prison. Since 95 percent of incarcerated individuals in the United States are released and re-enter communities, their ability to access opportunities and productivity is in everyone's best interest.[22]

Strategies for Inclusion

An inclusive approach to environmental education and action requires substantial effort and collaboration, but it is well worth the work. Among the strategies that will bring the greatest likelihood of success are:

- *Making programs accessible to new partners.* Think creatively and bravely about who has not been included, and how to best attract and welcome them as allies. Much more must be done to include underrepresented individuals and groups, particularly those from communities that are disproportionately affected by environmental problems.
- *Finding common ground.* Ask everyone in a group to share needs, interests, resources, and limitations. Ask lots of questions, and reflect back what others say. Point out areas where it sounds like interests and capabilities overlap—if any of those resonate in the group, then that is the topic to pursue further.
- *Considering how everyone can benefit.* How can programs improve partners' and students' lives? What do they want and need? Ask for their feedback in multiple ways—including anonymous submissions—and take action on their input. For example, visitors to prisons have learned that escaping the noise and constant crowding is an acute need for many incarcerated people. In Iowa's Correctional Institution for Women, landscape architects held workshops for incarcerated women and gained their input on the design for healing gardens at the prison. The gardens have become treasured areas of refuge for study and contemplation.[23]
- *Agreeing to shared values.* Motivation is based largely on values and beliefs. Discuss each partner's values, and look for areas of overlap. In SPP, partners accept a definition of sustainability that encompasses three spheres:

environmental, social, and economic. (See Figure 19–1.) While each part-ner may place greater or lesser emphasis on each sphere, all have agreed to respect and consider all three, as well as how they overlap and comple-ment each other.

- *Committing to shared principles.* Codify shared values to guide program planning and day-to-day operations. SPP's work is framed by five "essen-tial components" that are accepted by all partners and SPP Network sites;

Figure 19–1. The Three Spheres of Sustainability as Practiced by the Sustainability in Prisons Project

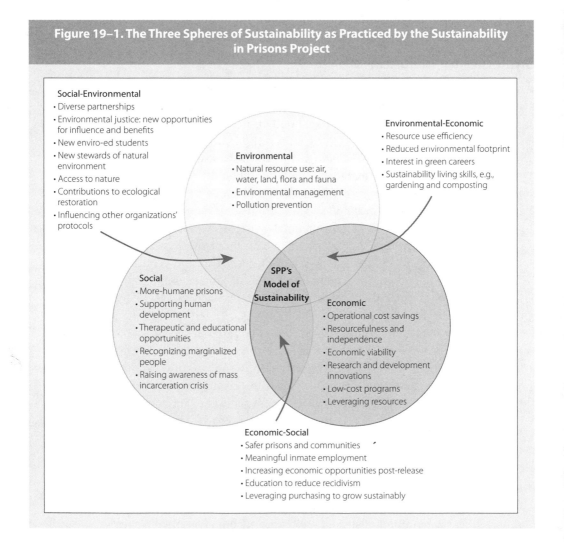

Social-Environmental
- Diverse partnerships
- Environmental justice: new opportunities for influence and benefits
- New enviro-ed students
- New stewards of natural environment
- Access to nature
- Contributions to ecological restoration
- Influencing other organizations' protocols

Environmental
- Natural resource use: air, water, land, flora and fauna
- Environmental management
- Pollution prevention

Environmental-Economic
- Resource use efficiency
- Reduced environmental footprint
- Interest in green careers
- Sustainability living skills, e.g., gardening and composting

Social
- More-humane prisons
- Supporting human development
- Therapeutic and educational opportunities
- Recognizing marginalized people
- Raising awareness of mass incarceration crisis

SPP's Model of Sustainability

Economic
- Operational cost savings
- Resourcefulness and independence
- Economic viability
- Research and development innovations
- Low-cost programs
- Leveraging resources

Economic-Social
- Safer prisons and communities
- Meaningful inmate employment
- Increasing economic opportunities post-release
- Education to reduce recidivism
- Leveraging purchasing to grow sustainably

briefly, they are: 1) partnerships and collaborations with multiple benefits, 2) bringing nature inside, 3) engagement and education, 4) safe and sustainable operations, and 5) evaluation, dissemination, and tracking. To make explicit that environmental education programs in prison are not to exploit or otherwise abuse incarcerated partners, consideration of their needs and wants is woven throughout the essential components.[24]

- *Asking for ideas and input.* Routinely ask all stakeholders for input to improve programs; every person has expertise and perspective that could advance the shared work. Since the first in-prison ecological conservation program was established as part of SPP in 2009, incarcerated technicians have helped develop and improve propagation protocols for dozens of native plants, many of which had never been cultivated previously. At Lovelock Corrections Center in Nevada, incarcerated men grow sagebrush in a multistate program managed by the Institute for Applied Ecology. One sagebrush technician refined techniques for transplanting and thinning the seedlings, carefully tracking and documenting results. His protocols show higher rates of survival and will translate into improved restoration of greater sage-grouse habitat.[25]
- *Recognizing and celebrating contributions.* In SPP programs, members of the staff cultivate a culture of thankfulness and mutual recognition. They frequently publish stories on individual and program successes. In the Roots of Success program at Stafford Creek Corrections Center in Washington State, every graduating class is invited to a graduation ceremony with speeches from class representatives, instructors, prison leadership, and special guests. These events are outpourings of admiration for individual and group accomplishments, commitment to the program, and environmental empowerment.[26]

There are numerous reasons to increase environmental education and SPP-type programs in prison settings. While a punitive model of incarceration used in the United States has proven ineffective at reducing crime or rates of imprisonment, education is rehabilitative. Environmental education programs have improved quality of life in prison and have energized environmental activism in people outside the environmental mainstream. In addition, these initiatives have significant environmental and economic benefits as correctional facilities reduce their resource use and produce environmentally friendly goods. At a time when the United States faces enormous environmental challenges and a criminal justice crisis, inmates and staff in prisons must be supported and empowered to play key roles in environmental and social change.

CHAPTER 20

Bringing the Earth Back into Economics

Joshua Farley

In 2014, the International Student Initiative for Pluralism in Economics, consisting of more than sixty-five associations of economics students from over thirty different countries, issued an open letter, with these opening lines: "It is not only the world economy that is in crisis. The teaching of economics is in crisis too, and this crisis has consequences far beyond the university walls. What is taught shapes the minds of the next generation of policymakers, and therefore shapes the societies we live in . . . . [I]t is time to reconsider the way economics is taught. We are dissatisfied with the dramatic narrowing of the curriculum that has taken place over the last couple of decades."[1]

What is this narrowing? Over recent decades, mainstream economics education has become increasingly obsessed with highly stylized mathematical models populated by rational, self-interested consumers. It has become less open to alternative methods and views, and less relevant to real-world events. Stylized models not only failed to predict the ongoing financial crisis, but, far more damning, they failed to see that it was even possible. These critiques are not new: for nearly two decades, economics students worldwide have been advocating for a "post-autistic economics" that focuses on concrete economic realities instead of imaginary mathematical worlds, and that provides a framework for critically debating economic issues.[2]

Even economics professors understand the need for a shift in how their discipline is being taught. Professors in France responded to their students'

Joshua Farley is an ecological economist, professor in community development and applied economics, fellow at the Gund Institute for Ecological Economics at the University of Vermont, and special visiting researcher at the Universidade Federal de Santa Catarina in Brazil. He is coauthor, with Herman Daly, of the first textbook in ecological economics.

manifesto on post-autistic economics by stating that, "Neoclassicalism's fiction of a 'rational' representative agent, its reliance on the notion of equilibrium, and its insistence that prices constitute the main (if not unique) determinant of market behavior are at odds with our own beliefs." Leading economists openly acknowledge that economics education "bears testimony to a triumph of ideology over science"; that "the economics profession went astray because economists, as a group, mistook beauty, clad in impressive-looking mathematics, for truth"; that "anyone who believes exponential growth can go on forever in a finite world is either a madman or an economist"; and that the depiction of a "rational" representative agent required to prove that market equilibriums are optimal is fatally flawed.[3]

At the same time, there is serious resistance to change within the discipline. Some 77 percent of economists believe that "economics is the most scientific of the social sciences." Economists earn higher salaries and find employment more easily than those in other disciplines; they have significant influence on policy decisions in government, business, and international organizations, often holding top political positions; and they are the only social science with their own "Nobel" Prize[†]—all of which are taken as validation of their theories. Furthermore, economists are unusually insular and hierarchical. As compared to other social sciences, citations in leading economics journals are less interdisciplinary and more concentrated among select journals, and economists profess significantly less regard for interdisciplinary research.[4]

Economists' influence greatly exacerbates the shortcomings of the field. Abundant evidence suggests that economics education is contributing not only to the world economic crisis, but also to a worsening ecological crisis. Since the core theories of modern economics emerged two hundred and forty years ago, the human population has increased at least ninefold, and per capita consumption, as measured by gross domestic product (GDP), has increased more than thirty-fivefold, with growth accelerating since 1950. The impact of market-driven economic activity on the global environment threatens to drive us across critical planetary boundaries, with potentially catastrophic outcomes for humans and other species. Human impacts now rival geological forces in their impacts on the planet, leading many scientists to assert that we have entered a new geological epoch, the Anthropocene.[5]

---

[*] Neoclassical economics is taught in the vast majority of universities and colleges.

[†] Actually, the Sveriges Riksbank Prize in Economic Sciences in Memory of Alfred Nobel.

Our economic system and economic education have failed, so far, to adapt to these challenges. We must create a resilient economic system that is capable of avoiding critical ecological thresholds while satisfying people's basic needs; we must design an economic system that is capable of balancing what is biophysically possible on a finite planet with what is socially, psychologically, and ethically desirable. The current approach to economics education blinds its students and practitioners to this reality. To create a resilient ecological economy, economics must abandon its narrow disciplinary focus and obsession with formal mathematical models to become a transdisciplinary field that is dedicated to addressing complex, real-world challenges.[6]

The Reigning Ideology: How Economics Is Currently Taught

The fundamental concept in mainstream economics for the past two hundred and forty years has been market equilibrium. In this perspective, competitive free markets are self-regulating systems that utilize free choice and the decentralized knowledge of rational, self-interested, and insatiable individuals to achieve an efficient and optimal allocation of resources. The fundamental goal, particularly since the mid-twentieth century, has been continuous economic growth.

For a market to be competitive, it must have a large number of buyers and sellers of nearly identical commodities, with free entry and exit for firms. Prices act as a fulcrum that balances the supply and demand of commodities, generating a "Pareto-efficient" equilibrium, meaning that there is no way for one individual to be made better off without making someone else worse off. Consumers are assumed to purchase the basket of goods that maximizes their individual utility, so the resulting equilibrium must maximize utility for society as a whole, subject to the initial distribution of wealth and resources. Firms are assumed to maximize their profits by reallocating resources from

Rational buyers and sellers at work in the Marrakesh, Morocco, *souk*, a traditional North African market.

Jorge Láscar

sectors earning lower profits to those earning higher ones, increasing supply in the latter and thus decreasing prices and profits.

In equilibrium, each firm earns a fair return on all its factors of production, but nothing more. Firms can earn profits temporarily by developing new innovations that lower their costs or improve their products, but this attracts new resources to the sector, driving profits back to fair levels. The result is a system that provides consumers with ever-better products at the lowest possible price that will entice producers to provide them. Economists incorporated these ideas into mathematical models in the nineteenth century, which showed that such market equilibria were both possible and optimal. These models now drive economic education, analysis, and policy.

Beginning in the 1940s and 1950s, economists and others began to worry about growing signs of resource scarcity. However, after careful study, economists concluded that in a free market, scarcity increases prices, which reduces demand and provides an incentive to innovate substitutes, thus alleviating scarcity in a negative feedback loop. Economists simply assumed that limitless innovation was possible. From the 1960s onward, society became increasingly worried about pollution, resource extraction, and land uses that degraded ecosystems and the life-sustaining functions that they provide. Economists diagnosed the problems as "externalities" of economic activity—unintended impacts on others for which no compensation occurred—caused by a lack of property rights. Such externalities could lead to suboptimal equilibriums.

Economists concluded that these external costs could be internalized into market prices, for example by creating private property rights or by taxing the undesired activities in proportion to the harm that they cause. These actions would internalize nature's values into the market economy, ensuring that the sole feedback loop of the price mechanism would be adequate for achieving a stable equilibrium for both the ecological and economic systems, rolled into one. Unquestionably, the price mechanism can be a powerful feedback loop that, if used correctly, can contribute to the solution of social and ecological challenges that we currently face. Unfortunately, by focusing obsessively on the equilibrium model of the economy, economics education teaches students that the market price mechanism is the only feedback loop that is necessary to solve our global problems, and that any outside intervention (such as government regulation) must be suboptimal.[7]

The conventional approach to university education is to teach students a set of disciplinary tools and theories that they can apply to any problem.

While disciplinary knowledge is important, economics education should be grounded in science, current events, and real-world problems. This approach allows students to empirically test the validity and limitations of the theories that they learn, exposes them to real-life complexity, forces them to learn and apply knowledge from other disciplines as necessary, and better prepares them to solve problems that urgently need solving. Grounding economics in science and real-world problems does not always provide the answers, but it can help identify false solutions, and where we should look for real ones.

Grounding Economic Programs in Reality: Economics for the Anthropocene

One model for the future of economics education is the Economics for the Anthropocene (E4A) program—a diverse partnership of academics, government, and nongovernmental organizations, led by McGill University, the University of Vermont, and York University—which "conducts problem-based, participatory research that combines academic and nonacademic partner knowledge with student training to create solutions to real-world challenges." The goal of the program is "to articulate, teach, and apply a new understanding of human-Earth relationships grounded in and informed by the insights of contemporary science." Broad, interdisciplinary coursework in ecological economics, "big history" (see Chapter 12), and integrative methods provides students with the necessary background for a capstone course focused on a specific problem, such as managing transboundary water resources, decarbonizing energy systems, or climate justice, all delivered with a variety of stakeholder partners.[8]

So what happens when we apply this approach? From physics, we know that it is impossible to make something from nothing (or *vice versa*), that doing work requires energy, and that entropy (which from an economic perspective is akin to disorder or uselessness) increases over time. Our finite planet is the only source of the raw material inputs into all economic production. If we continuously extract these raw materials faster than they can regenerate or we can produce substitutes, they must run out. Finite stocks of fossil fuels provide more than 86 percent of marketed energy supplies and generate two-thirds of greenhouse gas emissions. When the flow of any pollutant continuously exceeds the ecosystem's absorption capacity, it must accumulate as an ever-larger stock with ever-larger environmental impacts, from which the poorest populations often suffer the most. Although the use of renewable energy is growing rapidly, it is neither as energy-dense nor as versatile as fossil fuels, and, aside from hydropower, it is intermittent and difficult to store.[9]

From ecology, we know that the ecosphere is a highly complex, nonlinear, and evolving system in which everything is connected to everything else. Much of the raw material that we extract and convert into economic products alternatively serves as the structural building blocks of ecosystems. One particular configuration of ecosystem structure is capable of capturing solar energy flows that sustain vital ecological functions, including many that are essential for the survival of humans and all other species. Transforming ecosystem structure into economic products inevitably diminishes ecosystem services, including the capacity to regenerate the structure that we remove. Waste emissions that are in excess of absorption capacity further diminish the capacity of ecosystems to generate essential services.[10]

Sam Beebe

Clearcutting in progress on the Olympic Peninsula, Washington.

Ecosystems also face unknown thresholds beyond which a small change in some variable, such as resource extraction or waste emissions, can lead to potentially irreversible and catastrophic change. For example, rising concentrations of greenhouse gases can melt heat-reflecting ice pack, replacing it with heat-absorbing open ocean, leading to more ice melt in a positive feedback loop, or vicious circle, that threatens runaway climate change. Scientists do not know exactly when the climate system might tip or what the full repercussions might be, and they undoubtedly are unaware of many other serious ecological thresholds.[11]

This transdisciplinary, problem-based lens makes it obvious that economic production inevitably incurs serious ecological costs that are likely to worsen as the scale of the economy increases. When the worsening ecological costs of further growth exceed the economic benefits, growth becomes first uneconomic, then impossible. Fossil fuels and raw materials from nature can be privately owned and are depleted through use—the necessary conditions for markets to allocate them toward the economic products that maximize their monetary value. Building a resilient ecological economy confronts a distinctly different allocation problem: how much ecosystem structure should be left intact to provide life-sustaining ecological functions, and how much should be sacrificed for economic products necessary to meet essential human needs?[12]

Ecosystem services such as climate stability and protection from ultraviolet radiation cannot be privately owned and generate no price signal to reflect their scarcity, so markets cannot solve this problem. The provision of life-sustaining ecological functions can only be decided collectively, not by individual choice. Furthermore, since all economic production affects climate stability and other ecosystem services used by all, there is no way to improve the welfare of some individuals without harming others, and Pareto efficiency is a meaningless criteria.[13]

Looking toward solutions, maintaining ecological resilience will require dramatic reductions in resource extraction and waste emissions coupled with the innovation of green technologies that will allow us to satisfy basic human needs with these reduced resource flows. Since information is not depleted through use, making the information that underlies green technologies freely available to all maximizes the adoption of these technologies but provides no market incentive to supply them.[14]

Political science, game theory, and various other disciplines reveal that these challenges are best modeled as "prisoners' dilemmas," in which participants face two choices: cooperate with others or defect and act in one's own self-interest. Universal cooperation—for example, reducing greenhouse gas emissions, or contributing resources to the development of open-source sustainable technologies—generates the best outcome for society, and defection results in the worst. A self-interested individual will seek to free-ride on cooperators, reaping the benefits without making any sacrifice, resulting in widespread defection. Prisoners' dilemmas can be solved only through cooperation, not self-interest. Solving environmental problems therefore requires economic institutions that stimulate cooperation, which, in turn, requires insights into human behavior.[15]

Fortunately, psychology, behavioral economics, political science, and other fields provide important insights into economic institutions that inhibit or promote cooperation. Studies from these fields have shown that market economies stimulate self-interested and uncooperative behavior. Studying market economics or thinking within a market framework makes people, in comparison with the population as a whole, less cooperative, less compassionate, more corrupt, and more likely to free ride. In contrast, forging group ties, punishing defectors, generosity followed by reciprocity, and forgiving the occasional defection (which might be accidental) all promote cooperation.[16]

Recognition that the price mechanism alone is inadequate for achieving socioecological resilience does not mean that prices do not provide a powerful and useful feedback loop. Taxing or regulating activities that degrade the environment or otherwise reduce human welfare does raise their prices and reduce use. Similarly, we can cap resource extraction or waste emissions at levels that are likely to avoid critical ecological thresholds and unacceptable costs, and we can allow prices to adjust to ecological constraints. Both options provide incentives to develop and adopt greener technologies. Implementing price or quantity signals also requires collective action: they will not emerge spontaneously through market forces.

But market allocation weights peoples' preferences by their purchasing power. In a world with highly unequal income distribution, markets allocate resources to those who can afford them, not to those who need them. For example, when the prices of staple grains doubled during the food crisis of 2007–08, the malnourished population in poor countries increased by 40 million, while citizens of rich countries scarcely noticed. Internalizing ecological costs into food or energy prices would send these prices soaring, and would impose the greatest costs on those who have done the least to cause the problem. Even when markets work as intended, outcomes may be far from optimal.[17]

Economics education that is focused on real-life problems can help transform both the theory and practice of economics while training students to help solve real-world challenges. This model is common within the field of ecological economics, and numerous transdisciplinary programs in the environmental sciences integrate economics into the broader curriculum in this fashion. Few economics programs do so, however, although calls for change from students and the example of the E4A may help change this over time. What can be done in the meantime?[18]

Grounding Economics Courses in Reality: Testing Hypotheses with Current Events

Any economics professor can help transform economics education by grounding his or her individual courses in science and reality, as many economists from outside the mainstream already do. The scientific method is based on careful observation of a system, leading to hypotheses about how it works. These hypotheses are tested through rigorous experiments or continued observations of the system. It is impossible to prove a hypothesis true, so scientists strive instead to prove it false, and, if this fails after repeated attempts, the hypothesis becomes a theory. When a hypothesis or theory is proved wrong, new hypotheses must be formulated that better explain the system. In economics, planned experiments can be very difficult or even immoral, but unplanned experiments can be found everywhere. A scientific, reality-based approach to economics must continually check both assumptions and theory against current events and daily observations of how the world works.

Take, for example, the assumptions on which equilibrium theory is based. In the United States, weekly news stories of market concentration, mergers, and price setting illustrate how market forces can actually undermine competition, while soaring pharmaceutical prices illustrate how government-enforced patents and onerous licensing requirements do the same. Profit-maximizing firms are unlikely to invest in innovations when lobbying politicians earns far greater returns on investment. Recurring bubbles and busts in stocks, real estate, and commodities are clear evidence that rising prices can increase speculative demand, further driving up prices in a disequilibrium-generating positive feedback loop. Economists acknowledge that eliminating only some market distortions may not increase allocative efficiency if others remain. Since no society has ever eliminated all market distortions, the hypothesis that markets are somehow optimal is an untestable ideology—not science—raising the question of why perfectly competitive markets should be discussed at all.[19]

News stories, however, also show that some concepts from mainstream economics are both relevant and realistic. Take, for example, elasticity of demand. Resources that are essential and that have limited substitutes—such as food, water, energy, and health care—exhibit price-inelastic demand, which means that the amount that we wish to consume is insensitive to price. A large increase in price triggers only a small decrease in demand and hence a large increase in revenue for the firms providing them. The benefits of market power—the ability for firms to set prices—is obviously much greater for essential resources, motivating firms to seek market power in essential resources through mergers, collusion, or lobbying. This explains why market concentration tends to be greatest in these sectors. Paradoxically, reducing the output of food, energy, water, or prescription drugs will increase the contribution of these items to GDP, illustrating the absurdity of maximizing GDP as a goal.

These examples are intended only to illustrate the point that every concept taught in economics courses must be tested against reality. Theories that help students understand and explain what is going on in the world should be kept, but those that do not are useless or false, and in either case should be abandoned.

Toward a New Model for Economic Education

Economics can, and must, make important contributions to achieving a just and sustainable future, but this will require significant changes to both theory and education. The seriousness and urgency of the problems that we face leave little time to waste. Unfortunately, economics' outdated models of human behavior, free markets, and equilibrium outcomes, and its focus on abstract mathematical models instead of real-world problems, lead to policy recommendations that worsen our predicament.

Fortunately, there is a growing call to reform the field in recognition of its numerous serious flaws. Interdisciplinary and problem-based approaches to education can help identify when and where mainstream economic principles can be applied, when they should be rejected, and what better hypothesis might replace them. Problem-based, interdisciplinary approaches to education in general are increasingly common in universities and are actively promoted in the United States by the National Science Foundation and other funding agencies.[20]

Applied to economics education, this approach could train better economists and actively contribute to solving urgent problems and to building a new economic model. There is currently no complete, unified, and widely accepted alternative theory of economics, and many economists wrongly maintain that we cannot reject the current model until its critics have developed such an alternative model. However, there also is no widely accepted "Grand Unified Theory" in physics, yet this does not prevent physicists from rejecting incomplete or misguided theories and making important contributions to solving real-life problems.

Unlike the laws of physics, economic systems continually evolve in ways that profoundly affect the social and ecological systems in which they are nested. They must then adapt to the new environments they create, in a recursive process. A Grand Unified Theory in economics may be impossible, but, as in physics, economists should not let its absence prevent them from helping to solve real-life problems.[21]

CHAPTER 21

New Times, New Tools: Agricultural Education for the Twenty-First Century

Laura Lengnick

On a vacant lot in southeast Washington, D.C., just across the street from the Capitol Heights metro station, you can find the city's newest and largest urban farm. The 1.2 hectare East Capitol Urban Farm hosts research plots, a farmer's market, a mobile kitchen, a community garden, walking trails, and a playground, and soon it will produce fresh fish and vegetables in a high-tunnel aquaponics system. The farm is the first in a network of urban food hubs planned through a broad coalition of partnerships led by the University of the District of Columbia's College of Agriculture, Urban Sustainability and Environmental Sciences (CAUSES).[1]

"Being in Washington, D.C., in an exclusively urban environment and having the land-grant mission gives us a unique perspective on what we do and how we do it," said Dwane Jones, director of the Center for Sustainable Development and Resilience at CAUSES. "We want our students to understand that what we do locally in D.C. matters to people here, but it also has an impact on the world." CAUSES is the only publicly funded college of agriculture in the United States using experiential, "hands-on" learning as the core college-wide teaching strategy and using collaborative, community-based sustainable development projects as the primary learning platform.[2]

Widely recognized as a major factor pushing our planet beyond "safe operating space," the way that we eat fuels climate change, erodes community resilience, and contributes to many other twenty-first century challenges. Calls for

Laura Lengnick is a soil scientist and climate resilience consultant with Cultivating Resilience, LLC, in Asheville, North Carolina. She previously served as the director of sustainability education at Warren Wilson College and is the author of *Resilient Agriculture: Cultivating Food Systems for a Changing Climate* (New Society: 2015).

a transformation of the global industrial food system have grown over the last decade as it has become increasingly clear that sustainable food systems are the only way to achieve food security in a world of 9 billion people. This push for a fundamental shift in agriculture comes amid growing concern about the capacity of institutions of agricultural higher education to drive such a transformation.[3]

The Inadequacies of Modern Agricultural Education

As the twentieth century drew to a close, the U.S. land-grant university system prepared to celebrate one hundred and fifty years of public education. A profoundly radical innovation in higher education, the original mission of the land grants was to enhance public well-being through universal access to practical, place-based learning in agriculture within the context of a liberal education. This system of higher education is credited with the transformation of agricultural education worldwide through the development of agricultural science and technology, graduate training, and international development efforts.[4]

Despite its many successes, critics from within and outside the land grants expressed concerns about the capacity of the system to produce agriculturalists who could effectively address the sustainability challenges in agriculture and food systems. Although criticisms of the land-grant system varied—some focused on the limits of reductionist science, others on a lack of collaboration and problem-solving skills in graduates—critics agreed that the teacher-centered, content-driven strategies common in agricultural education had to change. Similar concerns about agricultural education were being raised at this time throughout the world.[5]

Although critics have been vocal and persistent over the last twenty years, the land-grant system has proven resistant to calls for change. The tendency of contemporary agricultural sciences toward increasing specialization into food system components—such as soil science, agribusiness, plant genetics, entomology, and food technology—has created both organizational and physical barriers to integration of the natural and social sciences. Other barriers to reinventing agricultural education in land-grant institutions include declining public support, which has compromised higher education budgets and given private funders more control of the research and education agenda, and agricultural faculty that typically have no formal training in teaching and little incentive to seek professional development opportunities.[6]

A New Purpose for Agriculture Education

Sustainable agriculture and food systems programs at community colleges, liberal arts colleges, and non-land-grant universities in the United States, many built around new student-run campus gardens and farms, have flourished to meet the public's growing interest in alternatives to industrial food. In addition, a proliferation of nonprofit organizations offers community-based, farmer-led education, research and extension programs, and farmer training programs in sustainable agriculture and food systems through local, national, and international networks. Faculty and staff at these institutions and organizations share the conviction that agricultural education, refocused on developing ecological, human, and social capacity for innovation, can put us on the path to a sustainable and resilient food future. (See Box 21–1.)[7]

Students in Florida International University's Agroecology Program work to separate and transplant pepper seedlings.

U.S. Department of Agriculture

Educators in sustainable agriculture, agroecology (the study of ecological processes applied to agriculture and food systems), and food systems have led the way in innovating student-centered teaching and learning strategies in higher education. At the heart of these innovations is a transformation of agricultural education that changes both what we teach (from industrial to ecological principles) and how we teach it (from traditional teacher-centered strategies to student-centered experiential learning strategies). A principle objective of these new methods of instruction is to prepare students to contribute responsibly to the sustainable development of farming and food systems in a world of unprecedented complexity and change.[8]

One of the earliest and most widely celebrated modern agricultural education programs using participatory on-farm methods was established in 1980 at Hawkesbury Agricultural College in Richmond, New South Wales, Australia. Viewed as a learning center situated within the national agricultural system, the college was tasked with creating opportunities for students to develop competencies through lived experience. Learner-centered teaching strategies

Box 21–1. TEMA: Training Farmers in the Field

The Turkish nonprofit organization TEMA (Turkish Foundation for Combating Soil Erosion, for Reforestation and the Protection of Natural Habitats) was founded in 1992 with the aim of creating effective and conscious public opinion about environmental problems, particularly soil erosion, deforestation, desertification, climate change, and biodiversity loss. TEMA's rural development work includes training Turkey's pistachio farmers on agricultural best practices as a means of improving crop production while reducing impacts on the environment and local communities.

Turkey ranks third in the world in pistachio production, and about two hundred thousand people across the country earn their living in this way. Most of the production occurs in the war-trodden provinces of Gaziantep and Şanlıurfa, on the border with Syria. Environmental challenges resulting from improper soil management and the widespread use of chemical fertilizers and pesticides are aggravated by the effects of climate change, which is hampering pistachio production and deepening the vulnerability of the region's population.

In 2011, TEMA initiated a rural development project aimed at tackling the diverse problems facing pistachio production, primarily through vocational training and peer learning. The organization trained and provided counseling services to more than five hundred farmers, disseminating practices such as farm management, pest and disease control, irrigation, and organic farming in villages throughout the region. TEMA also coordinated "experience sharing" days between leading farmers and their neighbors, emphasizing the importance of best practices and profitable production as well as the contributions of organized labor in industrialized countries. As a result of the project, more than one hundred farmers—many of them women—completed training and certification in master grafting and pruning, helping to boost their employment prospects.

In the first phase of the project (2011–13), the pistachio yield per tree increased by 49 percent, and average production (per 0.1 hectare) increased by 123 percent, compared to neighboring farms. Various pistachio pests and diseases decreased by 50 to 80 percent. Meanwhile, populations of beneficial insect species—especially those in the ladybug family (Coccinellidae)—increased as a direct consequence of the organic farming practices that were introduced. In addition, the project identified some two hundred older pistachio trees to be listed for protection. The second phase of the project (2014–19) aims to reach more farmers and to extend the participation of farmers geographically.

—Ali Değer Özbakır, TEMA

Source: See endnote 7.

were used to study systems and subsystems of agriculture rather than separate disciplines. Although the program inspired agricultural educators worldwide and graduated more than a thousand students prepared to lead systemic sustainable development in rural Australia, a lack of support by some faculty and students and declining resources ended the program in 1995.[9]

The Sustainable Agriculture Education Association (SAEA) is a professional association that serves sustainable agriculture and food systems educators, administrators, and students in the United States and abroad who are working to develop innovative approaches to sustainable agriculture education. "There is a relatively small group of people who are working to develop new ways to teach about sustainable agriculture and food systems," says Julie Cotton, current chair of the SAEA Steering Council. "SAEA is a really fruitful way for us to share ideas, learn from each other, and stay enthusiastic about what we are trying to do in the face of all of the challenges."[10]

Geir Lieblein, Tor Arvid Breland, and Charles Francis at the Norwegian University of Life Sciences (NMBU) have worked for almost twenty years to develop a program that represents a fundamental shift from traditional agricultural education. Their "critical learning landscape" integrates systemic and communicative learning through the use of open-ended cases and long-term, place-based experiential education designed to develop the systems thinking, collaboration, and leadership competencies that students need to contribute effectively to community-based sustainable development. "We give students a few tools for observation and interviewing, and then send them out onto farms and into community food systems right away," Francis says. "Then we all come back to the university and find out where the knowledge gaps are. It's 'just-in-time learning,' and we found out it really works. It is really amazing to see what happens when you give students some basic skills and turn them loose." Educational programs modeled on the NMBU program have been adopted in Ethiopia, India, Sweden, and Uganda.[11]

More recently, Francis and his colleagues in Norway and the United States proposed a new framework to guide future planning of agricultural research and education. They recommend a set of teaching and learning strategies designed to develop the competencies that agriculturalists need to effectively address sustainability challenges in farming and food systems, including an understanding of ecological design principles, the ability to integrate science and place-based knowledge, the use of adaptive management and integrated assessment, and capacities for observation, participation, reflection, dialogue, and visioning in community. (See Box 21–2.)[12]

With deep roots in ecology, psychology, complexity, and systems science, resilience science offers agricultural educators a useful new framework to integrate questions of sustainability, uncertainty, risk, and change in agriculture and food systems. Resilience theory situates food systems within ecosystems, adds new rigor to the definition of sustainability, extends sustainability concepts to include the explicit consideration of dynamic change, and encourages critical analysis of the structure, function, and purpose of agriculture and food systems.

Recent applications of resilience science to farming and food systems include behavior-based criteria useful in the visioning, design, assessment, and management of agroecosystems. Although resilience science is not currently integrated into agricultural higher education programs, resilience concepts are beginning to appear in some sustainable food systems and environmental studies curricula at the primary, secondary, and university levels. Community-based educational programs, such as the Transition Initiative, also have developed innovative teaching and learning strategies incorporating resilience concepts.

Source: See endnote 12.

Tools for Transformation

The unprecedented challenges of climate change; uncertain availability of soil, water, and energy resources; and demands of a growing and increasingly urban human population call into question deeply held beliefs about the capacity of science and technology to sustain the global industrial food system. Agricultural leaders must begin now to make the fundamental changes needed in higher education to produce agriculturalists capable of sustaining farming and food systems into the twenty-first century. "There is plenty of concern about educational reform at all levels and a growing recognition that this is the way to go," says Charles Francis. "I have a lot of hope. Will it happen? Yes. Will it happen easily? Probably not, but it is exciting to be working on a leading edge like this."[13]

Effective levers of change can be identified through the analysis of past transformations of higher education, by understanding the factors that contributed to the successes and failures of innovative programs, and by finding opportunity in change that may be unwelcome. "In many institutions, declining resources have forced departmental structures to become more integrated," says Julie Cotton. "Faculty are being challenged to think about how to integrate different disciplines, especially in undergraduate education. These efforts to create more integrated programs of education, while they may be more general, also allow for the incorporation of a broader spectrum of ideas."[14]

With the support and leadership of Sabine O'Hara, dean and director of land-grant programs at CAUSES, and through the development of strong relationships with local community organizations, Dwane Jones and his colleagues at the University of the District of Columbia have successfully navigated significant policy and process barriers to program innovations, such as the East Capitol Urban Farm. "One thing that has worked really well for us is to develop projects that address local needs," Jones says. "It is really important, especially for land-grant institutions, to be innovative, but also to remain relevant to the communities that they have a mission to serve."[15]

Farmer, scientist, student, educator, activist, administrator, business owner, policy maker, parent, or child: we are all teachers and learners. We all can play a role in transforming agricultural education. We can ask thoughtful questions about how the way that we eat fuels climate change, threatens public health, and degrades the natural resources on which agriculture and our communities depend. We can challenge the underlying assumptions of the global industrial food system and debate the wisdom of valuing yield, efficiency, and technological solutions over the regenerative potential of a sustainable food system. We can observe, participate in, reflect upon, dialogue about, and share our visions of twenty-first century food systems. In doing so, together we can weave a web of support for innovative agricultural education in service to a sustainable and resilient food future.

CHAPTER 22

Educating Engineers for the Anthropocene

Daniel Hoornweg, Nadine Ibrahim, and Chibulu Luo

John Smeaton was the first to declare himself a civil engineer, wanting to differentiate his work as non-military. Smeaton is best known for the lighthouse that he designed on the Eddystone Rocks, nineteen kilometers southwest of Plymouth Sound in the United Kingdom. The lighthouse, which started operation on October 16, 1759, was the third structure on the treacherous gneissic rocks (the first two were lost to waves and fire). Modeled after an oak tree and built from dovetailed blocks of granite, it marked a major step forward in building design and the use of concrete. Smeaton's Lighthouse ushered in an era of safer shipping and coincided with the start of the Industrial Revolution.[1]

In shipping, the nightmare scenario is the loss of clear indications of safe harbor; no captain wants to sail blind. Likewise, engineering is a cautious and conservative profession. The job of engineers is mostly to apply the discoveries of their scientific peers and of those that came before. The very basis of civil engineering design is based on the application of past experience as a reliable predictor of future conditions. Using clear signals of safety—the strength of concrete, the behavior of steel, the combustion of fuel—engineers learn about future possibilities based on past applications.

The principle of "stationarity" enables engineers to rely on historic formulas, tables, and standards to license and certify infrastructure and to manage risk and liability. Today, however, climate change and other risks are making this principle obsolete. Engineers cannot practice as in the past: new infrastructure likely will not be reliable or safe under future conditions if it is based on past

Daniel Hoornweg is Richard Marceau Research Chair at the University of Ontario Institute of Technology. **Nadine Ibrahim** is a postdoctoral researcher and lecturer, and **Chibulu Luo** is a PhD student, both in the Department of Civil Engineering at the University of Toronto.

practices. The profession is building blind. Lighthouses cannot be relied on to warn of all danger. Public health, safety, and welfare are at risk.

Engineers need new standards and processes that reflect today's changing world. Engineers need to better understand the new development paradigm and to partner with planners, community leaders, risk managers, and those who strive to deliver sustainable development in the present epoch, dubbed the "Anthropocene" because of the dominant influence of human activity on climate and the environment.[2]

Design conditions are no longer static and stationary. In this changing environment, infrastructure must be more sustainable as it meets the needs of people, equitably supports the fulfillment of human potential, and is appropriate for its specific location and culture. New resilient-system analysis must reinvent how things are built and the evolving standards to which they are built. A greater emphasis on lifecycle assessment and overall lifecycle cost analysis are required. A dynamic, systems approach is required.

Educating engineers to be more forward looking so as to develop new technologies and help build cities and society within this uncertainty is a crucial challenge. It is a challenge that the engineering profession is just starting to grapple with; however, the profession must tackle this head on, especially in an era of both climate change and continuing population growth. Even more important than graduating more engineers around the world is the need to educate better engineers. Throughout their careers, engineers need to learn how to work with the community to build and manage sustainable infrastructure and to apply and share the benefits of technology.

Compounding the Challenge of Engineering Education

The challenge of educating engineers to operate in the absence of stationarity is compounded by the growing demands on the profession. Around the world, the pace of city building and the development and application of new technologies is accelerating, yet the number of engineers graduating globally has remained relatively constant, at about 2 million. There is rising concern about the shortage of engineers in a carbon-constrained world that is urbanizing rapidly and where climate change is increasingly apparent. Sub-Saharan Africa, for example, requires at least an eightfold increase in the number of engineering students to meet current averages of the world's wealthier countries.[3]

Engineers continue to build the world's cities at a breakneck pace. Every year, more than 55 billion tons of material is extracted for mineral production and infrastructure construction, more than three times the amount carried by all the world's rivers. In the last two decades alone, the world's use of concrete has doubled, with some 3 billion tons a year going into the buildings and infrastructure bones of our cities. These new and bigger cities, which anchor today's civilizations, are hungry for resources such as food, water, materials, and energy. The demands on engineers will only increase. During the career spans of today's graduates, the world's cities, with their buildings, transportation infrastructure, and energy and water systems, will double in size. What took some two thousand years to build to this point will be doubled in just thirty-five years. More engineers are needed, and they also need to graduate with a new suite of tools.[4]

Unlike doctors, professional engineers cannot aspire to "do no harm." Engineers knowingly do great harm. Recognizing this impact, engineers use cost-benefit analysis, one of their more powerful tools, to attempt to ensure that the benefits outweigh the costs (or harm). Economist and engineer Jules Dupuit is thought to have exhibited the

The National Church of Nigeria, in Abuja, under construction in 2005.

first use of cost-benefit analysis in 1848 when evaluating a Paris bridge for President Louis Napoleon Bonaparte. Today, engineers must ask whether the bridge should be built at all, rather than just how to build it. Along with maximizing benefits and minimizing costs (financial, environmental, and social), engineers also must place public health and safety paramount in cost-benefit analysis.[5]

Engineers integrate technology with the behavior of materials and humans. This assessment is usually based on centuries of experience, and it usually includes a healthy factor of safety, or "fudge factor." Every now and then, engineers get it wrong. Catastrophes such as the gas leak at the Union Carbide plant in Bhopal, India; the nuclear disaster at Japan's Fukushima Daiichi plant; the *Challenger* space shuttle explosion, and the loss of the *Titanic* quickly come to mind. But so, too, should the compound problems that result from the aggregation of thousands of smaller failings: Beijing's air pollution, Toronto's traffic,

Lagos' inadequate supply of electricity, Jakarta's subsidence and flooding, and Atlanta's sprawl—these all are the failings of engineers.[6]

Every curve of the road, balcony railing, pipe sizing, and roof truss has a built-in factor of safety that balances user behavior, natural systems, and cost. To engineer is to define risk and then to apply a factor of safety against that risk. Engineers do this routinely for the stent in an aortic valve, a car's braking system, and in the bolts holding together the bicycle, baby's crib, space shuttle, or roller coaster. But for complex systems such as the climate, food and water supply, cities, and economies, these tools are no longer sufficient. The Anthropocene calls for new engineering tools. Systems such as climate, the nitrogen and phosphorous cycles, biodiversity, economies, and healthy communities are more complex than a building, road, or engine, yet engineering principles must still apply.[7]

Unlike for concrete and steel, there is no easy metric for the coefficient of public support, nor is there an angle of repose for civic action. No textbook provides the engineer with the tensile strength or modulus of elasticity of potential community conduct. Human strengths, such as our ability to adapt, innovate, and be resourceful, are traits that are not always easily measured. Neither is it easy to predict the impact of human frailties such as fragile egos, insecurities and fears, immediacy of the short term, and our selfishness as individuals that also manifests through our institutions. New and untested tools may be needed, but to not address these complexities is far riskier. Sustainable development, in many regards, seems akin to the factor of safety that every engineer incorporates into problem solving. But instead of determining how safe the building's foundation is from collapse, a "factor of sustainability" should also look at the activity in the context of what else is going on around the building—both locally and globally—for at least the next few decades.

Sustainability for Engineers

New education methods for engineers are needed that appreciate complex systems, continuously changing technologies, evidenced-based design, and the premium of good communication. Increasingly, these are the hallmarks of well-educated engineers. However, engineers still need to learn the fundamentals and to cope with an already crowded curriculum. Teaching the tenets of sustainability requires instilling an understanding of explicit and implicit values assessment; appreciation of the increasing lack of certainty and the quickening pace of change and need for faster reaction times; and integrating greater intensity and frequency of climate and social disruption. Controls on complex systems such as cities and global governance are often value-based and are not easily taught in an academic setting.[8]

Engineers must simplify and better communicate sustainability. For example, there are more than six hundred ways to measure the sustainability of infrastructure. Engineers, often with duplication or subtle efforts to promote one agency, country, or school over another, developed most of these. Engineers also are key in the development of standards and regulations that are integral to modern society. Along with the growing need for engineering research, the development of new codes and standards requires deeper public partnership. As medical doctors help us to monitor our personal health, community engineers need to help us monitor planetary health.[9]

Engineers are increasingly taught the fundamentals of sustainability, as most schools now include a course or two on sustainability in the curriculum. This trend is worldwide and growing. Several North American universities—such as Villanova University, Rochester Institute of Technology, and the School of Sustainable Design Engineering at the University of Prince Edward Island in Canada—offer innovative four-year sustainable engineering degrees. In Europe, many universities have a strong presence at the Engineering Education for Sustainable Development Conference, where they have met annually since 2002 to share experiences, incorporate disciplines of the social sciences and humanities, promote multidisciplinary teamwork, strengthen systematic thinking of a holistic approach, and raise awareness of the challenges posed by globalization.[10]

Sustainability is anchored in areas such as design and efficiency, systems analysis, and complex problem solving. However, limitations still arise because engineering practice often remains reductionist, and efforts toward greater evidence-based public policy are hampered by special interests and short-termism. Teaching any profession, engineering included, to embrace cooperation and the integration of nonquantifiable aspects such as social behavior remains a challenge.

Beyond updating the curriculum, engineering schools are developing novel ways of engaging engineering students. Many schools offer web-based curricula—including open courses for high schools and non-enrolled students, internships with a broad array of community groups, and engineering "change labs." These labs and internships expose students to the community early in their careers, providing better appreciation for the complexities of some problems and the application of technologies.

Yet the scale of the challenge requires that engineers graduating in the Anthropocene be equipped with a new set of tools and partners. The profession needs to have a greater appreciation of the demand for more engineering graduates, especially in low- and middle-income countries, and of the need for broad stakeholder involvement. Engineers in the Anthropocene are not just shaping the world around us; increasingly, they also need to work with the community as an integral part of the system. One way to advance this is through "cooperative education," which alternates teaching and employment terms and involves work in directly related areas. A degree usually takes about a year longer, but students gain exposure to relevant employment experience and are often cited as being more adaptable and innovative. The coop model was first developed in 1901 at the engineering faculty of Lehigh University and has since become commonplace at many universities.[11]

The University of Waterloo in Canada today hosts the world's largest coop program, with some seventy-five hundred undergraduate student placements per year. This program is regularly referred to as a key factor behind the university's consistent strength in reputational surveys. Worldwide, the high reputation and industry support associated with coop engineering programs is now being applied to sustainability objectives. A new era of cooperative education may be emerging where engineering students work with other professions and possibly engineers from other cultures and countries. The challenge of engineering in the Anthropocene will place a premium on a graduate's ability to cooperate with others and to problem-solve in a fluid and increasingly uncertain environment.[12]

Educating Engineers for a Broader Role?

Engineers can play a disproportionately small or large role in governance. Consider the examples of the United States and China. Of the five hundred and forty-one members in the 114th U.S. Congress, only eight were engineers, compared with two hundred and two from the legal profession, twenty-four journalists and media staffers, twenty-nine farmers, nineteen insurance agents, and seven ordained ministers. In China, on the other hand, eight out of nine top government officials were scientists or engineers in 2011.[13]

Organizations such as the United Nations and the World Bank typically have fewer than 5 percent of senior staff from the engineering profession. Perhaps not surprising, economists, lawyers, and journalists have a much stronger voice in public discourse. (See Chapter 20.) An antidote to nationalism and fear and growing parochialism is a sustained and honest (that is, professional) discussion of the costs and benefits of our collective actions. Engineers around the world need to better connect with people in communities and to work together to quantify, communicate, and strengthen actions of sustainability. Engineers must be integral members of the teams that search for, nurture, measure, and assist our better angels.[14]

The curriculum of most engineering programs is filled with calculus, problem solving, and design applications. Little room remains for communication, consensus building, and public debate. The aptitude of many engineers also suggests that these skills need to be acquired

Washington State University mechanical engineering students help install a telescoping camera platform they designed to remotely inspect underground storage tanks filled with radioactive waste at the Hanford Site in Washington.

through multidisciplinary teams. Broadening the education of engineers to operate with greater uncertainty is challenging but critical. Engineers also need to engage more fully with the public to better discern potential costs and benefits, and they need to be more forthright in outlining the growing degree of uncertainty incorporated within engineering approaches.[15]

Although efforts to cultivate this political and social engagement are rare, one inspiring effort is the student-driven Engineers Without Borders (EWB/ISF Ingénieurs Sans Frontières), which has more than three hundred chapters in over thirty-five countries. EWB raises awareness of development and sustainability issues on engineering campuses and mobilizes a cadre of passionate students and recent graduates.[16]

The Special Case of Africa

The growing shortage of graduate engineers that are able to operate in the uncertainty of the Anthropocene is most acute in Africa. Africa's rapid urban growth, together with the need for greater sustainability, creates considerable challenges and opportunities for engineering education. By 2100, Africa is expected to be home to five of the ten largest cities in the world by population. The number of African cities with populations over 5 million will likely grow from ten in 2010 to more than sixty in 2100 (representing over 40 percent of such cities globally). This will place significant demands on the continent, especially in light of increasing local and global resource constraints and environmental considerations. The potential demands on the engineering profession are likely so severe that new education methods, flexible design, and the establishment of multisector teams will originate here.[17]

A critical challenge is to ensure that new infrastructure enables African cities to meet population pressures and economic imperatives in a cost-effective and climate-conscious manner. To achieve these goals, African cities require comprehensive efforts to cultivate the engineering talent needed to build sustainable infrastructure at low costs. There also is a need to devise new approaches to engineering education and research to meet the required cadre of engineers and to fill skill gaps without compromising quality—even as global demand lures Africa's best and brightest away through "brain drain." Meeting these objectives will require a reimagined engineering education.[18]

Reliable and comprehensive data are lacking on enrollment numbers, graduation rates, and employment of engineering talent in sub-Saharan Africa. The United Nations Educational, Scientific and Cultural Organization (UNESCO) estimates a regional shortfall of 2.5 million engineers and technicians required to meet the recently adopted Sustainable Development Goals. Corroborated by the Royal Academy of Engineering, the lack of qualified engineers is attributed to diminished economic growth in the region. This contributes to persistent infrastructure gaps, reliance on costly foreign contractors, and diminished ability to use science and technology to address local needs and productivity improvements. The underlying causes of these factors include underinvestment in skills development, poorly resourced training institutions, underdeveloped regulatory regimes, and inadequate training of engineers.[19]

There also is an institutional tendency in sub-Saharan Africa, as in many places, to follow past practices and conventional educational approaches when training engineers. Reasons vary and undoubtedly are influenced by resource constraints. However, the educational models that predominate in older institutions, and that also largely shape the new schools, may be unduly restrictive. Old models are being challenged around the world. New demands are driven by emerging national and international realities, technological advances, rising costs of education, and a growing student population.

"Engineering for sustainability" needs better articulation, particularly in the context of resource-constrained settings and rapidly growing cities that are not seeing concomitant growth in wealth. Meeting the projected demands in Africa requires new engineering education models that are appropriate for rapid scale-up, with a view toward leveraging engineering for sustainable development. This exploration is part of a broader objective to examine new approaches to engineering education for sustainable cities.

The Role of Women in Engineering

Despite many efforts to encourage girls to enter technical careers, the number of women in engineering remains low. The profession remains heavily male-dominated in both industrial and non-industrialized countries. In Canada, for example, less than 12 percent of practicing engineers are women. In Australia, the share is less than 7 percent, and in Sudan, it is only 5 percent. Future capacity of the profession remains a major concern.[20]

The engineering profession, recognizing the benefits of better gender balance, is encouraging women in engineering, both as students and as faculty. A key reason for attracting women to engineering is their predilection to gravitate toward areas of study that place greater focus on sustainability. For example, recent estimates indicate that certain aspects of civil engineering—such as biosystems, environmental engineering, and pollution management—have greater than 40 percent female enrollment, whereas the disciplines of computer, mechanical, software, and electrical engineering and mining have less than 15 percent female enrollment.[21]

Organizations such as African Women in Science and Engineering provide mentorship to young women interested in the fields of science and engineering, while also working with various African universities to increase the number of female applicants. As Irina Bokova, director general of UNESCO, has observed: "No country today can afford to leave aside 50 percent of its creative genius, 50 percent of its innovation, 50 percent of its economic drivers. This is why gender equality in engineering is so important."[22]

More women are needed in engineering to augment the number of practitioners and to strengthen partnerships and linkages across technologies and social systems—not just in Africa, but around the world. The emerging challenges of the Anthropocene increase the urgency of attracting more women to the profession.

Looking Back from the Future

The future robustness of Earth's life-support systems is a big question. But when looking back from the future, a critical waypoint will be 2050. In 2050, when today's graduating engineers start retiring, the number of people living

---

Box 22–1. Updating Engineering Standards for the Anthropocene

Engineers need new standards and processes that can address today's changing world. Engineers need to better understand the new development paradigm and to rise to the challenge to partner with planners, community leaders, risk managers, and implementers of sustainable development.

Design conditions are no longer static and stationary. In this changing environment, all infrastructure must be sustainable so that it meets the needs of human welfare, equitably supports the fulfillment of human potential, and is appropriate for the specific location and culture.

Engineers must reinvent processes and redefine boundaries. They need to apply systems thinking with an expanded vision that looks at first principles, takes into account the root problems that they are trying to solve, and re-evaluates how best to apply new principles and standards to projects. Methods and processes must be advanced to meet future uncertainty and the evolving needs of human welfare.

To achieve this vision, engineers must focus on four key priorities to achieve a sustainable future:

- Focus on the actual needs of the community, and not simply on the project;
- Develop need-based standards for project development and promulgate them to the profession;
- Build the capacity of engineers so that engineers can confidently make the needed transformation; and
- Advocate and communicate with the public and key stakeholders to institutionalize sustainability as the new normal for infrastructure.

Engineers need both confidence and humility, and they need to expand their capacities to identify, understand, and manage risk and uncertainty. New process-based standards and protocols must be developed. Lifecycle assessment and lifecycle cost analysis are required to account for lifetime impacts as well as impacts beyond the useful life of proposed infrastructure. Engineers must be given the protocols that will help navigate unfamiliar waters where the old lighthouse is no longer able to guide.

—Michael Sanio, Director of Sustainability, American Society of Civil Engineers

in the world's cities will have doubled from today (during the course of their careers!). In 2050, there will likely be one hundred and twenty-two cities with populations exceeding 5 million, up from eighty cities today. This will come with a likely doubling of energy used, pollution generated, and global wealth, and a 30 percent rise in the total human population—to 9.7 billion by 2050.[23]

By 2050, humanity will have chosen, explicitly or implicitly, the overall degree of global sustainability. In 2050, the scale of planetary sustainability will largely have been locked in. "Peak" population, and corresponding peaks such as in oil and energy, waste, food, and water, might not yet have been reached, but their likely time frame and magnitude will be known. The next thirty-five years will determine much of the scope of sustainable development.

Engineers need new education methods to enable them to identify unseen dangers. (See Box 22-1.) They need to work more closely with society to discern where the hazards are, how to avoid them if possible, and how to work through them if they are encountered. Just as John Smeaton's lighthouse ushered in a new era, today's engineering graduates need to help signal the best path forward. Navigating the Anthropocene requires new tools. Old charts and past practices are no longer enough. The hazards are more severe, and the pace has quickened. Yet changes to education have not kept stride. Teaching new engineers to expect calamity is not common, but, given today's risk profile, this is prudent. Teaching tomorrow's engineers, as well as all members of society, to expect these disruptions requires much greater humility and openness than before.

CHAPTER 23

The Evolving Focus of Business Sustainability Education

Andrew J. Hoffman

In the mid-1990s, corporate social responsibility and environmental management emerged as small and somewhat peripheral considerations within business school education. In the ensuing two decades, they have grown to become a mainstream element of the curriculum under the broader subject heading of sustainable development or business sustainability. This is a good thing. And yet, for all the advances in curriculum and course content, a major shift in the focus of this teaching practice is beginning to emerge.[1]

Where the past incarnation of business sustainability education incorporated the issue within existing business logics and models, the next iteration focuses on changes within those logics and models themselves. The first mode of teaching, termed "enterprise integration," focuses on helping individual companies increase profits by translating sustainability into preexisting business considerations. The second mode, termed "market transformation," focuses on systemic changes in the business environment and prompts a reexamination of the role of the corporation in society. The first is focused on reducing *un*sustainability, the second is focused on creating sustainability.[2]

This transition presents business schools with a dilemma: they must teach both enterprise integration and market transformation simultaneously, even though these are fundamentally different approaches. The first will help business students get a job, the second will help them develop a focus for a lifelong career. Over time, the latter will eclipse the former as attention adjusts to addressing the root causes of unsustainability and not just its symptoms.[3]

Andrew J. Hoffman is Holcim (U.S.) Professor of Sustainable Enterprise at the University of Michigan Stephen M. Ross School of Business.

Business Sustainability 1.0: Enterprise Integration

Over the past half century, in order to meet the resource needs of growing human and livestock populations, humans have altered ecosystems "more rapidly and extensively" than in any comparable time in our history, according to the United Nations. How these resources are distributed, with the richest 20 percent of people now consuming 86 percent of all goods and services while the poorest 20 percent consume just 1.3 percent, is placing great strain on societal systems. In short, the historically exploitative relationship between the economy and the natural and social environments cannot be sustained. Into this emergent reality stepped business and business school education, starting in the mid-1990s and growing rapidly in the 2000s.[4]

During this time, sustainability programs grew within the corporate sector, from just under one hundred companies with dedicated programs in 2001 to more than three hundred and thirty in 2011. At the same time, companies increasingly published annual "sustainability reports," created positions such as the chief sustainability officer, and offered sustainability statements by senior executives. By 2010, surveys showed that more than 90 percent of CEOs believed that sustainability was important to a company's profits, and 72 percent of executives identified education as one of the critical development issues for the future success of their business sustainability efforts.[5]

Student demand for this education also has been strong, and is growing. Where students who wanted to change the world once turned to graduate schools of public policy and nonprofit management for their training, many now turn to schools of business management. Surveys show that 88 percent of business school students think that learning about social and environmental issues in business is a priority, 67 percent want to incorporate environmental sustainability considerations into whatever job they choose, and, when looking for full-time employment, 83 percent state that they are willing to take a salary cut for a job that makes a social or environmental difference in the world.[6]

To fill this demand, business schools responded. From 2001 to 2011, the number of sustainability-related courses available to Master of Business Administration (MBA) students increased more than fourfold. (See Figure 23–1.) Over the same time period, the share of business schools that required students to take a course dedicated to business and society increased from 34 percent to 79 percent, and specific academic programs on the topic now can be found in 46 percent of the top one hundred MBA programs in the United States.[7]

These programs can take multiple forms at the undergraduate, masters, and executive education levels, with content serving as part of either a standard business degree, dual degrees between business and environment schools, two-year and one-year specialization certificates, specialized business sustainability degrees, or schools dedicated to sustainable business. The central focus of these programs has been "enterprise integration": framing the issue as a market

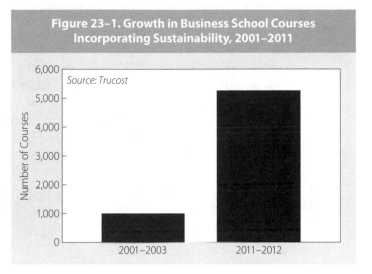

Figure 23-1. Growth in Business School Courses Incorporating Sustainability, 2001–2011

Source: Trucost

shift and fitting it within both the existing core disciplines of a business school (strategy, organizational behavior, marketing, operations, finance, and accounting) and the overriding objective of business education, namely, improving the competitive positioning of the firm and increasing its profits. Central to these programs has been a balance of the standard bottom line with the triple bottom line of the 3 P's: people, planet, and profit.

In this form of framing, coverage of sustainability within business school curricula can remain agnostic about the science of a particular issue (such as climate change) but still recognize its importance as a business issue. The full business scope is not an appeal to morals or to corporate social responsibility, but a response to key business constituents that are bringing these issues to the corporate agenda through core business channels. These constituent pressures can emerge from:

- *Coercive drivers*, in the form of domestic and international regulations and the courts;
- *Resource drivers*, emerging from suppliers, buyers, shareholders, investors, banks, and insurance companies;
- *Market drivers*, emerging from consumers, trade associations, competitors, and consultants; and
- *Social drivers*, emerging from environmental nonprofit organizations, the press, religious institutions, and academia.[8]

In this way, sustainability becomes much like any other business threat caused by a market shift. Market expectations change and technological developments advance, leaving certain industries to either adapt or face demise while others rise to fill their place in the long-accepted notion of "creative destruction." Put in such terms, much of the specific language of sustainability recedes, being replaced by the core language and framing of standard business education. Each frame has a preexisting repertoire with which to conceptualize its treatment within management education and practice, as shown in Figure 23–2.[9]

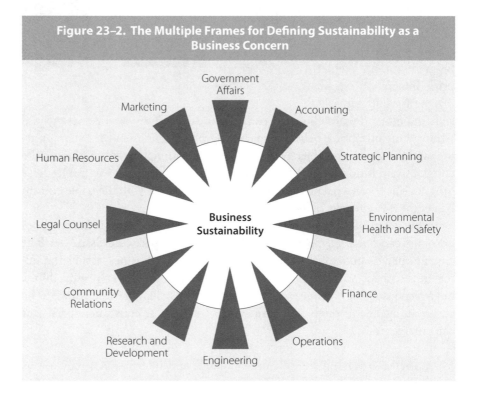

Figure 23–2. The Multiple Frames for Defining Sustainability as a Business Concern

As insurance companies apply sustainability pressures on the firm, the issue becomes one of risk management. From competitors, it becomes an issue of strategic direction. From investors and banks, it becomes an issue of capital acquisition and cost of capital. From suppliers and buyers, it becomes an issue of supply chain logistics. From consumers, it becomes an issue of market demand. Reflecting this translational framing, recruiters look less for narrow sustainability specialists and more for graduates who can find ways to merge business strategy with the objectives of sustainable development.[10]

But there is a problem. As promising as these developments are, our world continues to become less, not more, sustainable, and the problems we face are markedly different in nature than they were in the 1960s, when the "modern" environmental movement began. To mark this shift, scientists have proposed that we have left the Holocene and entered the Anthropocene, a new geologic epoch that acknowledges that humans are now a significant operating force within the Earth's ecosystems. Primarily through the market, "humans move more sediment than all the world's rivers combined. *Homo sapiens* has also warmed the planet, raised sea levels, eroded the ozone layer, and acidified the oceans," writes journalist Richard Monastersky.[11]

Recognition of the Anthropocene has broad implications for how we think about business sustainability. Rather than fitting sustainability into the logics of the market, we must now recognize that the market is taking control of natural systems, with potentially catastrophic consequences. Climate change, ozone depletion, droughts, wildfires, food insecurity, water scarcity, and the social unrest that results all point to a fundamental system failure created by our market and political structures. (See Chapter 20.) Where historic notions of business sustainability as "enterprise integration" have gained acceptance, they are inadequate for the scope of the issues we now face. By fitting sustainability into existing business logics, we are slowing the velocity at which we are approaching a system collapse, but we are not averting it by fully addressing the roots of the problem.

Business Sustainability 2.0: Market Transformation

In its next iteration, sustainable business education is moving beyond simply reducing unsustainability, and moving toward creating sustainability. Curtailing our impact on the environment is not enough. We must become a net positive influence on the environment to both ameliorate our legacy of harm and mitigate the impacts from a growing population that is expected to reach 9.7 billion by 2050. As such, business sustainability education as "market transformation" calls for a reexamination of the systemic aspects of the market, considering when they must be changed to properly address the issues we face. This reexamination takes multiple forms, as follows:[12]

*Sustainability 2.0 requires new conceptions of **market** parameters.*

Driven by concerns about the market failures around sustainability as well as those of the financial crisis of 2008, 57 percent of MBA students reported that they were rethinking their career objectives. In the course of that reexamination, there is a growing interest in courses that move beyond stale notions of a static free market in which companies serve only their shareholders and where government regulation is viewed as an unwarranted intrusion. Courses that teach about the malleability and multiple forms of capitalism (for example, Scandinavian, Japanese, and American capitalism differ markedly on their rules of the market and on the role of government) help business school students understand how the market can change to better serve society. A popular course at the Harvard Business School, "Reexamining Capitalism," explores "the evolution, power and limitations of our current capitalist systems" and "how the 'rules of the game' by which capitalism is structured should change" to address the social and environmental issues of our day.[13]

One area in which market rules must change lies in the urgency of addressing climate change. The immediate task is to constrain the emission of greenhouse gases through a set of regulatory policies and business responses ("enterprise integration"). However, the ultimate solution is to become carbon-neutral and eventually carbon-negative. This cannot be accomplished by one firm or one product competing in the market as it presently exists. The Rewiring the Economy project at the University of Cambridge Institute for Sustainability Leadership calls for ways to "lay the foundations for a sustainable economy" and to "lift and 'tilt' the playing field for business, such that, over time, the economy generates positive outcomes for people within safe environmental boundaries." The output of this project is used to inform the school's Master of Studies in Sustainability Leadership program, as well as its postgraduate certificate programs in both Sustainable Business and Sustainable Supply Chains.[14]

*Sustainability 2.0 requires new conceptions of **systems** parameters.*

Sustainability solutions are a property of the system as a whole, not of one company. The notion of an energy company installing a wind farm and calling itself sustainable makes no empirical sense. A more sustainable energy system incorporates the whole electricity grid, encompassing generation, transmission, distribution, use, and mobility. As such, new courses and programs are integrating more systems-focused approaches. (See Chapter 12.) For example, the MIT Sloan School of Management's sustainable business program (S-Lab) relies heavily on system dynamics modeling in its curriculum. The Presidio Graduate School focuses its curricula on helping its graduates "see connections and make better decisions in the context of a whole system." Surveys of corporate executives foretell this trend. In defining what companies want from business school sustainability education, surveys show that executives want more skills in systems thinking and its application to business goals and operations, as well as knowledge of how to create or manage social networks that are directly relevant to a company's business objectives and processes.[15]

Sustainability 2.0 requires new conceptions of **operational** parameters.

System approaches to business sustainability require special consideration of operations and the design and optimization of supply chain logistics. This is a rapidly growing domain of business sustainability education, with programs, such as the Yale School of Management's Center for Business and the Environment, relying heavily on courses in sustainable supply chain management, lifecycle analysis, and industrial ecology. Other programs offer courses in the area of the "circular economy," which focuses on closed loops in industrial systems that dematerialize and reduce energy use within both global and local supply chains.[16]

Sustainability 2.0 requires new conceptions of **organizational** parameters.

Business school education traditionally has taught three types of organizational form: for-profit, nonprofit, and government. But as business sustainability education enters into new systemic approaches, new forms of organization also emerge or become more relevant, such as hybrid organizations, benefit corporations, cooperatives, employee-owned companies, and networked organizational forms. These forms offer new challenges and new opportunities in governance and employee engagement that the standard models of the public corporation do not fully cover. For example, the Weatherhead School of Management at Case Western Reserve University offers an Appreciative Inquiry Certificate in Positive Business and Society Change, which teaches students how to draw out "the best in people, their organizations, and the relevant world around them" by considering "what gives 'life' to a living system when it is most alive, most effective, and most constructively capable in economic, ecological, and human terms."[17]

*Sustainability 2.0 requires new conceptions of **business metrics and models**.*

New forms of business sustainability education compel questions around the underlying theories and models used to understand and explain the market. Pushing management education further into new domains, courses are emerging to explore new models for market exchange and service provisioning in multiple areas, from regenerative capitalism and collaborative consumption to conflict-free sourcing and environmental finance. The Stanford Graduate School of Business, for example, offers an executive education course on Business Model Analysis and Design, which includes content on the emerging phenomena of the "sharing economy."[18]

To aid this diffusion, Net Impact, an organization that supports students and professionals in developing sustainability careers, provides training modules on the sharing economy, the circular economy, and others. And the Erb Institute for Global Sustainable Enterprise at the University of Michigan is developing new teaching tools to help bring new content into the business curricula, such as constructive lobbying, business and human rights, impact assessment, stakeholder engagement, and materiality assessment.[19]

*Sustainability 2.0 redefines **the role of the corporation in society**.*

At the root of market transformation lies a reexamination of the role of the corporation in society. Many business schools are redirecting their programmatic focus to reflect the positive role that business can and must play in solving the great challenges that society faces. Students who will become tomorrow's business leaders are demanding this shift. For example, the Ross School of Business, with the help of its Center for Positive Organizations, has adopted four pillars to brand its focus, one of which is "positive business, to develop leaders who make a positive difference in the world."[20]

Part of this redirection has been a response to critics of the unsustainable foundations of business school education, such as neoclassical economics and principal-agent theory, both of which are built on rather dismal simplifications of human beings as largely untrustworthy and driven by avarice, greed, and selfishness. (See Chapter 20.) Additionally, there is growing attention to the limitations of sacrosanct metrics, such as discount rates and gross domestic product, and the ways in which these metrics limit efforts at addressing social and environmental sustainability. For example, students at the Ivey School of Business at Canada's Western University are exposed to course content that examines the costs to businesses and society of short-term thinking (such as quarterly earnings) and the innovative opportunities that are exposed by thinking in the long term. Other programs are questioning the taken-for-granted assumption that business serves the shareholder to the exclusion of other critical stakeholders.[21]

Inspiring Students Through the Transition

Advertisment for coworking space in New Farm, Brisbane, Australia.

There is great interest and energy in bringing sustainability more deeply into the norms of business education. For example, the Association to Advance Collegiate Schools of Business (AACSB), the premier accrediting body for business programs, created new standards for social and environmental sustainability that institutions were required to adopt by the 2016–17 school year. In the near term, compliance will focus on Sustainability 1.0. But eventually, if business schools are to properly address our sustainability challenges, they also will have to develop the skills and knowledge to teach Sustainability 2.0.

This shift is critical. If society is to adequately address sustainability, the solutions must come from the market, and, more specifically, from the corporate sector. The market (comprising corporations, the government, nongovernmental organizations, as well as the many stakeholders in market transactions, such as consumers, suppliers, buyers, insurance companies, banks, etc.) is the most powerful organizing institution on Earth, and corporations are the most powerful organizations within it. Without business, there will be no scalable solutions. That does not mean that only business can generate solutions, but rather that the powers of innovation, production, and distribution that business possesses must play an essential role in making the necessary changes in our lifestyles. Business will develop the buildings we live and work in, the clothes we wear, the food we eat, the forms of mobility we employ, and the energy systems that propel them. If there are no solutions coming from the market, there will be no solutions.

But for all the advances in business sustainability education, more needs to be done to meet both the challenges of the Anthropocene and the growing demands of business students. Surveys find that while MBA students overwhelmingly believe that business must play a role in addressing environmental and social issues, only 31 percent of MBA students "think that corporations are working towards the betterment of society," and 79 percent of students feel that they are not receiving adequate training in "how to make business more environmentally sustainable."[22]

Despite this, students continue to enroll in business sustainability courses and programs. They are driven by a personal motivation to solve these problems through business. To fully serve what this demographic is seeking, there is one more component of the "market transformation" approach that must be developed. Future business sustainability education must move beyond just teaching skills and models. It also will require the cultivation of vocations or callings in management, helping students connect to a larger purpose of bringing about a sustainable world through the power of business. This emphasis is, at present, beyond the standard domain of traditional business education. But it represents the hope of a future that fully responds to the burdens of the unprecedented sustainability challenges that we are placing on the next generation.[23]

CHAPTER 24

Teaching Doctors to Care for Patient and Planet

Jessica Pierce

In April 2016, the U.S. government released a three hundred and sixteen-page report titled *The Impacts of Climate Change on Human Health in the United States*. The assessment, mandated by the President's Climate Action Plan, aims to provide a comprehensive and qualitative overview of "observed and projected climate change-related health impacts" and warns of serious and sustained risks from elevated temperatures, extreme weather events, degraded water and air quality, and infectious disease. During the report's official unveiling, U.S. Surgeon General Vivek Murthy warned that climate change poses "a serious, immediate, and global threat to human health." He went on to add, "As far as history is concerned, this is a new kind of threat we are facing."[1]

As Murthy's words vividly suggest, our future will be shaped, whether we like it or not, by the realities of living with a disrupted climate, and in ecosystems already strained to the brink of collapse by growing numbers of humans and unprecedented levels of pollution, resource consumption, and biodiversity loss. Those thinking broadly about the protection and maintenance of human health have identified the critical need to reshape our healthcare systems to respond to these environmental challenges. But this evolution is still in its earliest stages, and, for the most part, doctors and other health professionals work as if climate change and other environmental challenges are unrelated to their professional lives. This disconnect is rooted in a system of medical education that has not yet integrated climate change, sustainability, or ecosystems thinking.

Jessica Pierce is a bioethicist and faculty affiliate at the Center for Bioethics and Humanities at the University of Colorado Anschutz Medical School.

Why Medicine Needs to Take the Environment Seriously

Scientists and public health experts have long understood the crucial links between the natural environment and human health, but only in the past several decades has the true significance of these interconnections become apparent. The enormous literature on these links continues to grow, and environmental degradation—particularly climate change—has become a top priority for all of the world's major health organizations. The World Health Organization has expressed profound concern about a looming crisis brought on by climate change—a threat that is not merely speculative, but that is affecting health right now. The United Nations, for its part, urges immediate action. "Climate change," the UN argues, "is now affecting every country on every continent. It is disrupting national economies and affecting lives, costing people, communities, and countries dearly today and even more tomorrow."[2]

Climate change is perhaps the most catastrophic threat to human health. But other environmental health hazards such as indoor and outdoor air pollution, lack of clean water, the presence of pesticides in soils and foods, and the accumulation of toxic chemicals such as lead and dioxins in people's bodies also pose significant health risks. Poor environmental quality is already estimated to account for at least 25 percent of the global burden of disease, and toxic agents are ranked as the fifth most important underlying cause of death in the United States.[3]

Public health experts also are concerned about the degradation and increasing fragility of "ecosystem services," such as nutrient cycles, crop pollination, regulation of the water cycle, and regulation of climate. These services form the basic and necessary substrate for human well-being, and if this substrate becomes too unstable and too polluted, all the pills and scalpels in the world will not save us.

With climate change and other environmental threats taking center stage in discussions of future health and survival, it is perhaps surprising that these issues remain largely absent from medical curricula. Medicine will need to adapt to an environmentally challenged planet and to learn to protect and sustain health without causing further damage to overstressed ecosystems. The medical educators of today and the doctors of tomorrow need to foster a revolution in how medicine is practiced, how hospitals and clinics are designed, and how patients are treated and counseled. And this revolution needs to begin now.

First Steps: Making Environmental Medicine Mandatory

How might we integrate environmental concerns into the current medical curriculum? Let's begin with what we might call "light green" curricular reform, involving the most immediately achievable and superficial changes to the curriculum.

One baby step is to improve knowledge of "environmental medicine"—the diagnosis, treatment, and prevention of disease and ill health that originate from the environment. There is agreement among leading health institutions that the lack of training in environmental medicine is a key weakness within our current medical corps. A survey of U.S. medical schools conducted in the mid-1990s found that one-quarter of medical schools have no environmental medicine curriculum requirements whatsoever, and the remaining 75 percent require, on average, only seven hours of study in environmental medicine over four years. This survey is outdated, but little has changed in the intervening years, and the data remain relevant. Those working to improve the teaching of environmental medicine focus on competencies such as understanding the influence of environmental agents (such as lead or radon) on different organ systems, recognizing signs and symptoms of potential environmental exposures, and taking patients' environmental histories.[4]

Second Steps: Adding Environmental Awareness and Ecosystems Thinking

Learning how to diagnose, treat, and prevent health problems related to environmental exposures is vital, but it is only a first step. Medical students need to have a deeper understanding of how climate change and the other environmental challenges of the coming decades may influence human health, how the provision of health services can inadvertently cause environmental damage, and how environmental sustainability can be aligned with healthy lifestyles. In contrast to environmental medicine, we might call these broader competencies "sustainable health" and might think of them as "medium green."

Although there has been no formal survey of what is being taught (or not taught) about sustainability in U.S. medical schools, an informal review suggests that the answer is "very little to nothing." That the doctors of tomorrow are not learning about our current environmental situation—even about climate change—is shocking. Small studies from other countries suggest that things may be only slightly better outside the United States. In the United

Kingdom, fewer than half of the schools surveyed provided any teaching on climate change and/or sustainability. A survey of Australian schools found that only 10 percent included climate change in the core curriculum.[5]

It is unclear exactly how sustainability can best be integrated into health-care practices, thinking, and education, and considerable work remains to be done in defining learning goals and content. In one early effort, the General Medical Council in the United Kingdom asked the Sustainable Healthcare Education Network in 2011 to recommend consensus learning objectives for medical curricula in the area of sustainability. Three priority outcomes were identified:

1) Understanding how the environment and human health interact at different levels;
2) Demonstrating the knowledge and skills needed to improve the environmental sustainability of health systems; and
3) Exploring how the duty of a doctor to protect and promote health is shaped by the dependence of human health on the local and global environment.[6]

Some of the subthemes identified as content additions include defining sustainability, understanding the social and economic dimensions of sustainability, population growth and control, nutrition, sustainable procurement, resource use within health care, resilience of health services to environmental perturbations and challenges, waste management, impacts of long-term healthcare trends on sustainability, and conflicting priorities (for example, is environmental sustainability the responsibility of doctors or of citizens, and can these two roles be separated?). Although there was disagreement about exactly what content should take priority, there was consensus among all participants that doctors have a duty to protect and promote health, and that sustainability is fundamentally linked to this duty.[7]

A related but distinct question is, what do physicians need to know about ecosystems and ecosystem services? Sarah Walpole and her colleagues at two medical schools in the United Kingdom conducted an interesting review of this question. As with environmental sustainability, ecosystems and ecosystem services are not yet taught at medical schools, and questions remain about how, when, and where to best integrate this material into the curriculum. There was broad consensus that doctors and medical students need to understand the links between ecosystems and human health, how environmental change can impact health, and how human activities can impact the environment, both positively and negatively.[8]

As Walpole and her colleagues note, where knowledge about ecosystems is lacking, "health professionals are less likely to enact environmentally sustainable practices" such as energy efficiency, waste reduction, and careful segregation of waste, nor will they speak readily with patients, colleagues, or members of their commu-

A U.S. Navy Hospital Corpsman sprays insecticide around a hut in Vanuatu to control the mosquito population and prevent the spread of malaria.

nity about environmental threats to health—such as local air pollution—or opportunities to promote health through protecting the environment.[9]

Some of the specific competencies that emerge as important include communicating environmental issues to the public and contributing to the sustainability of health systems by implementing positive changes—which rests on an understanding of health care's ecological footprint and how it can be managed. The researchers admit that open questions remain regarding the ideal breadth and depth of knowledge to convey, and what teaching practices would be most effective in transmitting this theoretical knowledge and practical competencies.[10]

Third Steps: Changing the Lens

Although researchers have identified opportunities to integrate ecoliteracy and discussion of climate change into the existing curriculum, the "eco" would still be peripheral to medical training—an add-on or specialization. And this, according to many who have studied the issue, is simply not enough. Rather than thinking about specific content points to be added to existing curricula, it might be far better to focus on integrating an ecosystems view of health and a commitment to sustainability across the curriculum. There are a variety of ways in which the curriculum could expand the way that students think about health and sustainability.

For example, climate change and other environmental concerns could be integrated seamlessly across the existing curriculum, as could learning about the capacity of ecosystems to support and sustain human health through ecosystem services (such as regulation of the climate and provision of food and clean water). "Dark green" change requires more than fiddling with the details of a catastrophically unsustainable system. Sustainability cannot simply be an add-on, but must become the central ethos of medical education and medical care. And sustainability must be understood in deep ways: protecting and promoting health without undercutting the ability of future generations to be healthy and without threatening the vast web of life with which we share this planet.

Medical students can be encouraged to become advocates for sustainability at the university medical centers where they train, an important step in transitioning medicine from being part of the problem to being part of the solution. Student groups, for example, could partner with "green healthcare" initiatives aimed to make healthcare delivery less damaging to the environment. Specific initiatives adopted by medical centers in the United States include reducing meat purchasing and insisting on antibiotic-free meat, buying locally grown foods and composting food waste, requiring that all newly built structures be green building-certified, replacing toxic cleaners with greener alternatives, eliminating polyvinyl chloride (PVC) plastics and phthalates from hospitals, and buying reprocessed medical equipment instead of new. Medical students can then carry these and other initiatives over to the clinics and hospitals where they eventually work. The ultimate goal, of course, is not merely to make health care less *un*sustainable, but to make it sustainable. But these initiatives are an important stopgap and can help with the transition.[11]

Students also can compare existing U.S. healthcare practices with the ideals of a sustainable health system: to be economically and ecologically sustainable over the long term, to be prepared for both increased and novel health challenges ushered in by climate disruption, to do more with less, and to be more energy-efficient, less toxic, and less expensive. Health systems also must be finely adapted to the specific needs and opportunities of each community, since each place faces unique health challenges and has a unique set of resources at hand. Because different countries have evolved different ways to address health care (see Box 24–1), they can offer diverse examples of doing more with less. In Singapore, India, and Costa Rica, for example, good health is achieved at much lower cost than in the United States.[12]

The links between poverty and ill health are clear and are part of the current

Box 24–1. Maya Social-Natural Medicine

Maya medicine has been practiced continuously for more than two thousand years in Mesoamerica and is still the primary source of health care for most of the indigenous Maya population in Guatemala (about 6 million people).

From 2011 to 2015, sixty-seven traditional healers from five ethnolinguistic Maya groups, called *Ajq'ij* or *Ajq'omanel*, engaged in joint research with scientists and medical doctors representing Western modern biomedicine. The goal was to understand cancer better by looking at knowledge systems based on completely different assumptions about how a person maintains health and overcomes disease—in other words, to engage in mutual learning by exchanging medical knowledge.

Maya medicine is based on the principle of *Ixbisb'al li wan*, understood as the ability of a person to keep his or her physical body, mind (thoughts), feelings, and spiritual expression in balance. Disease is explained as the consequence of losing balance in any of these four bases of human existence, and treatment therefore is oriented toward finding and treating the cause of this imbalance.

Whereas biomedical physicians gear "healing" toward treating the physical condition (the organ or body part affected), an *Ajq'omanel* regards the physical manifestation of sickness as a mere "alarm signal" that points to a need to uncover and treat the real cause behind the illness. This includes looking at the recent behavior of the patient and whether he or she is living the principles of respect (*nimb'el*), coexistence (*sahil wanq*), and harmony (*tzalajb'il*) with other people, animals, nature, and the spiritual world. So while it is important to treat the symptoms, relieve pain, and restore the body, a Maya *Ajq'omanel* also will treat the person holistically, understanding the relational aspects that embed the patient in his or her larger family and societal circle, but also in the natural surroundings.

Involving family and social networks in Maya medicine is considered a key aspect for restoring health in a sick individual, as relationships in the therapeutic encounter transcend the typical doctor-patient dyad in Western biomedicine. The goal of the *Ajq'omanel* is to restore the health of the entire system surrounding the patient; thus, treatment includes assigning roles to family members that include emotional, spiritual, informational, and instrumental support (such as caretaking or economic support). A distinct "therapeutic unit" emerges that includes the patient or wellness-seeker, his or her family with its larger societal circle, and the healer—all of whom have distinct yet interdependent roles. Maya belief includes the concept of living energies (*Rajawal*) in nature, so healers also are mindful of whether a patient is having an adequate relationship with the natural environment.

There are many categories of disease that originate in breaking the balance with *Nantat Ix* (Mother Earth). For example, one man from a Mam Maya family suffered from strong headaches and epileptic attacks, the cause of which medical tests could not identify. After consultation with an *Ajq'omanel*, the patient revealed that he had cut down two hundred

continued on next page

Box 24–1. continued

trees illegally from a mountain. The *Ajq'omanel* confirmed that the man was being affected by the mountain's guardian spirit (*Rajawal*) and gave him herbal teas and medicinal baths, conducted two fire ceremonies to ask for forgiveness and offer compensation, and required him—with the help of his family—to replant seven times the number of trees cut and to nurture them for four years. Shortly after the patient started planting trees, his symptoms stopped completely, and he has been in good health since. This traditional practice clashes with the biomedical tradition in most hospitals and clinics in the region that separates the patient from the context, strips his or her identity, and annuls sociocultural support networks.

The level of engagement that an *Ajq'omanel* typically has with the patient and his or her social network is one of the most striking characteristics to biomedical observers. Rather than having a utilitarian encounter, bonds form that support the well-being of a patient by acknowledging the configuration of the patient's particular belief system. There is no fear to address emotional preferences, mental attitudes, or spiritual concerns. On the contrary, Maya healing asks that health practitioners acknowledge the multidimensional aspects of a patient, understood in a continuum that links the patient to the surrounding social and natural world. Maya healing is, in essence, a call back to a symbiotic medicine that understands how we are linked to each other and to the world around us.

—*Monica Berger-González, Center for Health Studies, Universidad del Valle de Guatemala*

Source: See endnote 12.

medical paradigm. But important links also exist between poverty and environmental deterioration, and likewise between the pursuit of wealth and unsustainability, and these connections can be discussed within the curriculum. To understand why social justice is part of the solution to our environmental crisis, medical students might study areas of the world that are poor but that nevertheless achieve high levels of health, such as Costa Rica, Sri Lanka, and the Indian state of Kerala. What do these areas share? They share a commitment to women's autonomy, education, universal access to primary healthcare services, and egalitarianism.[13]

Students also might be encouraged to study countries like Cuba, which has achieved a "post-petroleum" health system that is about as close to sustainable health care as we have seen. Is it perfect? No—and the problems are worth exploring. But Cuba offers several important lessons: health care can shed its heavy reliance on petroleum and find alternative sources of energy, such as the solar panels that grace the roofs of Cuban health clinics; health care should be free and universally available; medical education also should be free; and primary care providers can be, to very good effect, "embedded" into communities, so that nearly everyone has a neighbor who is either a doctor or a nurse.

Cuba also offers a vivid reminder that money spent does not necessarily equal gains in health. Despite spending just $817 per capita on health care in 2014—compared to $9,403 per capita in the United States—Cuba achieves health outcomes every bit as good as the United States. Life expectancy in the two countries is the same, at about seventy-eight years, and Cuba has lower infant mortality rates (four deaths per thousand live births in 2015, compared to six per thousand in the United States). It is essential that medical students study the economics of health care and recognize that profit-driven health systems are unlikely ever to be sustainable.[14]

Much of the emphasis falls within the realm of ethics and values. Part of rebooting medical education for the Anthropocene—or the "age of humans" that we have now entered—will involve new ways of thinking about bioethics. For example, students might explore how the four principles of medical ethics (respect for autonomy, nonmaleficence, beneficence, and justice) might look through a green lens, such as in the expansion of "do no harm" to include avoiding harm to the broader community or ecosystem through toxic or unsustainable practices. Ethics classes might place additional emphasis on justice, going beyond the standard discussions about rationing healthcare resources and encouraging future doctors to think about the role of poverty and wealth acquisition in driving environmental damage, and about the need for greater economic equity. Ecological stability and intergenerational justice might be added as fifth and sixth core principles of medical ethics. Many of these topics could be framed as discussions in bioethics courses. (See Box 24–2.)[15]

The Path Forward: From Light Green to Dark Green

As medical students finish their training and transition into the healthcare workforce, they will be entering into an environment ripe for change. Yet their mindset must be primed for thinking sustainably about health and health care. This is why it is critical that medical education evolve, too.

Two interconnected threads need to be woven through the medical school curriculum. First, the doctors of tomorrow need to have the knowledge and practical skills to diagnose and treat patients under a changing ecological and

---

Box 24–2. Discussion Questions for Future Bioethics Courses

• Does the Hippocratic Oath, and particularly the core ethical principle of "do no harm," apply only to an individual doctor in relationship with his or her patient within the context of each unique medical encounter? If individuals and entire communities and ecosystems are being harmed by the provision of health services, is this a violation of the Hippocratic Oath?

• How should doctors understand demographic trends, as related to ecological sustainability? Is there a problem with human "overpopulation" and, if so, should it influence how doctors counsel patients about reproductive services?

• Must doctors "believe" in the consensus opinion of climate scientists? Can you be a responsible physician and yet also be a climate denier? How, if at all, should climate change influence counseling of patients about lifestyle choices?

• Since reducing consumption of red meat would lead to decreased greenhouse gas emissions and better public health, do all doctors have a responsibility to counsel patients toward vegetarianism or at least less frequent meat consumption?

• In a world of limits, what happens to patient autonomy? Should patient choice be curtailed (for example, no, you do not get to choose to have that elective and unnecessary cosmetic surgery procedure)?

• Who is the patient? Is it just the person in front of the doctor, or is it the broader community, the Earth?

• With so many healthcare resources being funneled into care for those who are very aged or ill, would it make sense to think about embracing death and refusing aggressive or expensive care at the end of life, as an act of ecological activism? Should there be greater acceptance of assisted suicide, or fewer resources dedicated to the very sick or elderly?

---

economic regime, adapting as well as possible to a hotter, more unstable, and less resilient world. Second, tomorrow's doctors need to erase the artificial line between working to keep people healthy and working to keep the environment healthy. Being an advocate for human health means, necessarily, also being an advocate for ecological health. The ecological is not personal, it is professional. Medical education should mobilize students and doctors to commit to sustainability and ecological stewardship, so that they can help reshape medicine in ways that are less damaging and more sustainable.

Sustainability is not a goal. It is a core value that shapes choices, behaviors, and even thought patterns across all areas of our lives. It is, as Douglas Klahr writes, "an endless process of constant implementation, assessment, and readjustment." At this point in time, it is not a matter of averting crisis—it is too late for that—but of slowing its pace and adapting to new ways of life as wisely and compassionately as possible. Every choice about sustainable practices is a compromise, and every good we pursue involves tradeoffs. From where we sit now, it is hard even to envision what a truly sustainable approach to human health might look like because we are working from inside the box. "A true paradigm shift," says Klahr, "means that, within a realm of life, everything is subject to change, and this encompasses change that is often unforeseen and unimaginable at the outset."[16]

The Anthropocene will require, perhaps ironically, that we break free from a human-centered worldview. This is essential to learning to live differently. The challenge for medicine is to become "post-anthropocentric," to evolve past our current small view to a more encompassing and long-range perspective. Although we likely cannot even envision what truly sustainable health will look like, a variety of elements may emerge within the new paradigm of post-anthropocentric medicine. These include:

- *An Earth-centered Worldview* (in contrast to our current human-centered one). Humans are part of a wider community of beings.
- *One Health.* The health of ecosystems and their inhabitants are interconnected. A fundamental tenet of post-anthropocentric medicine is that humans pay attention to the health of ecosystems as a whole, as well as to the viability of the multitude of creatures who depend on these ecosystems for survival.
- *The Planet as Patient.* Caring for human patients dovetails with caring for the planet; healthcare systems do not undercut human health by damaging the environment.
- *Treating the Whole Person.* Each individual patient is treated as a whole person, which includes their embeddedness within their community, society, and natural environment. The social and ecological context of care is relevant to medicine.
- *A Commitment to Life.* All life is valued, not just human life (and especially not just certain human lives). The survival of other living beings is intimately tied to our own survival.
- *Violence as Pathogen.* Violence toward each other and toward other forms of life must be addressed as interlinked problems with interlinked solutions, and both have profound implications for health.
- *Health Decoupled from Profit.* As long as health systems and medical care are profit-driven, sustainability will remain elusive.

- *Think Globally, Treat Locally.* Health is local and community-based, but the global situation informs all aspects of healthcare delivery.
- *Health as Social Justice.* Poverty and inequity drive ill health and also environmental destruction. One cannot be addressed without the other.

Given the immediacy and seriousness of our planetary situation, it is critical that medical education begin preparing doctors for the Anthropocene. If we start work now, the doctors of tomorrow can become the visionaries who help create health systems that are environmentally sustainable and economically viable over the long term and that can sustain human flourishing under increasingly challenging conditions.

Conclusion

CHAPTER 25

The Future of Education: A Glimpse from 2030

Erik Assadourian

It's 2030. The world's population has now grown to 8.5 billion people. Global temperatures are now an average of 1.2 degrees Celsius higher than in 1880. Seas have already risen forty centimeters since 2016, suggesting that the models of that year that projected a rise of two meters by 2100 were likely significant underestimates. The Arctic Ocean is now consistently ice-free every summer. And several countries have lost a primary source of fresh water and freshwater storage as glaciers grow smaller and smaller each year.[1]

Many cities have been damaged significantly by climate change, and more than 40 million environmental refugees have fled their homelands and settled elsewhere, triggering sometimes violent backlash in their host countries. Flooding and disasters routinely cost tens of billions of dollars a year in damages, which has depleted the coffers of many national governments and diverted spending away from critical social investments, including schools.[2]

In many places, this has caused education to take a major step backward, with governments shuttering schools, laying off teachers, cutting instructional hours, and even reducing total years of schooling. In other places, however, the convergence of economic, environmental, and social crises has led to a flurry of educational innovation: new programs, new curricula, new priorities, and new types of schools, perhaps revealing the first steps on a new educational path that is better adapted to life on a changing planet.

Many pioneering schools saw the writing on the wall back in 2017 (and some even earlier), when climate change and other neon signs of a changing planet became too bright to screen out. Creative school leaders started to understand

Erik Assadourian is a senior fellow at the Worldwatch Institute and director of *State of the World 2017* and Worldwatch's EarthEd Project.

that "all education must be environmental education," to paraphrase environmental educator David Orr. On a changing planet, these pioneers realized, education must prepare children—and students at all levels—with the skills, knowledge, and wisdom necessary to navigate the turbulent future ahead.[3]

These schools integrated ecoliteracy and spending time in nature as a means to reconnect children to the Earth on which their lives depend. They taught systems thinking and critical thinking so that children would be prepared intellectually to deal with the rapid changes that they would face. They taught character education and social and emotional skills so that children would be prepared socially. They taught home economics and traditional and Indigenous skills so that children would be prepared physically. And they taught activism, social entrepreneurship, and leadership so that children would be prepared politically. Professional schools also evolved, integrating the environment directly into the teaching practices and curricula of economics and engineering programs, into the agricultural sciences, and into business and medical schools.

By today—2030—some of the most innovative schools have become global "franchises," making them as recognized now as Catholic schools, Waldorf schools, and Montessori schools were in earlier decades. For example, nature-based preschools are now present in nearly every country of the world. Many are located in forests, but others are situated on farms, beaches, deserts, mountains, and even rivers. The beauty of these nature schools is that they have integrated vocational training into the core of their curricula, providing students with a means not only to generate a livelihood later in life, but also to help the school sustain itself—whether through farming, tending livestock, hunting and trapping, or foraging forest crops such as acorns, fruit, maple sap, and lumber. This has enabled many nature schools to thrive even as neighboring conventional schools have been forced to shut their doors.

The chain of social entrepreneurship schools, founded by Mechai Viravaidya in 2009, also has expanded globally. Starting with just one high school in the Thai province of Buriram (see Chapter 14), in just two decades this movement has grown to more than three hundred and sixty "Bamboo Schools" spread over forty-three countries. Their model of empowering both students and the communities of which they are a part has proved highly attractive, as well as successful. A study from 2027 found that the presence of Bamboo Schools enabled those communities to maintain a decent quality of life, even as national economic trends declined, and helped them better resist the shocks of widespread climate disasters, as a result of the strong levels of social capital and local ecological knowledge built up over the years.[4]

Other efforts are still in their startup stages but show great promise. One need only glimpse at a "typical" day at three of the most innovative schools of 2030—a river-based middle school in The Gambia, an ecoengineering magnet high school in Singapore, and an activist high school in Brazil—to see this potential. As these examples reveal, wide differences remain not just in the level of resources available to schools, but in the direct effects that school systems have suffered from a rapidly changing planet. In places where flooding, droughts, and other climate disasters have become omnipresent challenges, these experiences and the strategies used to deal with them have become part of the core curriculum, even the school design. In places where stability has remained, these concerns have tended to be more "academic," with activity focused on how students can prepare both themselves and global society for thriving in a changing world.

Hawa's Day Aboard the Tigerfish Floating School

Today, Hawa arrived early at the dock where her middle school will pick her up for the day. Yes, instead of taking the bus, she will actually be picked up by her school, as the Tigerfish Float-ing School is a large boat. Over the course of the school day, Tigerfish meanders up and down a few kilo-meters of the Gambia River, just downstream of the city of Bansang, picking up the school's one hundred and sixty students.

Forgemind ArchiMedia

Makoko Floating School in Nigeria, an early prototype of the Tigerfish Floating School in The Gambia.

Today, it is Hawa's turn—along with her other classmates who meet at this dock—to harvest the day's catch from the school's dockside fish farm. Lunch each day consists of a mix of vegetables, rice, and the fish that the school raises in a series of small tanks, located on each of the docks. This ensures that even the poorest of Tigerfish's students get adequate protein every day. The fish farms also are an integral part of the curriculum: all students will gradu-ate middle school with a comprehensive knowledge of fish farming—from growing the insects that the fish eat, to proper harvesting and management of the tanks, to basic veterinary training.

After feeding the fish and harvesting today's catch, Hawa and her classmates see their school approaching. The floating school is a large, pyramid-shaped boat—designed not for speed but for stability, even in the worst weather. The school consists of several well-lit classrooms, a kitchen, and two science labs, as science is a priority at Tigerfish. Today, the first- and second-year students (sixth and seventh graders) are learning about circuits and solar electricity. With a solar array covering the boat, the students have the chance not only to learn about photovoltaics in the abstract, but also to take part in maintaining the boat's electrical system.

In the other science lab, the eighth graders, including Hawa, have been spending the day dealing with a problem. The tilapia in one of the school's fish farms have developed some sort of disease, with many of the fry dying and many of the adult fish developing skin lesions and rubbing themselves raw against the sides of the tanks. Over the course of the day, the students have dissected several fish to explore internal symptoms, examined fish cells under the microscope, and conducted online research—first on Googlepedia and then in academic journals—to assess the problem. Their hypothesis: the fish are suffering from Trichodina, tiny parasites that attach to the gills, skin, and fins and cause irritation.

The teacher, who has been quietly nudging the process along—helping with the equipment, engaging those who get left out, settling down those who get too excited—now makes a video call to the local veterinarian and allows the students to present their case that the tilapia are suffering from Trichodina. The vet, seeing the evidence, supports their conclusion and agrees to come by the next day to give the fish a potassium salt bath to kill the parasites. After the call, the teacher praises the excellent work of the class, although it is the success of correctly identifying and dealing with the problem that is most rewarding to Hawa and many of the other students.

Not every day does such a perfect project manifest at Tigerfish, offering the students an opportunity to expand their vocational knowledge, research skills, critical thinking, and ability to work together. However, routinely integrating river life into the school curriculum tends to offer more opportunities than otherwise would exist. Biology, chemistry, climatology, ecology, and physics are all naturally a part of life on a river—a river that most of these students will live along their entire lives.

Having a deep knowledge of and connection to the Gambia River is perhaps the most valuable aspect of Tigerfish, although gaining an understanding of the many changes occurring in the ecosystem also is very valuable. As climate change and population pressures have reduced wild fish stocks to endangered levels, farmed fish have largely replaced wild fish. And after several serious floods made schooling impossible for tens of thousands of children living along the riverbank, the idea of floating schools was embraced, with one-quarter of The Gambia's students now spending at least some of their school years at a floating river school.

Recognizing the high risk for future climate-related changes—including the potential submersion of vast areas of the country—certain skills are an integral part of the curriculum: the ability to swim well, disaster education (how to respond effectively in a crisis), and, most importantly, multilingualism. Although English is the primary school language, all students also learn French and Mandinka. The hope is that knowing two global languages will increase students' employment opportunities in good times, and, if a large share of the population eventually becomes climate refugees (a possibility that the government now openly acknowledges), knowing both English and French will help people better integrate into other countries.

While floating schools are not solving the climate crisis in The Gambia and the other coastal areas where they have emerged, they have proven to be an ingenious adaptation—one that not only prepares students for life on a changing planet, but that has made them into vocal advocates for dealing proactively with the climate crisis.

Arivan's Day at the Garden City Eco-engineering Academy

Arivan has just stepped off Singapore's Mass Rapid Transit train and is now walking his last few blocks to his high school, the Garden City Eco-engineering Academy, along a pedestrian and bicycle-only street. This is his favorite part of the commute. Even though school starts later in Singapore to avoid the worst of the morning rush hour, the train ride is still chaotic. But these last few blocks along Agnes Avenue—with its lush tree canopy, birdsong, and verdant sidewalk cafés—is more park than street. Of course, not all roads in the city are so picturesque. But Arivan is proud that the students of Garden City Academy have played an important role over the years in helping to make Singapore one of the greenest cities on the planet.

Even as a second-year student, Arivan is still orienting himself to the possibilities—and responsibilities—that come with being a student at Garden City.

Students at Garden City Eco-engineering Academy examine specimens from their latest experiment.

Ever since he was little, Arivan's education has been centered on character education. Having a strong moral character, or, simply put, "being good," has been deeply integrated into every aspect of Singapore's educational system—from teaching empathy early on to exploring the moral complexities of modern life as children mature.

With Arivan having passed both his character education and science exams at the top of his class, he has earned a coveted spot at Garden City (although, naturally, he accepted it with humility). Students here gain access to some of the most controversial environmental technologies on the planet, from genetically modified organisms (GMOs) and nanotechnology, to geoengineering and carbon capture and storage (CCS). They are expected to graduate not only with in-depth knowledge of these technologies, but also having helped advance humanity's understanding of them, and of how to use them responsibly (if at all).

Garden City's philosophy is that as the planet's ecological crises have accelerated, the need for eco-engineering has too. Not only are governments working diligently to make their cities and industries more sustainable, but widespread ecological disruptions are requiring us to make our key systems— water, electricity, transportation, agriculture, and coastal infrastructure— more resilient. As the planet continues to heat up, there will be greater and greater pressure to try more controversial technologies in an attempt to dig humanity's way out of crisis.

Rather than ignoring or banning these technologies outright, the Singapore government feels that it is better to train the next generation of scientists to be morally developed leaders who can analyze rationally whether the sacrifices that come with using the technologies merit the benefits. As an added bonus, many of the eco-engineers that have graduated from Garden City have become a valuable asset to Singapore, bringing significant global leadership in filing patents and generating abundant journal citations, royalties, and remittances.

When Arivan started at Garden City, he was invited to participate in a longitudinal study on the health and environmental implications of saltwater-tolerant perennial rice. This GMO crop was designed back in 2019 and has been undergoing long-term testing to ensure that, along with being safe, it is productive, palatable, and profitable. During the early years of the study, students were involved in growing and harvesting the rice and feeding rice meal to rats. When the field tests found no adverse health effects, students moved on to feeding the rice to chickens, then dogs and cats.

Two years ago, the students (under the close supervision of scientists at the school and at Singapore's Agri-Food & Veterinary Authority) declared the rice safe for human testing. The rice is now served in the school cafeteria. After all, one of the mottos of Garden City is, "What we expect of others, we must expect of ourselves." Arivan now plays an important role in collecting and analyzing health data. As a student in the school's GMO track, he routinely reads the latest journal studies and has even joined his team to present at the prestigious International Congress of Agricultural Biotechnologies. He and others in his track also regularly present updates of their work as well as broader "field briefs" to students in the other three tracks at Garden City: Geoengineering & CCS, Nanotechnology & Biomimicry, and Urban & Civil Design.

This morning, the day is starting with a presentation from a Civil Design team working on growing Singapore's first living house. Planted nine years ago, the trees and grasses that make up the house have now fused to a point where the interior can be built. As the lead presenter explains, "If all goes well, we'll have our first resident by the end of the school year." When not attending full-school presentations, most students are either taking core courses or participating in their teams' studies.

But it is not all science and moral education at Garden City. Languages—particularly English and Mandarin (the top two scientific languages), along with Malay—are a required part of the curriculum. The arts and other means to cultivate creativity and critical thinking are encouraged, as are opportunities to get outdoors in Singapore's many managed natural spaces. Arivan is part of the wilderness skills club, where he is currently learning how to make fire using a bow drill. As Garden City administrators often state, "Connecting students to the eco-social-technical organism that our city-state has become reveals the mysteries of our living planet and their duty as stewards." While that might sound like jargon to the outsider, it is music to the students of the Garden City Eco-engineering Academy.

Beatriz's Day at the Freire School of Activism

It is hard even to see the Freire School of Activism in São Paulo, Brazil, as a school. The two hundred high school students spend far more of their time directly engaged in activist campaigns than in anything resembling traditional academic work. Of course, there are classes—two days a week—where all the basics are covered: Portuguese and English, math, history, systems thinking, sustainability sciences, persuasive speech, advocacy, law, and ethics. But unlike other high schools in São Paulo, all coursework is regularly oriented toward how students can use this knowledge to make their communities, their city, their country, and their world more just, more sustainable, and healthier places to live.

To keep costs to a minimum, class days take place at a few local churches, and the tutorials and campaign meetings that dominate the week typically rotate among students' homes, public libraries, and cafés. Most meetings consist of updates and strategic planning for the dozen or so campaigns that the students have chosen to initiate and participate in—from efforts to clean up abandoned brownfields and establish new health clinics and bike paths, to campaigns to reduce air pollution and increase national carbon tax rates. Perhaps most ambitious is the students' ongoing campaign to persuade city officials to pass a law requiring green roofs and rainwater catchment systems on all new or rehabilitated buildings—essential infrastructure as climate change and a growing population make access to fresh water a dire challenge in the city.

What has been most powerful about the activist educational model is that the students, like any good campaigners, learn to reach out to a broad range of constituencies to build strong coalitions. Students at Freire learn to identify potential partners, communicate strategically, engage groups with varying interests, and apply a whole host of skills such as running meetings, management, and organizing. Inevitably, they also end up spreading a philosophy of empowerment to their families and communities as they seek out broader support, often inspiring others to join the campaigns.

Beatriz, now a teacher at Freire, graduated from the very first class in 2024. One of the school's earliest campaign successes was an effort to expand the Clean City Law—a 2007 law that banned billboard advertising across the city of São Paulo—to the entire state. The removal of billboards from the city had significant impacts on revealing social inequities and also reducing materialism and unhealthful consumption patterns. Although the campaign succeeded a few years after Beatriz graduated, her role in organizing nonviolent civil disobedience actions—including a relentless "ad-jamming" campaign to replace billboard ads with public service announcements and artwork—had a major impact in exhausting the opposition and persuading the populace and the state to pass the new law. Today, the state of São Paulo is the largest ad-free area in the world.[5]

Now, Beatriz teaches Portuguese, persuasive speech, and civil disobedience and serves as a mentor and adviser for students as they run their campaigns. It is part of the philosophy of the school that students always lead the campaigns (including organizing the spokespersons, community liaisons, lobbyists, and other leadership) and that teachers take only an advisory role. A recent survey of the school's first decade found that the majority of Freire graduates have continued to be socially and politically active, and many have gone on to be leaders in local and state government, in education, and in socially responsible business.

Beyond activism, complementary skills such as conflict mediation, debate, and management are deeply integrated into the curriculum at Freire. And everyone is encouraged to participate in physical activities—particularly aikido, a martial art that encourages exploration of

Beatriz speaking with students at a demonstration in support of raising Brazil's carbon tax.

both how to de-escalate conflict and, when that fails, how to use the attacker's energy against himself, a skill regularly put into service in the students' activism. While there are many more campaigns to wage, the Freire School has been instrumental in making the city and state of São Paulo healthier, more sustainable, and more livable places to reside.

How Can We Get Here?

Forest schools? Schools on boats? Activist, social entrepreneur, eco-engineering, and character education high schools? These may sound fantastical, but, even in 2017, models exist for the schools described above. Admittedly, such robust forms as these are few and far between, but the foundational stones from which similar schools could be built are already in place. Consider the more than one thousand forest schools now found around the world, the Makoko Floating School in Nigeria and the successful Semester at Sea program in the United States, Barefoot College's Night School, or the Mechai Pattana "Bamboo" School for social entrepreneurs and philanthropists in Thailand.[6]

Yet the question remains: how can we transition away from schools that are based on outmoded ideas or even rote memorization; that feed children unhealthy foods and give them just twenty minutes a day to be outside and active; that overwhelm them with technologies for which they are not ready; and that "teach to the test" rather than offer creative opportunities to learn cooperatively, connect with nature, and "learn how to learn"?

While the scenarios presented above cannot offer answers on how to get there, one can hope that success will breed success and that the challenges faced will breed a resolve to innovate, not capitulate. As more children attend—and thrive in—forest schools, and as parents demand spots for their own children or even start their own forest schools, the movement will grow. As climate change-induced flooding inundates school after school, more designers and governments will give floating schools a try, and—if the schools prove not just to be safer and unsinkable, but also to accelerate learning opportunities—these too may proliferate. As social entrepreneurship schools create new opportunities for students as well as for the villages, towns, and cities of which they are a part, communities may clamor for these. And as more social change leaders recognize that students are natural activists, activism schools may go from being a rare phenomenon to a leading trend of the future.

The key, as always, will be cultural pioneers—whether in the role of educators, students, administrators, policy makers, or parents—who refuse to take no for an answer. These pioneers are instrumental in creating new programs, working within educational systems to adapt them to the changing century, passing new laws, funding new projects—and never giving up, even as resources grow more scarce and as the challenges thrown at them from a growing population and a changing planet intensify.[7]

Regardless of what, exactly, education in the future looks like, if it is designed to be Earth-centric—teaching students to understand their dependence on a living planet and providing them with the skills that they need to live restoratively and to navigate the conflict that life on a changing planet will bring—then they will be better prepared for the tumultuous future ahead. Perhaps that is the most we can hope that education can provide.

Notes

Foreword

1. Donella Meadows, *Thinking in Systems* (White River Junction, VT: Chelsea Green, 2008).

2. Edward O. Wilson, *Half-Earth: Our Planet's Fight for Life* (New York: W. W. Norton & Company, 2016).

3. William Ophuls, *Ecology and the Politics of Scarcity Revisited* (New York: W. H. Freeman, 1977/1992).

4. Joachim Radkau, *The Age of Ecology* (Malden, MA: Polity Press, 2011).

Chapter 1. EarthEd: Rethinking Education on a Changing Planet

1. Bijal Trivedi, "Chimps Shown Using Not Just a Tool but a 'Tool Kit,'" *National Geographic News*, October 6, 2004; John Downer Productions Ltd., "Dolphins: Spy in the Pod," Episode 1, 2014; Charles I. Abramson and Ana M. Chicas-Mosier, "Learning in Plants: Lessons from *Mimosa pudica*," *Frontiers in Psychology*, March 31, 2016; "Mimosa Plants Have Long Term Memory, Can Learn, Biologists Say," Sci-News.com, January 16, 2014; Michael Hopkin, "Bacteria 'Can Learn,'" *Nature*, May 8, 2008.

2. David F. Lancy, John Bock, and Suzanne Gaskins, "Putting Learning in Context," in David F. Lancy, John Bock, and Suzanne Gaskins, eds., *The Anthropology of Learning in Childhood* (Plymouth, U.K.: AltaMira Press, 2012), 5.

3. Erik Assadourian, "The Rise and Fall of Consumer Cultures," in Worldwatch Institute, *State of the World 2010: Transforming Cultures* (New York: W. W. Norton & Company, 2010), 3–20.

4. Michon Scott, "What's the Hottest Earth Has Been 'Lately'?" Climate.gov, September 17, 2014; Millennium Ecosystem Assessment, *Ecosystems and Human Well-being: Synthesis* (Washington, DC: Island Press, 2005); John S. Watson, "Why I Think Oil and Natural Gas Are Indispensable for the Foreseeable Future," LinkedIn, August 30, 2016.

5. Leo Barraclough, "Global Advertising Spend to Rise 4.6% to $579 Billion in 2016," *Variety*, March 21, 2016; Assadourian, "The Rise and Fall of Consumer Cultures"; "The New Class War," *The Economist*, July 9, 2016.

6. Holli Riebeek, "Global Warming," Earth Observatory, June 3, 2010; Brady Dennis and Chris Mooney, "Scientists Nearly Double Sea Level Rise Projections for 2100, Because of Antarctica," *Washington Post*, March 30, 2016.

7. Carl Folke, "Respecting Planetary Boundaries and Reconnecting to the Biosphere," in Worldwatch Institute, *State of the World 2013: Is Sustainability Still Possible?* (Washington, DC: Island Press, 2013), 19–27; United Nations Population Division, *World Population Prospects: The 2015 Revision* (New York: 2015); Rakesh Kochhar, "A Global Middle Class Is More Promise Than Reality," Pew Research Center, July 8, 2015.

8. Oliver Milman, "Rate of Environmental Degradation Puts Life on Earth at Risk, Say Scientists," *The Guardian* (U.K.), January 15, 2015.

9. John Taylor Gatto, "Against School," *Harper's Magazine*, September 2003, 38; Lancy, Bock, and Gaskins, "Putting Learning in Context."

10. Jeanna Bryner, "Teaching Evolution Just Got Tougher in Tennessee," LiveScience, April 12, 2012; Deborah Zabarenko, "Tennessee Teacher Law Could Boost Creationism, Climate Denial," *Reuters*, April 13, 2012; Michael Smyth, "Smyth: Vision Vancouver Refusal to Take Corporate Cash for Schools 'Classic Case of Ideology Trumping Reality,' Says LaPointe," *The Province*, November 1, 2014; Josh Golin and Melissa Campbell, "Reining in the Commercialization of Childhood," in Worldwatch Institute, *State of the World 2017: Rethinking Education on a Changing Planet* (Washington, DC: Island Press, 2017); Common Sense Media Inc., *The Common Sense Census: Media Use by Tweens and Teens* (San Francisco, CA: 2015).

11. Erik Assadourian, "The Path to Degrowth in Overdeveloped Countries," in Worldwatch Institute, *State of the World 2012: Moving Toward Sustainable Prosperity* (Washington, DC: Island Press, 2012), 22–37; Harald Welzer, *Climate Wars* (Cambridge, U.K.: Polity Press, 2012); Potsdam Institute for Climate Impact Research and Climate Analytics, *Turn Down the Heat: Why a 4°C Warmer World Must Be Avoided* (Washington, DC: World Bank, 2012); John Schwartz, "Paris Climate Deal Is Too Weak to Meet Goals, Report Finds," *New York Times*, November 18, 2016.

12. Michael K. Stone, "Ecoliteracy and Schooling for Sustainability," in Worldwatch Institute, *State of the World 2017*; Luis González Reyes, "Growing a New School Food Culture," in idem.

13. Buffy Cushman-Patz, School Leader, School for Examining Essential Questions of Sustainability (SEEQS), Honolulu, HI, personal communication with author, October 18, 2016; SEEQS, *Examining Our World: 2016 Yearbook* (Honolulu, HI: 2016).

14. David Sobel, "Beyond Ecophobia," *Yes! Magazine*, November 2, 1998; David Sobel, "You Can't Bounce Off the Walls If There Are No Walls: Outdoor Schools Make Kids Happier—and Smarter," *Yes! Magazine,* March 28, 2014; Timothy D. Walker, "Kindergarten, Naturally," *The Atlantic*, September 15, 2016; David Sobel, "Outdoor School for All: Reconnecting Children to Nature," in Worldwatch Institute, *State of the World 2017*; Melissa K. Nelson, "Education for the Eighth Fire: Indigeneity and Native Ways of Learning," in idem; "School's Out: Lessons from a Forest Kindergarten," a film directed by Lisa Molomot and produced by Rona Richter, 2014, www.bullfrogfims.com/catalog/school.

15. Jacob Rodenburg and Nicole Bell, "Pathway to Stewardship: A Framework for Children and Youth," in Worldwatch Institute, *State of the World 2017*.

16. Marvin W. Berkowitz, "The Centrality of Character Education for Creating and Sustaining a Just World," in Worldwatch Institute, *State of the World 2017*.

17. Pamela Barker and Amy McConnell Franklin, "Social and Emotional Learning for a Challenging Century," in Worldwatch Institute, *State of the World 2017*; Jessica Alexander, "America's Insensitive Children?" *The Atlantic*, August 9, 2016; James Gaines, "This School Replaced Detention with Meditation. The Results Are Stunning," Upworthy, September 22, 2016.

18. Anya Kamenetz, Steve Drummond, and Sami Yenigun, "The One-Room Schoolhouse That's a Model for the World," NPR Morning Edition, June 9, 2016; Center for Education Innovations, "Escuela Nueva," www.educationinnovations.org/program/escuela-nueva; Fundacion Escuela Nueva, "Escuela Nueva Pedagogical Concept in 3 Minutes!" video, May 7, 2013, https://www.youtube.com/watch?v=x1uMr4hI6I8; Emily Gustafsson-Wright and Eileen McGivney, "Fundación Escuela Nueva: Changing the Way Children Learn from Colombia to Southeast Asia," Brookings, April 23, 2014.

19. David Whitebread, "Crisis in Childhood: The Loss of Play," *Cambridge Primary Review*, November 16, 2015.

20. David Whitebread, "Prioritizing Play," in Worldwatch Institute, *State of the World 2017*; Timothy D. Walker, "The Junk Playground of New York City," *The Atlantic*, August 11, 2016; Claire Duffin, "Streets Are Alive with

the Sound of Children Playing," *The Telegraph* (U.K.), February 22, 2014; Sheila Wayman, "Let the Children Play: The Secret to Finnish Education," *The Irish Times*, October 4, 2016; Timothy D. Walker, "The Joyful, Illiterate Kindergartners of Finland," *The Atlantic*, October 1, 2015; Ariana Eunjung Cha, "The U.S. Recess Predicament," *Washington Post*, October 2, 2015; Adam Taylor, "Why Finland's Unorthodox Education System Is the Best in the World," *Business Insider*, November 27, 2012.

21. Marilyn Mehlmann with Esbjörn Jorsäter, Alexander Mehlmann, and Olena Pometun, "Looking the Monster in the Eye: Drawing Comics for Sustainability," in Worldwatch Institute, *State of the World 2017*.

22. "Rise of the Machines," *The Economist*, May 9, 2015.

23. Linda Booth Sweeney, "All Systems Go! Developing a Generation of 'Systems-Smart' Kids," in Worldwatch Institute, *State of the World 2017*.

24. Dennis McGrath and Monica M. Martinez, "Deeper Learning and the Future of Education," in Worldwatch Institute, *State of the World 2017*.

25. Edward O. Wilson, *Consilience: The Unity of Knowledge* (New York: Alfred A. Knopf, Inc., 1998), 269.

26. Helen Maguire and Amanda McCloat, "Home Economics Education: Preparation for a Sustainable and Healthy Future," in Worldwatch Institute, *State of the World 2017*.

27. Yudhijit Bhattacharjee, "Why Bilinguals Are Smarter," *New York Times*, March 17, 2012; Claudia Dreifus, "The Bilingual Advantage," *New York Times*, May 30, 2011; Naja Ferjan Ramírez and Patricia K. Kuhl, *Bilingual Language Learning in Children* (Seattle, WA: Institute for Learning & Brain Sciences, University of Washington, June 2016); Melinda D. Anderson, "The Economic Imperative of Bilingual Education," *The Atlantic*, November 10, 2015.

28. Mona Kaidbey and Robert Engelman, "Our Bodies, Our Future: Expanding Comprehensive Sexuality Education," in Worldwatch Institute, *State of the World 2017*.

29. Nancy Lee Wood, "Preparing Vocational Training for the Eco-Technical Transition," in Worldwatch Institute, *State of the World 2017*; Kei Franklin, "Providing Environmental Consciousness Through Life Skills Training," in idem.

30. Paulo Freire, *Pedagogy of the Oppressed* (London: Bloomsbury Academic, 2000); Paulo Freire, *Education for Critical Consciousness* (New York: Continuum, 2005).

31. The Grove Community School, "Our Core Values," http://thegrovecommunityschool.ca/#core-values; Wilma Verhagen, Principal, The Grove Community School, Toronto, ON, personal communication with author, December 2, 2016.

32. John Hardy, "My Green School Dream," TED Talk, July 2010; Melati Wijsen and Isabel Wijsen, "Our Campaign to Ban Plastic Bags in Bali," TED Talk, January 2016; Bye Bye Plastic Bags website, www.byebyeplasticbags.org, viewed November 29, 2016.

33. Celia Oyler, *Actions Speak Louder Than Words: Community Activism as Curriculum* (New York: Routledge: 2012), 12–24.

34. Sobel, "Outdoor School for All"; Mehlmann with Jorsäter, Mehlmann, and Pometun, "Looking the Monster in the Eye"; Sweeney, "All Systems Go!"; Maguire and McCloat, "Home Economics Education."

35. Wood, "Preparing Vocational Training for the Eco-Technical Transition"; Joslyn Rose Trivett et al., "Sustainability Education in Prisons: Transforming Lives, Transforming the World," in Worldwatch Institute, *State of the World 2017*; Bunker Roy, "The Barefoot Model," in idem; "What Is a Danish Folk High School," a film by denmark.dk, http://danishfolkhighschools.com/about.

36. College of the Atlantic, "Areas of Study," www.coa.edu/academics/areas-of-study.

37. American Association for the Advancement of Sustainability in Higher Education (AASHE), *Sustainable Campus Index: 2016 Top Performers & Highlights* (Philadelphia, PA: October 2016); Elisabeth B. Wall, "Appalachian Earns AASHE's Top Overall Sustainability Ranking," *Appalachian Magazine*, November 16, 2016; Appalachian State University, "Student Learning Outcomes," https://sd.appstate.edu/academics/internships/student-learning-outcomes, viewed November 29, 2016.

38. Jennifer Washburn, *University Inc.: The Corporate Corruption of Higher Education* (New York: Basic Books, 2005); Michael Maniates, "Suddenly More Than Academic: Higher Education for a Post-Growth World," in Worldwatch Institute, *State of the World 2017*.

39. Jonathan Dawson and Hugo Oliveira, "Bringing the Classroom Back to Life," in Worldwatch Institute, *State of the World 2017*.

40. Daniel Hoornweg, Nadine Ibrahim, and Chibulu Luo, "Educating Engineers for the Anthropocene," in Worldwatch Institute, *State of the World 2017*.

41. Laura Lengnick, "New Times, New Tools: Agricultural Education for the Twenty-First Century," in Worldwatch Institute, *State of the World 2017*.

42. Joshua Farley, "Bringing the Earth Back into Economics," in Worldwatch Institute, *State of the World 2017*.

43. Andrew J. Hoffman, "The Evolving Focus of Business Sustainability Education," in Worldwatch Institute, *State of the World 2017*; Jessica Pierce, "Teaching Doctors to Care for Patient and Planet," in idem.

44. Michael K. Stone, "Ecoliteracy and Schooling for Sustainability," in Worldwatch Institute, *State of the World 2017*.

45. The Annie E. Casey Foundation, *Theory of Change: A Practical Tool for Action, Results and Learning* (Baltimore, MD: 2004).

46. Barbara Chow, "Policy as Opportunity – Your Best and Worst Friend," video, February 15, 2012, https://www.youtube.com/watch?time_continue=17&v=zMfmvJ1feCw; United Nations Educational, Scientific and Cultural Organization, *National Journeys Towards Education for Sustainable Development 2013* (Paris: 2013).

47. United Nations Sustainable Development Goals, "Goal 4: Ensure Inclusive and Quality Education for All and Promote Lifelong Learning," www.un.org/sustainabledevelopment/education.

48. Erik Assadourian, "The Future of Education: A Glimpse from 2030," in Worldwatch Institute, *State of the World 2017*.

Chapter 2. Outdoor School for All: Reconnecting Children to Nature

1. Victoria J. Rideout, Ulla G. Foehr, and Donald F. Roberts, *Generation M²: Media in the Lives of 8- to 18-Year-Olds* (Menlo Park, CA: Kaiser Family Foundation, January 2010), 2.

2. David Sobel, *Childhood and Nature: Design Principles for Educators* (Portland, ME: Stenhouse Publishers, 2008), 114.

3. Louise Chawla and Victoria Derr, "The Development of Conservation Behaviors in Childhood and Youth," in Susan Clayton, ed., *Oxford Handbook of Environmental and Conservation Psychology* (Oxford, U.K.: Oxford University Press, 2012), 30–31.

4. Ibid.

5. "School's Out: Lessons from a Forest Kindergarten," a film directed by Lisa Molomot and produced by Rona Richter, 2014, www.bullfrogfilms.com/catalog/school; Rosemary Bennett, "If You Go Down to the Woods Today,"

The Times (London), October 6, 2009.

6. David Sobel and Rachel Larimore, *Nature Cements the New Learning: A Case Study of Expanding a Nature-based Early Childhood Program from Preschool into the K–5 Curriculum in Public Schools in Midland, Michigan* (Keene, NH: Antioch University New England, June 2016).

7. David Sobel, *Place-based Education: Connecting Classrooms and Communities*, Nature Literacy Monograph Series #4 (Great Barrington, MA: The Orion Society, 2013), 30–31.

8. Niki Buchan, "Bush School—Nature Education in Australia," Precious Childhood Blog, May 1, 2012.

9. "Acknowledgement of Traditional Aboriginal Territory in British Columbia," Safe Harbor—Respect for All Blog, April 18, 2014; STAR School website, www.starschool.org.

10. Green School website, www.greenschool.org.

11. The College School, "Adventure Education," www.thecollegeschool.org/about-tcs/forward-thinking-education/adventure.

12. Appalachian Mountain Club, "Youth Opportunities Program," www.outdoors.org/youth-programs/youth-opportunities-program.

13. Kroka Expeditions, "Welcome to the Kroka Semester Program," www.kroka.org/semester/semester_overview.shtml.

14. Rediscovery website, http://rediscovery.org.

15. Ming Wei Koh, "Place-based Education *Is* Education for Sustainability," in *Educating for Sustainability: Case Studies from the Field PreK-12* (Shelburne, VT: Shelburne Farms Sustainable School Project, 2016), 1.

16. Ibid., 40.

17. Lilian Mongeau, "Oregon Asks, What If Camp Were Part of School?" *Christian Science Monitor*, August 6, 2016.

18. Ibid.

19. Children & Nature Network website, www.childrenandnature.org; National League of Cities, "Cities Connecting Children to Nature," www.nlc.org/find-city-solutions/institute-for-youth-education-and-families/youth-and-young-adult-connections/cities-connecting-children-to-nature; Lauren Fay Carlson, "A Breath of Fresh Air: City of Grand Rapids Aims to Reconnect Children with Nature," Rapid Growth, October 6, 2016.

20. North American Association for Environmental Education website, https://naaee.org; U.S. Department of Education, "Every Student Succeeds Act (ESSA)," www.ed.gov/essa?src=rn; International Association of Nature Pedagogy, "Summary of Erasmus Project," www.naturepedagogy.com/erasmus.

Chapter 3. Ecoliteracy and Schooling for Sustainability

1. Michael K. Stone, "Solving for Pattern: The STRAW Project," *Whole Earth* (Spring 2001): 78.

2. Ibid.

3. Quoted in Laurette Rogers, *The California Freshwater Shrimp Project: An Example of Environmental Project-Based Learning* (Berkeley, CA: Heyday Books, 1996), 31.

4. As Michael Pollan has observed: "The word 'sustainability' has gotten such a workout lately that the whole concept is in danger of floating away on a sea of inoffensiveness. Everybody, it seems, is for it—whatever 'it' means" (Michael Pollan, "Our Decrepit Food Factories," *New York Times Magazine,* December 16, 2007). For a discussion of ongoing efforts to develop standards for education for sustainability, see the *Journal of Sustainability*

Education, especially Jaimie Cloud, "Education for a Sustainable Future: Benchmarks for Individual and Social Learning," *Journal of Sustainability Education,* April 14, 2016. For more on the Center for Ecoliteracy, see www .ecoliteracy.org.

5. Fritjof Capra, "Preface: How Nature Sustains the Web of Life," in Michael K. Stone and Zenobia Barlow, eds., *Ecological Literacy: Educating Our Children for a Sustainable World* (San Francisco, CA: Sierra Club Books, 2005), xiii.

6. Cited in Michael K. Stone, "Applying Ecological Principles," www.ecoliteracy.org/article/applying-eco logical-principles.

7. David W. Orr, *Earth in Mind: On Education, Environment, and the Human Prospect,* tenth anniversary edition (Washington, DC: Island Press, 2004), 94–96.

8. Wendell Berry, *The Gift of Good Land* (New York: North Point Press, 1982), 134–48; National Farm to School Network website, www.farmtoschool.org.

9. London Sustainable Development Commission, *Children and Nature: A Quasi-Systematic Review of the Empirical Evidence* (London: Greater London Authority, 2011).

10. "Meet Anna and Carolina from Nature Schools in Sweden," Inspiring School Grounds brochure from 2015 International School Grounds Alliance conference, available at www.inspiringschoolgrounds.com.

11. Michael K. Stone, *Smart by Nature: Schooling for Sustainability* (Healdsburg, CA: Watershed Media, 2009), 123.

12. Sustainability Academy at Lawrence Barnes website, http://sa.bsdvt.org.

13. Stone, *Smart by Nature,* 122–27.

14. Stone, "Solving for Pattern: The STRAW Project."

15. Zenobia Barlow, "Confluence of Streams," *Resurgence,* no. 226 (September/October 2004): 6.

16. Center for Ecoliteracy, *Cultivating 20 Years of Ecoliteracy* (Berkeley, CA: 2015), 7.

17. David A. Gruenewald and Gregory A. Smith, "Creating a Movement to Ground Learning in Place," in David A. Gruenewald and Gregory A. Smith, eds., *Place-Based Education in the Global Age* (New York: Lawrence Erlbaum Associates, 2008), 347.

18. Quoted in Center for Ecoliteracy, "Forging the Justice Path," www.ecoliteracy.org/download/forging -justice-path.

19. Michael K. Stone and Zenobia Barlow, "Living Systems and Leadership: Cultivating Conditions for Institutional Change," *Journal of Sustainability Education* 2 (March 2011).

20. Center for Ecoliteracy, *Cultivating 20 Years of Ecoliteracy,* 19–21.

21. International School Grounds Alliance, "About," www.internationalschoolgrounds.org/about.

22. "Chapter 36: Promoting Education, Public Awareness and Training," in United Nations Conference on Environment & Development, *Agenda 21* (Rio de Janeiro, Brazil: June 1992); Jennifer Seydel, Executive Director, Green Schools National Network, personal communication with author, July 21, 2016.

23. The ten themes are biodiversity and nature, climate change, energy, global citizenship, health and well-being, litter, marine and coast, school grounds, transport, waste, and water. Eco-Schools, "Seven Steps Toward an Eco-School," www.ecoschools.global/seven-steps. Figure 3–1 from Brid Conneely, Director, International Eco-Schools, personal communication with author, July 1, 2016.

24. Sustainable Caerphilly County Borough, "Eco School Case Studies: Ynysddu Primary School–Incredible Edible Project," http://your.caerphilly.gov.uk/sustainablecaerphilly/schools-and-esdgc/eco-schools/eco-school-case-studies; Eco-Schools Indian Ocean, "Mauritius Hosts Its First Eco-Schools Award Ceremony," eco-schools.io/news/mauritius-hosts-its-first-eco-schools-award-ceremony.

25. Box 3–2 from the following sources: Bruce Stokes, Richard Wike, and Jill Carle, "Global Concern About Climate Change, Broad Support for Limiting Emissions," Pew Research Center, November 5, 2015; Richard Wike, "What the World Thinks About Climate Change in 7 Numbers," Pew Research Center, April 18, 2016; Diego Román and K. C. Busch, "Textbooks of Doubt: Using Systemic Functional Analysis to Explore the Framing of Climate Change in Middle-school Science Textbooks," *Environmental Education Research* (September 2015): 1–23; Roger Beck et al., *Modern World History: Patterns of Interaction* (Austin, TX: Holt McDougal, 2007) (the 2012 edition eliminates the line "Not all scientists agree with the theory of the greenhouse effect"; however, many schools continue to use older editions of the textbook); Portland Public Schools, "Resolution to Develop an Implementation Plan for Climate Literacy," Board of Education Resolution No. 5272 (Portland, OR: May 17, 2016); Bill Bigelow, "Nation's Largest Teachers Union Endorses Teaching 'Climate Justice,'" *Common Dreams*, July 19, 2016; Our Children's Trust, "Landmark U.S. Federal Climate Lawsuit," www.ourchildrenstrust.org/us/federal-lawsuit; classroom activities from Bill Bigelow and Tim Swinehart, *A People's Curriculum for the Earth: Teaching Climate Change and the Environmental Crisis* (Milwaukee, WI: Rethinking Schools, 2015).

26. Global Action Plan International, "ESD in Action, Ukraine: Phase 2," www.globalactionplan.com/esda2.

27. Biomimicry 3.8, "Life's Principles," http://biomimicry.net/about/biomimicry/biomimicry-designlens/lifes-principles.

Chapter 4. Education for the Eighth Fire: Indigeneity and Native Ways of Learning

1. Jon Rehyer et al., eds., *Honoring Our Children: Culturally Appropriate Approaches for Teaching Indigenous Students* (Flagstaff, AZ: Northern Arizona University, 2013); Kelsey Sheehy, "Graduation Rates Dropping Among Native American Students," *U.S. News & World Report*, June 6, 2013; Education Week, *Diplomas Count 2016: High School Redesign* (Bethesda, MD: 2016).

2. Etymology Online, www.etymonline.com; Paulo Freire, *Pedagogy of the Oppressed* (New York: Continuum, 1970); Md. Mahbubul Alam, "Banking Model of Education in Teacher-Centered Class: A Critical Assessment," *Research on Humanities and Social Sciences* 3, no. 15 (2013): 27–31.

3. Greg Cajete, *Native Science: Natural Laws of Interdependence* (Santa Fe, NM: Clear Light Publishing, 1999), x; Rose von Thater-Braan, "What Is the Native American Learning Bundle," 2014, http://silverbuffalo.org/NSA-ScienceOfLearning.html.

4. Marie Battiste, *Decolonizing Education: Nourishing the Learning Spirit* (Saskatoon, SK: Purich Publishing, 2013); Rose von Thater-Braan and Melissa K. Nelson, "Grandfather How Do I Learn? Exploring the Foundations of Diversity," documentary film, 2013, https://vimeo.com/71449994.

5. See Joan Roach, "By 2050 Warming to Doom Millions of Species," *National Geographic News*, July 12, 2004.

6. Adam Hadhazy, "'Second Brain' Influences Mood and Well-being," *Scientific American*, February 12, 2010; Joseph Ciarrochi et al., *Emotional Intelligence in Everyday Life: A Scientific Inquiry* (New York: Psychology Press, 2001); Andy Fischer, *Radical Ecopsychology: Psychology in Service of Life* (New York: State University of New York, 2013); Catherine Baumgartner, Embodied Ecologies website, http://embodiedecologies.moonfruit.com; Lynne C. Manzo and Patrick Devine-Wright, *Place Attachment: Advances in Theory, Methods, and Applications* (New York: Routledge, 2014).

7. Robin Kimmerer, "*Mishkos Kenomagwen*, 'The Lessons of Grass': Restoring Reciprocity with the Good Green Earth," in Melissa K. Nelson and Dan Shilling, eds., *Keepers of the Green World: Traditional Ecological Knowledge*

and Sustainability (Cambridge, U.K.: Cambridge University Press, forthcoming 2017); Leanne Simpson, "Land as Pedagogy: Nishnaabeg Intelligence and Rebellious Transformation," *Decolonization: Indigeneity, Education & Society* 3, no. 3 (2014).

8. Dan Moonhawk Alford, *A Report on the Fetzer Institute-sponsored Dialogues Between Western and Indigenous Scientists*, presented at the annual meeting of the Society for the Anthropology of Consciousness, April 11, 1993, http://hilgart.org/enformy/dma-b.htm; David Bohm, *On Dialogue* (New York: Routledge, 2004).

9. Sage LaPena, presentation at annual Bioneers Conference, San Rafael, CA, October 24, 2016.

10. Melissa K. Nelson and Nicola Wagenberg. "Linking Ancestral Seeds and Waters to the Indigenous Places We Inhabit," in Jeannine M. Canty, ed., *Ecological and Social Healing: Multicultural Women's Voices* (New York: Routledge, 2016); Winona LaDuke, "Slow, Clean, Good Food," *Indian Country Today Media Network*, September 30, 2016.

11. Daniel R. Wildcat, *Red Alert! Saving the Planet with Indigenous Knowledge* (Golden, CO: Fulcrum Publishing, 2009).

12. The Cultural Conservancy, "Guardians of the Waters: Preserving and Regenerating Water Guardianship," www.nativeland.org/guardians-of-the-waters.

13. The Cultural Conservancy, "Native Foodways: Protecting the Sanctity of Native Foods," www.nativeland.org /native-foodways; Wendy Johnson and Melissa K. Nelson, "Wild and Cultivated Foods: A Collaboration," *Whole Thinking Journal* (Center for Whole Communities), no. 8 (Spring 2013): 30–33; Jane Brody, "Babies Know: A Little Dirt Is Good for You," *New York Times*, January 26, 2009.

14. Melissa K. Nelson, "Mending the Split-Head Society with Trickster Consciousness," in Melissa K. Nelson, ed., *Original Instructions: Indigenous Teachings for a Sustainable Future* (VT: Inner Traditions, 2008); Leanne Simpson, *Lighting the Eighth Fire: The Liberation, Resurgence, and Protection of Indigenous Nations* (Winnipeg, ON: Arbeiter Ring Publishing, 2008).

15. Jeannette Armstrong, *Constructing Indigeneity: Syilx Okanagan Oraliture and Tmix Centrism*, doctoral dissertation (Griefswald, Germany: Ernst-Moritz-Arndt University, 2009); Rowen White, presentation at Cultural Conservancy Harvest Festival, Indian Valley Organic Farm and Garden, College of Marin, CA, October 8, 2016; Wes Jackson, *Becoming Native to This Place* (Berkeley, CA: Counterpoint, 1996).

16. Enrique Salmon, "Kincentric Ecology: Indigenous Perceptions of the Human-Nature Relationship," *Ecological Applications* 10, no. 5 (October 2000): 1,327–32; Dan Hart, "White Shamans and Plastic Medicine Men," film, 1995, www.iupui.edu/~mstd/e320/white%20shamans%20video.html; Toby McLeod, "Pilgrims and Tourists," film, 2014, http://standingonsacredground.org/sites/default/files/Pilgrims-Tourists_SOSG-teachers_book marked.pdf.

Chapter 5. Pathway to Stewardship: A Framework for Children and Youth

1. David Sobel, *Beyond Ecophobia, Reclaiming the Heart in Nature Education*, second edition (Great Barrington, MA: The Orion Society, June 2013).

2. Simeon Ogonda, "What Kind of Children Are We Leaving Behind for Our Planet," *Huffington Post*, April 21, 2015.

3. Mark Olfson, Benjamin G. Druss, and Steven C. Marcus, "Trends in Mental Health Care Among Children and Adolescents," *New England Journal of Medicine* 372 (May 21, 2015): 2,029–38; Jean T. Twenge et al., "Birth Cohort Increases in Psychopathology Among Young Americans, 1938–2007: A Cross-temporal Meta-analysis of the MMPI," *Clinical Psychology Review* 30, no. 2 (March 2010): 145–54; U.S. Centers for Disease Control and Prevention, "Increasing Prevalence of Parent-Reported Attention-Deficit/Hyperactivity Disorder Among Children,

United States, 2003 and 2007," *Morbidity and Mortality Weekly Report*, November 12, 2010; E. J. Costello et al., "Prevalence and Development of Psychiatric Disorders in Childhood and Adolescence," *Archives of General Psychiatry* 60, no. 8 (August 2003): 837–44; Parliament of Canada, *Healthy Weights for Healthy Kids*, Report of the Standing Committee on Health, 39th Parliament, 1st Session (Ottawa, ON: March 2007).

4. Richard Louv, *Last Child in the Woods: Saving Our Children from Nature-Deficit Disorder* (New York: Workman Publishing, 2005); Richard Louv, *Vitamin N: The Essential Guide to a Nature-Rich Life* (Chapel Hill, NC: Algonquin Books, 2012); Joy A. Palmer et al., "Significant Life Experiences and Formative Influences on the Development of Adults' Environmental Awareness in the UK, Australia and Canada," *Journal of Environmental Research* 5, no. 2 (1999): 181–200.

5. Cathy Dueck, *Pathway to Stewardship: A Framework for Children and Youth*, draft (Peterborough, ON: Pathway to Stewardship Framework Committee, March 2016).

6. Louise Chawla, "Significant Life Experiences Revisited: A Review of Research on Sources of Environmental Sensitivity," *Journal of Environmental Research* 4, no. 4 (1998): 369–82.

7. Box 5–1 from the following sources: Glen Aikenhead and Herman Michell, *Bridging Cultures: Indigenous and Scientific Ways of Knowing Nature* (Toronto, ON: Pearson Canada Inc., 2011), 98; Melissa K. Nelson, *Original Instructions: Indigenous Teachings for a Sustainable Future* (Rochester, VT: Bear & Company, 2008), 3; Joyce Green, *The Three Rs of Relationship, Respect and Responsibility: Contributions of Aboriginal Political Thought for Ecologism and Decolonization in Canada*, paper presented to the Canadian Political Science Association (Regina, SK: University of Regina, 2007), 1; Peter Keith Kulchyski, *Like the Sound of a Drum: Aboriginal Cultural Politics in Denendeh and Nunavut* (Winnipeg, MB: University of Manitoba Press, 2005), 78.

8. Drew Monkman is a retired teacher, author, and newspaper columnist from Peterborough, ON. He is the coauthor with Jacob Rodenburg of *The Big Book of Nature Activities* (Gabriola Island, BC: New Society Publishers, 2016).

9. Camp Kawartha Outdoor Education Centre, Douro-Dummer, ON.

10. Dueck, *Pathway to Stewardship*.

11. Camp Kawartha Environment Centre, Peterborough, ON. Box 5–2 from the following sources: Lesley Le Grange, "Ubuntu, Ukama, Environment and Moral Education," *Journal of Moral Education* 41, no. 3 (2012): 329–40; Lesley Le Grange, "Ubuntu, Ukama and the Healing of Nature, Self and Society," *Educational Philosophy and Theory* 44, no. S2, special issue on Africa (September 2012): 56–67; Lesley Le Grange, "Ubuntu as Architechtonic Capability," *Indilinga* 11, no. 2 (2012): 139–45; Lesley Le Grange, "Ubuntu as Ecosophy and Ecophilosophy," *Journal of Human Ecology* 49, no. 3 (2015): 301–08; Raèssa Pather, "Garden Project Feeds on Fresh Ideas," *Mail & Guardian* (Cape Town), November 27, 2015; Wildlife and Environment Society of South Africa, "WESSA Eco-schools," www.wessa.org.za/what-we-do/schools-programme/eco-schools.htm, viewed October 14, 2016.

12. Hilary Inwood and Susan Jagger, *DEEPER, Deepening Environmental Education in Pre-Service Education Resource* (Toronto, ON: University of Toronto, Ontario Institute for Studies in Education, 2014), 44–45.

13. ParticipACTION, *The Biggest Risk Is Keeping Kids Indoors: The 2015 ParticipACTION Report Card on Physical Activity for Children and Youth* (Toronto, ON: 2015); Camp Kawartha Environment Centre, Peterborough, ON.

14. Stan Kozak and Susan Elliott, *Connecting the Dots: Key Strategies That Transform Learning for Environmental Education, Citizenship, and Sustainability* (North York, ON: Learning for a Sustainable Future, 2014).

Chapter 6. Growing a New School Food Culture

1. Ramón Fernández Durán and Luis González Reyes, *En la espiral de la energía* (Madrid: Libros en Acción y Baladre, 2014).

2. United Nations General Assembly, "Report Submitted by the Special Rapporteur on the Right to Food, Olivier De Schutter," Human Rights Council, Sixteenth session, Agenda item 3, Promotion and protection of all human rights, civil, political, economic, social and cultural rights, including the right to development (New York: December 20, 2010).

3. For more on re-ruralization, see Fernández Durán and González Reyes, *En la espiral de la energía*.

4. Foodlinks, *Revaluing Public Sector Food Procurement in Europe: An Action Plan for Sustainability* (Brussels: 2010), 14; FUHEM Foundation, "Nuestros comedores ecológicos en 'Comando Actualidad' de TVE," www.fuhem.es /educacion/noticias.aspx?v=10022&n=0; Menjadors Ecològics, "Qui som i què fem," www.menjadorsecologics .cat/associacio/el-projecte/; Georgia Organics website, http://georgiaorganics.org; Seven Generations Ahead website, http://sevengenerationsahead.org.

5. Ecocentral, "Ecomenja," http://ecocentral.cat/empreses-gestores-menjadors-escolars/ecomenja-gestor-de -menjadors-escolars/; "Nation's First School District to Serve 100% Organic, Non-GMO Meals," EcoWatch, August 18, 2015; National Farm to School Network website, www.farmtoschool.org.

6. AMPA Gómez Moreno, "Comedor ecológico," https://ampagomezmoreno.wordpress.com/servicios/come dor; Chris Hardman, "How One Visionary Changed School Food in Detroit," Civil Eats, April 6, 2015; Tracie McMillan, "Detroit's Rebel Lunch Lady Wants to Fix More Than Food," *National Geographic*, March 4, 2016.

7. Bridget Huber, "Welcome to Brazil, Where a Food Revolution Is Changing the Way People Eat," *The Nation*, July 28, 2016; Foodlinks, *Revaluing Public Sector Food Procurement in Europe*.

8. Menjadors Ecològics, "Qui som i què fem."

9. FUHEM Foundation, "Nuestros comedores ecológicos en Comando Actualidad de TVE."

10. Ibid.

11. Ibid.

12. Cidades Sem Fome, "School Gardens Project," https://cidadessemfome.org/en/#projekt_sg; Jane Goodall Institute China, "Roots and Shoots," www.jgichina.org/en/rootsandshoots/environment/index.aspx?PartNod eId=13; Kitchen Garden Foundation, "Pleasurable Food Education," www.kitchengardenfoundation.org.au /content/pleasurable-food-education; Edible Garden City website, www.ediblegardencity.com; Slow Food Foundation for Biodiversity, "Mbuyuni School Garden," www.fondazioneslowfood.com/en/slow-food-gardens-africa /mbuyuni-school-garden.

13. Danielle Nierenberg and Allyn Rosenberger, "35 Food Education Organizations," Food Tank, September 13, 2016; The Edible Schoolyard Project, "Train with Us!" http://edibleschoolyard.org/training.

14. GrowNYC, "Learn It Grow It Eat It," www.grownyc.org/learn-it-grow-it-eat-it.

15. The Independent School Food Plan, "Cooking in the Curriculum," www.schoolfoodplan.com/actions /cooking-in-the-curriculum.

16. Curriculum of Cuisine website, http://thecurriculumofcuisine.org; Alimentando Otros Modelos, "Materiales didácticos Alimentando Otros Modelos," August 1, 2016, https://alimentarotrosmodelos.wordpress .com/2016/08/01/materiales-didacticos-alimentando-otros-modelos.

17. SEED, "Schools Education," http://seed.org.za/schools-education.

18. INCLUD-ED website, http://creaub.info/included.

19. Information based on the author's experience in these movements. See also MST, "Educação," www.mst.org .br/educacao, and Raúl Zibechi, *Descolonizar la rebeldía. (Des)colonialismo del pensamiento crítico y de las prácticas emnacipatorias* (Málaga, Spain: Baladre y Zambra, 2014).

Chapter 7. The Centrality of Character Education for Creating and Sustaining a Just World

1. Charles C. Haynes and Marvin W. Berkowitz, "What Can Schools Do?" *USA Today,* February 20, 2007, 13a.

2. Marvin W. Berkowitz, Kristen Pelster, and Amy Johnston, "Leading in the Middle: A Tale of Pro-social Education Reform in Two Principals and Two Middle Schools," in Philip Brown, Michael W. Corrigan, and Ann Higgins-D'Alessandro, eds., *The Handbook of Prosocial Education: Volume 2* (Lanham, MD: Rowman & Littlefield, 2012), 619–26; Haynes and Berkowitz, 13a.

3. Berkowitz, Pelster, and Johnston, "Leading in the Middle."

4. Character.org website, www.character.org; Kathy Beland, *Eleven Principles Sourcebook: How to Achieve Quality Character Education in K–12 Schools* (Washington, DC: Character.org, 2003).

5. Marvin W. Berkowitz and Melinda C. Bier, *What Works in Character Education: A Research-driven Guide for Educators* (Washington, DC: Character Education Partnership, 2005).

6. Joseph A. Durlak et al., "The Impact of Enhancing Students' Social and Emotional Learning: A Meta-analysis of School-based Universal Interventions," *Child Development* 82, no. 1 (2011): 405–32; Jon C. Marshall, Sarah D. Caldwell, and Jeanne Foster, "Moral Education the CHARACTER*plus* Way'," *Journal of Moral Education* 40, no. 1 (2011): 51–72.

7. Marvin W. Berkowitz, "What Works in Values Education," *International Journal of Educational Research* 50, no. 3 (2011): 153–58; Berkowitz and Bier, *What Works in Character Education*; Marvin W. Berkowitz, Melinda C. Bier, and Brian McCauley, *Effective Features and Practices That Support Character Development*, paper presented at National Academies of Sciences, Engineering, and Medicine Workshop on Approaches to the Development of Character, Washington, DC, July 26–27, 2016.

8. Marvin W. Berkowitz, *You Can't Teach Through a Rat: And Other Epiphanies for Educators* (Chapel Hill, NC: Character Development Publishing, 2012).

9. See Charles F. Elbot and David Fulton, *Building an Intentional School Culture: Excellence in Academics and Character* (Thousand Oaks, CA: Corwin, 2008).

10. Built on the work of Bob Hassinger, former principal of Halifax Middle School in Halifax, PA, another National School of Character.

11. David W. Johnson and Roger T. Johnson, *Learning Together and Alone: Cooperative, Competitive, and Individualistic Learning* (Upper Saddle River, NJ: Prentice-Hall, 1987).

12. Marilyn Watson, *Learning to Trust: Transforming Difficult Elementary Classrooms Through Developmental Discipline* (San Francisco, CA: Jossey-Bass, 2003).

13. Cathryn Berger Kaye, *The Complete Guide to Service Learning: Proven, Practical Ways to Engage Students in Civic Responsibility, Academic Curriculum, & Social Action* (Minneapolis, MN: Free Spirit Publishing, 2003); Mary O'Brien, *Making Better Environmental Decisions: An Alternative to Risk Assessment* (Cambridge, MA: The MIT Press, 2002); Ron Berger, *An Ethic of Excellence: Building a Culture of Craftsmanship with Students* (Portsmouth, NH: Heinemann, 2003); EL Education website, www.elschools.org.

14. Giraffe Heroes Project website, www.giraffe.org.

15. Center for the Collaborative Classroom website, www.collaborativeclassroom.org.

16. The Jubilee Centre for Character & Virtues, University of Birmingham website, www.jubileecentre.ac.uk.

17. Berger, *An Ethic of Excellence*; Inspire>Aspire website, www.inspiringpurpose.org.uk; John Templeton Foundation website, www.templeton.org.

18. See, for example, Berkowitz and Bier, *What Works in Character Education*; Collaborative for Academic Social and Emotional Learning (CASE), *Safe and Sound: An Educational Leader's Guide to Evidence-based Social and Emotional (SEL) Programs* (Chicago, IL: 2005); National School Climate Center website, www.schoolclimate.org; EL Education website; Collaborative for Academic, Social, and Emotional Learning (CASEL) website, www.casel.org.

19. Berkowitz, Bier, and McCauley, *Effective Features and Practices That Support Character Development*.

Chapter 8. Social and Emotional Learning for a Challenging Century

1. Alice Fothergill and Lori Peek, *Children of Katrina* (Austin, TX: University of Texas Press, 2015).

2. Ibid.; Amy McConnell Franklin, "The Integrated Model: A Framework for Emotional Intelligence: Awareness, Intention, and Choice in Decision-making and Relationship," 2014, www.amymcconnellfranklin.com/the-integrated-model-of-emotional-intelligence-3.

3. Collaborative for Academic, Social, and Emotional Learning (CASEL) website, www.casel.org.

4. Joseph A. Durlak et al., "The Impact of Enhancing Students' Social and Emotional Learning: A Meta-analysis of School-based Universal Interventions," *Child Development* 82, no. 1 (2011): 405–32; Roger P. Weissberg et al., "Social and Emotional Learning: Past, Present, and Future," *Handbook of Social and Emotional Learning: Research and Practice* (New York: Guilford, 2015); Damon E. Jones, Mark Greenberg, and Max Crowley, "Early Social emotional Functioning and Public Health: The Relationship Between Kindergarten Social Competence and Future Wellness," *American Journal of Public Health* 105, no. 11 (2015): 2,283–90; Weissberg et al., "Social and Emotional Learning: Past, Present, and Future"; Ann S. Masten, "Resilience in Developing Systems: Progress and Promise as the Fourth Wave Rises," *Development and Psychopathology* 19, no. 3 (2007): 921–30; James J. Heckman, Jora Stixrud, and Sergio Urzua, *The Effects of Cognitive and Noncognitive Abilities on Labor Market Outcomes and Social Behavior* (Washington, DC: National Bureau of Economic Research, 2006).

5. Clive Belfield et al., "The Economic Value of Social and Emotional Learning," *Journal of Benefit-Cost Analysis* 6, no. 3 (2015): 508–44. Box 8–1 from the following sources: Thomas J. Doherty and Susan Clayton, "The Psychological Impacts of Global Climate Change," *American Psychologist* 66, no. 4 (2011): 26; Janet Swim et al., *Psychology and Global Climate Change: Addressing a Multi-faceted Phenomenon and Set of Challenges* (Washington, DC: American Psychological Association, 2009); Tom Crompton and Tim Kasser, *Meeting Environmental Challenges: The Role of Human Identity* (Godalming, U.K.: WWF-UK, 2009).

6. Alistair Woodward, Simon Hales, and Philip Weinstein, "Climate Change and Human Health in the Asia Pacific Region: Who Will Be Most Vulnerable," *Climate Research* 11, no. 1 (1998): 31–38; Karen L. O'Brien and Robin M. Leichenko, "Double Exposure: Assessing the Impacts of Climate Change Within the Context of Economic Globalization," *Global Environmental Change* 10, no. 3 (2000): 221–32.

7. Katie Parsanko-Malone, Poudre High School, Fort Collins, CO, personal communication with Pamela Barker, July 22, 2016.

8. Cynthia Frantz et al., "There Is No 'I' in Nature: The Influence of Self-Awareness on Connectedness to Nature," *Journal of Environmental Psychology* 25, no. 4 (2005): 427–36.

9. Weissberg et al., "Social and Emotional Learning: Past, Present, and Future." Figure 8–1 adapted from Ibid., 7.

10. CASEL, "Inclusion Criteria for SELect Programs," www.casel.org/guide/criteria, viewed July 6, 2016; CASEL, "Selecting Evidence-Based SEL Programs," www.casel.org/guideselecting-programs, viewed July 6, 2016; Dean L. Fixsen et al., *Implementation Drivers: Assessing Best Practices* (Chapel Hill, NC: Frank Porter Graham Child Development Institute, University of North Carolina Chapel Hill, 2013).

11. Lori Nathanson et al., "Creating Emotionally Intelligent Schools with RULER," *Emotion Review*, August 18, 2016; Yale Center for Emotional Intelligence, RULER website, http://ei.yale.edu/ruler-schools/, viewed November 14, 2016.

12. Ibid.; Marc A. Brackett and Susan E. Rivers, "Transforming Students' Lives with Social and Emotional Learning," in Reinhard Pekrun and Lisa Linnenbrink-Garcia, eds., *International Handbook of Emotions in Education* (London: Routledge, 2014), 368–88.

13. Brian D. Ostafin, Michael D. Robinson, and Brian P. Meier, eds., *Handbook of Mindfulness and Self-regulation* (Berlin: Springer, 2015); Tamar Mendelson et al., "Feasibility and Preliminary Outcomes of a School-based Mindfulness Intervention for Urban Youth," *Journal of Abnormal Child Psychology* 38, no. 7 (2010): 985–94; Kimberly A. Schonert-Reichl and Molly Stewart Lawlor, "The Effects of a Mindfulness-based Education Program on Pre-and Early Adolescents' Well-being and Social and Emotional Competence," *Mindfulness* 1, no. 3 (2010): 137–51; Emily Deruy, "Does Mindfulness Actually Work in Schools?" *The Atlantic*, May 20, 2016; Tara Garcia Mathewson, "Where Mindfulness Education Fits in Schools," Education Dive, August 31, 2016. Box 8–2 from UWC Thailand website, www.uwcthailand.net, viewed October 14, 2016, and from UWC International website, www.uwc.org, viewed October 14, 2016.

14. Abbey J. Porter, "Restorative Practices in Schools: Research Reveals Power of Restorative Approach, Part II," *Restorative Practices E-Forum* 3 (2007).

15. Eric Rasmussen, "Connection Circles: How to Establish a Restorative Circle Practice," *Kaleidoscope* (Knowles Science Teaching Foundation), Spring 2016, 3–7.

16. Barrowford Primary School website, www.barrowford.lancs.sch.uk.

17. Box 8–3 from the following sources: Hearts and Horses Therapeutic Riding Center website, www.heartsandhorses.org; Linda Kohanov, *The Power of the Herd: A Nonpredatory Approach to Social Intelligence, Leadership and Innovation* (Novato, CA: New World Library, 2013); Pamela Barker and Tamara Merritt, *Final Report: TSD at Hearts and Horses Spring 2016*, unpublished, 2016.

18. CASEL, "State Scan Scorecard Project," www.casel.org/state-scan-scorecard-project, viewed August 27, 2016; CASEL, "Partner Districts," www.casel.org/partner-districts, viewed August 27, 2016.

19. Organisation for Economic Co-operation and Development, *Skills for Social Progress: The Power of Social and Emotional Skills* (Paris: 2015); Catalina Torrente, Anjali Alimchandani, and J. Lawrence Aber, "International Perspectives on SEL," *Handbook of Social and Emotional Learning: Research and Practice* (New York: Guilford, 2015).

20. Fothergill and Peek, *Children of Katrina*.

Chapter 9. Prioritizing Play

1. Timothy D. Walker, "The Junk Playground of New York City," *The Atlantic*, August 11, 2016.

2. Ibid.

3. David Whitebread et al., *The Importance of Play: A Report on the Value of Children's Play with a Series of Policy Recommendations* (Brussels: Toys Industries for Europe, 2012).

4. "School Testing Regime," www.politics.co.uk/reference/school-testing-regime; Michael Gove, U.K. Department for Education, "The Purpose of Our School Reforms," speech delivered at Policy Exchange, June 7, 2014; Marcus Winters, "Standardized Tests Are Costly, But Worth It," *New York Times*, July 23, 2012; Latasha Gandy, "Don't Believe the Hype: Standardized Tests Are Good for Children, Families and Schools," Education Post, January 11, 2016; Douglas Holtz-Eakin, "A Hidden Benefit to Common Core: High Education Standards Prevent Unprepared College Students and Help the Economy," *U.S. News & World Report*, April 27, 2016.

5. Pasi Sahlberg, *Finnish Lessons: What Can the World Learn About Educational Change in Finland?* (New York: Teachers College Press, 2012).

6. David Whitebread and Sue Bingham, "School Readiness in Europe: Issues and Evidence," in Marilyn Fleer and Bert van Oers, eds., *International Handbook on Early Childhood Education and Development. Volume II: Western-Europe and UK* (Dordrecht, The Netherlands: Springer, 2017).

7. Denis Campbell and Sarah Marsh, "Quarter of a Million Children Receiving Mental Health Care in England," *The Guardian* (U.K.), October 3, 2016.

8. Peter Gray, "The Decline of Play and the Rise of Psychopathology," *American Journal of Play* 3, no. 4 (2011): 443–63; David Whitebread, "Play, Metacognition & Self-regulation," in Pat Broadhead, Justine Howard, and Elizabeth Wood, eds., *Play and Learning in the Early Years* (London: Sage, 2010).

9. Laura E. Berk, Trisha D. Mann, and Amy T. Ogan, "Make-Believe Play: Wellspring for Development of Self-Regulation," in Dorothy G. Singer, Roberta Michnick Golinkoff, and Kathy Hirsh-Pasek, eds., *Play = Learning: How Play Motivates and Enhances Children's Cognitive and Social-Emotional Growth* (Oxford, U.K.: Oxford University Press, 2006), 74–100; Kristin Valentino et al., "Mother–Child Play and Maltreatment: A Longitudinal Analysis of Emerging Social Behavior from Infancy to Toddlerhood," *Developmental Psychology* 47, no. 5 (2011): 1,280–94; Insa Weber-Börgmann et al., "Associations with ADHD and Parental Distress Within Play in Early Childhood," *Zeitschrift fur Kinder-und Jugendpsychiatrie und Psychotherapie* 42, no. 3 (2014): 147–55; V. Taneja et al., "'Not by Bread Alone': Impact of a Structured 90-minute Play Session on Development of Children in an Orphanage," *Child: Care, Health and Development* 28, no. 1 (2002), 95–100; Sue Bratton et al., "Head Start Early Mental Health Intervention: Effects of Child-centered Play Therapy on Disruptive Behaviors," *International Journal of Play Therapy* 22, no. 1 (2013): 28; Maggie Fearn and Justine Howard, "Play as a Resource for Children Facing Adversity: An Exploration of Indicative Case Studies," *Children & Society* 26, no. 6 (2012): 456–68.

10. Kathy Sylva et al., *The Effective Provision of Pre-School Education (EPPE) Project: Technical Paper 12 – The Final Report* (London: Department for Education and Skills and Institute of Education, University of London, 2004); Sebastian P. Suggate, Elizabeth A. Schaughency, and Elaine Reese, "Children Learning to Read Later Catch Up to Children Reading Earlier," *Early Childhood Research Quarterly* 28, no. 1 (2012): 33–48; Sebastian P. Suggate, "School Entry Age and Reading Achievement in the 2006 Programme for International Student Assessment (PISA)," *International Journal of Educational Research* 48, no. 3 (2009): 151–61.

11. J. S. Bruner, "Nature & Uses of Immaturity," *American Psychologist* 27 (1972): 687–708; Peter K. Smith, "Evolutionary Foundations and Functions of Play: An Overview," in Artin Göncü and Suzanne Gaskins, eds., *Play & Development: Evolutionary, Sociocultural and Functional Perspectives* (Mahwah, NJ: Lawrence Erlbaum, 2012), 21–49; Peter Gray, "Play as a Foundation for Hunter-Gatherer Social Existence," *American Journal of Play* 1, no. 4 (2009): 476–522; Sergio Pellis and Vivien Pellis, *The Playful Brain: Venturing to the Limits of Neuroscience* (Oxford, U.K.: Oneworld Publications, 2009); Kathy Sylva, Jerome S. Bruner, and Paul Genova, "The Role of Play in the Problem-Solving of Children 3–5 Years Old," in Jerome S. Bruner, Alison Jolly, and Kathy Sylva, eds., *Play: Its Role in Development and Evolution* (Harmondsworth, U.K.: Penguin Books, 1976), 55–67; C. S. Tamis-LeMonda and M. H. Bornstein, "Habituation and Maternal Encouragement of Attention in Infancy as Predictors of Toddler Language, Play and Representational Competence," *Child Development* 60, no. 3 (1989): 738–51; Rebecca A. Marcon, "Moving Up the Grades: Relationship Between Pre-school Model and Later School Success," *Early Childhood Research & Practice* 4, no. 1 (2002): 517–30; Anthony D. Pellegrini and Kathy Gustafson, "Boys' and Girls' Uses of Objects for Exploration, Play and Tools in Early Childhood," in Anthony D. Pellegrini and Peter K. Smith, eds., *The Nature of Play: Great Apes and Humans* (New York: Guilford Press, 2005), 113–35; David Whitebread, Helen Jameson, and Marisol Basilio, "Play Beyond the Foundation Stage: Play, Self-regulation and Narrative Skills," in Janet Moyles, ed., *The Excellence of Play, Fourth Edition* (Maidenhead, U.K.: Open University Press, 2015), 84–93.

12. Jane E. Barker et al., "Less-structured Time in Children's Daily Lives Predicts Self-directed Executive

Functioning," *Frontiers in Psychology* 5 (2014); C. Hughes et al., "Measuring the Foundations of School Readiness: Introducing a New Questionnaire for Teachers – The Brief Early Skills and Support Index (BESSI)," *British Journal of Educational Psychology* 85, no. 3 (2015): 332–56.

13. James F. Christie and Kathleen A. Roskos, "Standards, Science, and the Role of Play in Early Literacy Education," in Singer, Golinkoff, and Hirsh-Pasek, eds., *Play = Learning.*

14. Whitebread, "Play, Metacognition & Self-regulation"; Marilou Hyson, Carol Copple, and Jacqueline Jones, "Early Childhood Development and Education," in K. Ann Renninger and Irving E. Sigel, eds., *Handbook of Child Psychology: Volume 4. Child Psychology in Practice* (New York: Wiley, 2006), 3–47; Marcel V. J. Veenman and Marleen A. Spaans, "Relation Between Intellectual and Metacognitive Skills: Age and Task Differences," *Learning and Individual Differences* 15, no. 2 (2005): 159–76; John Hattie, *Visible Learning: A Synthesis of over 800 Meta-analyses Relating to Achievement* (London: Routledge, 2009); Megan M. McClelland et al., "Relations Between Preschool Attention Span-Persistence and Age 25 Educational Outcomes," *Early Childhood Research Quarterly* 28, no. 2 (2013): 314–24.

15. Judy S. DeLoache, Susan Sugarman, and Ann L. Brown, "The Development of Error Correction Strategies in Young Children's Manipulative Play," *Child Development* 56, no. 4 (1985): 928–39; Claire Cook, Noah D. Goodman, and Laura E. Schulz, "Where Science Starts: Spontaneous Experiments in Preschoolers' Exploratory Play," *Cognition* 120 (2011): 341–49.

16. Angeline Lillard, "Guided Participation: How Mothers Structure and Children Understand Pretend Play," in Göncü, and Gaskins, eds., *Play & Development*, 131–53; Mark Nielsen, and Tamara Christie, "Adult Modelling Facilitates Young Children's Generation of Novel Pretend Acts," *Infant and Child Development* 17, no. 2 (2008), 151–62.

17. Deena Skolnick Weisberg, Kathy Hirsh-Pasek, and Roberta Michnick Golinkoff, "Guided Play: Where Curricular Goals Meet a Playful Pedagogy," *Mind, Brain, and Education* 7, no. 2 (2013): 104–12.

18. Doris Pronin Fromberg and Doris Bergen, *Play from Birth to Twelve: Contexts, Perspectives, and Meanings, Third Edition* (New York: Routledge, 2015); Anthony D. Pellegrini, *The Oxford Handbook of the Development of Play* (Oxford, U.K.: Oxford University Press, 2015); LEGO Foundation, "Building Children's Writing Skills Through Learning Through Play," video, October 15, 2015, https://vimeo.com/142506730.

19. David Whitebread, Martina Kuvalja, and Aileen O'Connor, *Quality in Early Childhood Education: An International Review and Guide for Policy Makers* (Doha, Qatar: World Innovation Summit for Education, 2015); Government of Ireland, *Ready, Steady, Play! A National Play Policy* (Dublin: Department of Childhood and Youth Affairs, 2004); Government of Newfoundland, *The Power of Play* (St. John's, NL: Department of Education and Early Childhood Development, 2011); Claire Duffin, "Streets Are Alive with the Sound of Children Playing," *The Telegraph* (U.K.), February 22, 2014; "AnjiPlay," https://www.facebook.com/AnjiPlayWorld.

Chapter 10. Looking the Monster in the Eye: Drawing Comics for Sustainability

1. Shankar Musafir, *To The Principal* (Mumbai: Leadstart Publishing Pvt Ltd, 2011).

2. Osamu Tezuka, *Apollo's Song* (New York: Vertical Inc., 2007); originally published in Japanese in 1970. Box 10–1 based on the following sources: Jing Shi et al., "Knowledge as a Driver of Public Perceptions About Climate Change Reassessed," *Nature Climate Change Letters* 6 (April 25, 2016): 759–62; Ariel Malka, Jon A. Krosnick, and Gary Langer, "The Association of Knowledge with Concern About Global Warming: Trusted Information Sources Shape Public Thinking," *Risk Analysis* 29, no. 5 (May 2009): 633–47; Dan M. Kahan et al., "The Polarizing Impact of Science Literacy and Numeracy on Perceived Climate Change Risks," *Nature Climate Change Letters* 2 (May 27, 2012): 732–35; Kathryn T. Stevenson et al., "Overcoming Skepticism with Education: Interacting Influences of Worldview and Climate Change Knowledge on Perceived Climate Change Risk Among Adolescents," *Climate Change* 126, no. 3 (October 2014): 293–304; Asahi Glass Foundation, "Gring and Woodin's

Environmental Doomsday Clock," www.af-info.or.jp/en/gw_clock/index.html.

3. Guity Novin, *A History of Graphic Design*, online textbook, 2010, http://guity-novin.blogspot.com; United Nations Educational, Scientific and Cultural Organization (UNESCO), *2016 GEM Report: Education for People & Planet* (Paris: 2016).

4. Frans Lenglet, "ESD and Assessing the Quality of Education and Learning," in Victoria W. Thoresen et al., eds., *Responsible Living: Concepts, Education and Future Perspectives* (Berlin: Springer, 2015), 57–72; Marilyn Mehlmann and Olena Pometun, *ESD Dialogues: Practical Approaches to Education for Sustainable Development by and for Educators* (Books on Demand, 2014); Farid Gardizi, "Interview with Charles Hopkins, UNESCO Chair in Education for Sustainable Development, York University, Canada" (Paris: UNESCO, 2009). See also The Learning Teacher Network, *The Learning Teacher Journal*, various issues.

5. Susana Gonçalves and Suzanne Majhanovich, *Art and Intercultural Dialogue* (Rotterdam: Sense Publishers, 2016).

6. Jeff Creswell, *Creating Worlds, Constructing Meaning: The Scottish Storyline Method* (Portsmouth, NH: Heinemann, 1997); Elizabeth Grugeon and Paul Garder, *The Art of Storytelling for Teachers and Pupils* (London: Routledge, 2013).

7. Figure 10–1 from Esbjörn Jorsäter, "The Hero's Journey," Wikimedia Commons, https://commons.wikimedia.org/wiki/File:Hero´s_Journey.jpg.

8. Quote from a student of Drawing for Life in Kosovo, 2012.

9. United Nations, "Sustainable Development Goals," www.un.org/sustainabledevelopment/sustainable-development-goals; William Rees, Mathis Wackernagel, and Phil Testemale, *Our Ecological Footprint: Reducing Human Impact on the Earth* (Gsbriola Island, BC: New Society Publishers, 1998); Esbjörn Jorsäter et al., *Drawing for Life Teacher's Manual* (in several languages) and resulting comics, available at http://comics.globalactionplan.com.

10. John Haworth and Graham Hart, eds., *Well-Being: Individual, Community and Social Perspectives* (London: Palgrave Macmillan, 2007).

11. Jorsäter et al., *Drawing for Life Teacher's Manual*, 19–20.

12. Ibid., 31.

13. Arthur Koestler, *The Act of Creation* (New York: Penguin Books, 1964).

14. Lenglet, "ESD and Assessing the Quality of Education and Learning."

15. Åse Eliason Bjurström, "Trying to Make 'Reality' Appear Different: Working with Drama in an Intercultural ESD Setting," *Social Alternatives* 31, no. 4 (2012): 18–23.

Chapter 11. Deeper Learning and the Future of Education

1. Author field notes, October, 25, 2011; author interview with teacher team, June 2, 2012; King Middle School, "Past Expeditions," http://king.portlandschools.org/news___events/on_expedition.

2. Ibid.

3. Ibid.

4. Organisation for Economic Co-operation and Development, *Education at a Glance* (Paris: 2012); Ken Robinson, *Out of Our Minds: Learning To Be Creative* (Chichester, U.K.: Capstone Publishing, 2011), 49.

5. American Institutes for Research, *Does Deeper Learning Improve Student Outcomes?* (Washington, DC:

2016); Pedro Noguera, Linda Darling-Hammond, and Diane Friedlander, *Equal Opportunity for Deeper Learning* (Washington, DC: Jobs for the Future, October 2015).

6. Author field notes, November 30, 2011.

7. King Middle School, "Past Expeditions."

8. Larry Rosenstock, Chief Executive Officer, High Tech High, San Diego, CA, personal communication with author, January 15, 2013.

9. Author field notes, October 27 and 28, 2011; Casco Bay High School, "Academics," https://cbhs.portland schools.org/academics.

10. Author field notes, November 20, 2011; author interview, November 15, 2012.

11. Author field notes, September 28 and October 24, 2011; Avalon School, "What Is Project-Based Learning?" www.avalonschool.org/pbl.

12. Author field notes, November 24 and 25, 2011.

13. Michael Fullan and Maria Langworthy, *A Rich Seam: How New Pedagogies Find Deeper Learning* (London: Pearson, 2014), 2.

14. Ibid., 16.

15. Ibid., 19; Richard Elmore, "What Happens When Learning Breaks Out in Rural Mexico?" Education Week Blog, May 18, 2011.

16. European Schoolnet website, www.eun.org; Monica Martinez, Dennis McGrath, and Elizabeth Foster, *How Deeper Learning Can Create a New Vision for Teaching* (Washington, DC: National Commission on Teaching & America's Future (NCTAF), 2016); iTEC: Designing the Future Classroom website, http://itec.eun.org.

17. Common Core State Standards Initiative website, www.corestandards.org.

18. Sarah Jenkins et al., *The Shifting Paradigm of Teaching: Personalized Learning According to Teachers* (Washington, DC: NCTAF and Knowledgeworks, 2016).

19. XQ: The Super School Project, "Ten U.S. 'Super Schools' Awarded $10M Each for Reimagining Education," https://xqsuperschool.org/news/3150; Next Generational Learning Challenges website, http://nextgenlearning .org.

20. Box 11–1 from the following sources: John Dewey, "Self-realization as the Moral Ideal," *The Philosophical Review* 2, no. 6 (1893): 652–64; Joe Bryan, "Participatory Mapping," in Tom Perreault, Gavin Bridge, and James McCarthy, eds., *The Routledge Handbook of Political Ecology* (New York: Routledge, 2015), 249–62; TEMA, "Türkiye Su Varliklarina Yönelik Tehditler Haritasi," http://sertifika.tema.org.tr/_Ki/SuTehditleriHaritasi/.

Chapter 12. All Systems Go! Developing a Generation of "Systems-Smart" Kids

1. Thanks to Frank Draper, former field science teacher at Catalina Foothills High School in Tucson, AZ, for this question.

2. Douglas Smith, *Decade of the Wolf: Returning the Wild to Yellowstone*, revised and updated edition (Guilford, CT: Lyons Press, 2012).

3. John Sterman, *Business Dynamics: Systems Thinking and Modeling for a Complex World* (Boston: Irwin/McGraw-Hill, 2000), 5–13.

4. Tina A. Grotzer and Belinda Bell Basca, *Helping Students to Grasp the Underlying Causal Structures When Learning About Ecosystems: How Does It Impact Understanding?* paper presented at the National Association for

Research in Science Teaching Annual Conference, Atlanta, GA, 2000; David W. Green, Explaining and Envisaging an Ecological Phenomenon," *British Journal of Psychology* 88, no. 2 (1997): 199–217; Peter A. White, "Naive Ecology: Causal Judgements About a Simple Ecosystem," *British Journal of Psychology* 88, no. 2 (1997): 219–233; Linda Booth Sweeney, *Thinking About Systems: An Empirical Investigation of Middle School Students' and Their Teachers' Conceptions of Natural and Social Systems*, unpublished doctoral dissertation (Cambridge, MA: Harvard University, 2004); Linda Booth Sweeney and John D. Sterman, "Thinking About Systems: Student and Teacher Conceptions of Natural and Social Systems," *System Dynamics Review* 23, no. 2–3 (2007): 285–311.

5. Figure 12–1 from PBS Learning Media, "Systems Literacy: Teaching About Systems," STEM professional development module for teachers, www.pbslearningmedia.org/collection/systemsliteracy.

6. Linda Booth Sweeney and Dennis Meadows, *The Systems Thinking Playbook: Exercises to Stretch and Build Learning and Systems Thinking Capabilities* (White River Junction, VT: Chelsea Green, 1995/2010); World Commission on Environment and Development, *Our Common Future*, Annex to United Nations General Assembly document A/42/427, Section IV-1 (New York: 1987).

7. Clive Hamilton, "Define the Anthropocene in Terms of the Whole Earth, *Nature*, August 17, 2016.

8. Martos Hoffman and Daniel Barstow, *Revolutionizing Earth System Science Education for the 21st Century, Report and Recommendations from a 50-State Analysis of Earth Science Education Standards* (Cambridge, MA: TERC Center for Earth and Space Science Education, 2007), 5.

9. Ibid. Box 12–2 from Melia Robinson, "Bill Gates Is Revolutionizing How History Is Taught, and We Went to a Poor NYC High School to See It in Action," *Business Insider*, January 7, 2015.

10. Jaimie Cloud, "Education for a Sustainable Future: Benchmarks for Individual and Social Learning," *Journal of Sustainability Education*, June 2016; The Center for Green Schools at the U.S. Green Building Council and Houghton Mifflin Harcourt, *National Action Plan for Educating for Sustainability* (Boston and Washington, DC: 2014), 4.

11. The Center for Green Schools at the U.S. Green Building Council and Houghton Mifflin Harcourt, *National Action Plan for Educating for Sustainability*.

12. Putnam/Northern Westchester BOCES Curriculum Center, Education for Sustainability website, www.pnwboces.org/EFS/efs_home.htm.

13. Next Generation Science Standards website, www.nextgenscience.org.

14. PBS Learning Media, "Westward Expansion: A Systems Approach to History," http://rmpbs.pbslearningmedia.org/resource/syslit14-ush-sys-westexp/westward-expansion-a-systems-approach-to-history.

15. Mary Catherine Bateson, *Willing to Learn: Passages of Personal Discovery* (Hanover, NH: Steerforth Press, 2004), 290.

16. Tracy Benson, James K. Doyle, and Frank Draper, *Tracing Connections: Voices of Systems Thinkers* (Lebanon, NH and Acton, MA: isee systems, Inc. and Creative Learning Exchange, 2010), 54–58.

17. Ibid., 54–55.

18. Ibid., 61.

19. Waters Foundation, Systems Thinking in Education website, www.watersfoundation.org.

20. Linda Booth Sweeney, *Healthy Chickens, Healthy Pastures: Making Connections at Drumlin Farm and Beyond* (Lincoln, MA: Creative Learning Exchange and Mass Audubon Drumlin Farm Wildlife Sanctuary, 2010).

21. Washington University in St. Louis, Social System Design Lab (SSDL) website, https://sites.wustl.edu/ssdl; Washington University in St. Louis, SSDL, "Changing Systems Gun Violence," https://sites.wustl.edu/ssdl

/changing-systems-youth-summit/changing-systems-gun-violence-2.

22. Steph Grimes, "3D Printing for Solving Baltimore's Problems," Digital Harbor Foundation, May 25, 2016; Digital Oyster Foundation website, http://digitaloysterfoundation.weebly.com.

23. AZ Quotes, "Joseph Campbell Quotes," www.azquotes.com/quote/796947.

24. Dr. Martin Luther King, Jr., "Christmas Sermon on Peace," delivered to the congregation of Ebenezer Baptist Church, Atlanta, GA, December 24, 1967.

Chapter 13. Reining in the Commercialization of Childhood

1. Susan Linn and Josh Golin, "Save the Lorax, Shun the Stuff," *Huffington Post*, March 1, 2012.

2. Emma Brown, "The Lorax Helps Market Mazda SUVs to Elementary School Children Nationwide," *Washington Post*, February 28, 2016.

3. Susan Gregory Thomas, *Buy Buy Baby: How Consumer Culture Manipulates Parents and Harms Young Minds* (Boston: Houghton Mifflin, 2004), 109–35; Duncan Hood, "Is Advertising to Kids Wrong? Marketers Respond," *Kidscreen*, November 1, 2000, 15.

4. Dale Kunkel, "Children and Television Advertising," in Dorothy G. Singer and Jerome L. Singer, eds., *The Handbook of Children and Media* (Thousand Oaks, CA: Sage, 2001), 375–93; James Vincent, "Teens Can't Tell the Difference Between Google Ads and Search Results," The Verge, November 20, 2015.

5. Anup Shah, "Children as Consumers," Global Issues Blog, November 21, 2010; Juliet Schor, *Born to Buy: The Commercialized Child and the New Consumer Culture* (New York: Scribner, 2004), 21; Bruce Horovitz, "Marketing to Kids Gets More Savvy with New Technologies," *USA Today*, August 14, 2011, B1.

6. American Academy of Pediatrics Council on Communications and Media, "Media Use by Children Under the Age of Two," *Pediatrics* 128, no. 5 (October 2011): 1,040–45; Victoria Rideout, *Zero to Eight: Children's Media Use in America* (San Francisco, CA: Common Sense Media Inc., 2011), 44; Pooja S. Tandon et al., "Preschoolers' Total Daily Screen Time at Home and by Type of Child Care," *Journal of Pediatrics* 158, no. 2 (February 2011): 297–300; Common Sense Media Inc., *The Common Sense Census: Media Use by Tweens and Teens* (San Francisco, CA: 2015).

7. Debra Holt et al., *Children's Exposure to TV Advertising in 1977 and 2004* (Washington, DC: Federal Trade Commission, 2007); Jane Wakefield, "Children Spend Six Hours or More a Day on Screens," *BBC News*, March 27, 2015; Francisco Lupiáñez-Villanueva et al., *Study on the Impact of Marketing Through Social Media, Online Games and Mobile Applications on Children's Behavior* (Brussels: European Commission Consumers, Health, Agriculture and Food Executive Agency, 2016), 66.

8. Patrick Callan, "Fast Forward," *Kidscreen*, June 16, 2016.

9. Lily Hay Newman, "NY Cracks Down on Mattel and Hasbro for Tracking Kids Online," *Wired*, September 13, 2016; John Herrman, "Who's Too Young for an App? Musical.ly Tests the Limits," *New York Times*, September 17, 2016.

10. Campaign for a Commercial-Free Childhood (CCFC), "International Coalition to McDonald's: Keep Your Word and Ban Ronald McDonald from Schools," July 29, 2014.

11. T. Lobstein and S. Dibb, "Evidence of a Possible Link Between Obesogenic Food Advertising and Child Overweight," *Obesity Reviews* 6, no. 3 (2005): 203–08; Marjorie Hogan and Victor Strasburger, "Body Image, Eating Disorders, and the Media," *Adolescent Medicine: State of the Art Reviews* 19, no. 3 (2008): 521–46; Dimitri Christakis, "Virtual Violence: Council on Communications and Media," *Pediatrics* 138, no. 1 (July 2016); American Psychological Association, Task Force on the Sexualization of Girls, *Report of the APA Task Force*

on the Sexualization of Girls (Washington, DC: 2007); Moniek Buijzen and Patti Valkenburg, "The Effects of Television Advertising on Materialism, Parent–Child Conflict, and Unhappiness: A Review of Research," *Journal of Applied Developmental Psychology* 24, no. 4 (2003): 437–56.

12. Brad Tuttle, "Why We Are Eating Fewer Happy Meals," *Time*, April 23, 2012.

13. Tim Kasser, *The High Price of Materialism* (Cambridge, MA: The MIT Press, 2002).

14. Jean Twenge and Tim Kasser, "Generational Changes in Materialism and Work Centrality, 1976–2007: Associations with Temporal Changes in Societal Insecurity and Materialistic Role Modeling," *Personality and Social Psychology Bulletin* 39, no. 7 (July 2013): 883–97; Jeffrey Brand and Bradley Greenberg, "Commercials in the Classroom: The Impact of Channel One Advertising," *Journal of Advertising* 34, no. 1 (1994): 18–21; Vandana and Usha Lenka, "A Review on the Role of Media in Increasing Materialism Among Children," *Journal of Business Research* 133 (May 15, 2014): 456–64; Sara Kamal, Shu-Chuan Chu, and Mahmood Pedram, "Materialism, Attitudes, and Social Media Usage and Their Impact on Purchase Intention of Luxury Fashion Goods Among American and Arab Young Generations," *Journal of Interactive Advertising* 13, no. 1 (2013): 27–40.

15. William Kilbourne and Gregory Pickett, "How Materialism Affects Environmental Beliefs, Concern, and Environmentally Responsible Behavior," *Journal of Business Research* 61, no. 9 (September 2008): 885–93; Gregory Maio et al., "Changing, Priming, and Acting on Values: Effects via Motivational Relations in a Circular Model," *Journal of Personality and Social Psychology* 97, no. 4 (2009): 699–715.

16. CCFC, "Stop McTeacher's Nights," http://commercialfreechildhood.org/action/stop-mcteachers-nights.

17. Consumers Union, *Captive Kids: A Report on Commercial Pressure on Kids in School* (Yonkers, NY: 1998); Michele Simon, *Best Public Relations That Money Can Buy: A Guide to Food Industry Front Groups* (Washington, DC: Center for Food Safety, 2013); Sourcewatch (The Center for Media and Democracy), "American Farm Bureau Federation," www.sourcewatch.org/index.php/American_Farm_Bureau_Federation, viewed September 22, 2016; Timna Jacks and Henrietta Cook, "Big Business Muscles into Schools," *The Age*, September 17, 2016.

18. Kunkel, "Children and Television Advertising," 375–93.

19. Consumers International, "Advertising to Children Is Now Illegal in Brazil," April 10, 2014; Daybreak Kamloops, "Senator Nancy Green Raine Wants to Ban Junk Food Ads Directed at Kids," *CBC News*, September 29, 2016.

20. Sacha Pfeiffer, "An Ad or a Show? Some Say YouTube Kids Blurs the Line. Group Wants Rules for TV Ads Applied to Children's Video App," *Boston Globe*, April 21, 2015.

21. Farida Shaheed, *Report of the Special Rapporteur in the Field of Cultural Rights* (Boulder, CO: National Education Policy Center, University of Colorado, 2014); Center for a New American Dream, *Analysis Report: New American Dream Survey 2014* (Charlottesville, VA: 2014).

22. Bill Bigelow, "Scholastic Inc. Pushing Coal: A 4th Grade Curriculum Lies Through Omission," *Rethinking Schools* (Summer 2011): 30–33.

23. Bill Bigelow, "This Is What Happened When Scholastic Tried to Bring Pro-coal Propaganda into Schools," *Yes Magazine*, March 6, 2014; Tamar Lewin, "Scholastic InSchool Backing Off Corporate Ties," *New York Times*, August 1, 2011, A10.

24. Common Sense Media Inc., *The Common Sense Census.*

25. M. Marinelli et al., "Hours of Television Viewing and Sleep Duration," *JAMA Pediatrics* 168, no. 5 (May 2014): 458–64; Erik C. Landhuis et al., "Programming Obesity and Poor Fitness: The Long-term Impact of Childhood Television," *Obesity* 16, no. 6 (2008): 1,457–59; Yalda T. Uhls et al., "Five Days at Outdoor Education Camp Without Screens Improves Preteen Skills with Nonverbal Emotion Cues," *Computers in Human Behavior* 39

(October 2014): 387–92. Box 13–1 from "McDonald's Japan Posts Profit Boosted by Pokémon Go Partnership," *Advertising Age*, August 9, 2016.

Chapter 14. Home Economics Education: Preparation for a Sustainable and Healthy Future

1. Juliet Schor, "Foreword," in Arjen E. J. Wals and Peter Blaze Corcoran, eds., *Learning for Sustainability in Times of Accelerating Change* (Wageningen, The Netherlands: Wageningen Academic Publishers, 2012), 15–18; Damian Carrington, "The Anthropocene Epoch: Scientists Declare Dawn of Human-influenced Age," *The Guardian* (U.K.), August 29, 2016.

2. Amanda McCloat and Helen Maguire, "Reorienting Home Economics Teacher Education to Address Education for Sustainable Development," in Miriam O'Donoghue, *Global Sustainable Development: A Challenge for Consumer Citizens*, e-book, 2008; Eleanore Vaines, "Wholeness, Transforming Practices and Everyday Life," in Mary Gale Smith, Linda Peterat, and Mary Leah de Zwart, eds., *Home Economics Now: Transformative Practice, Ecology and Everyday Life* (Vancouver, BC: Pacific Educational Press, 2004), 133–65.

3. Sue L. T. McGregor, "Everyday Life: A Home Economics Concept," *Kappa Omicron Nu FORUM* (National Honor Society for the Human Sciences) 19, no. 1 (2012); Vaines, "Wholeness, Transforming Practices and Everyday Life."

4. Sue L. T. McGregor, *Locating the Human Condition Concept Within Home Economics*, McGregor Monograph Series No. 201002 (Halifax, NS, Canada: 2010), 240; Irish Department of Education and Skills, *Leaving Certificate: Home Economics Scientific & Social Syllabus* (Dublin: The Stationery Office, 2001), 2; Irish Department of Education and Skills, *The Junior Certificate Home Economics Syllabus* (Dublin: The Stationery Office, 2002).

5. Roland Tormey et al., "Working in the Action/Research Nexus for ESD: Two Case Studies from Ireland," *International Journal of Sustainability in Higher Education* 9, no. 4 (2008): 428–40; International Federation for Home Economics, "IFHE Position Statement: Home Economics in the 21st Century" (Bonn, Germany: 2008); Helen Maguire et al., "Images and Objects: A Tool for Teaching Education for Sustainable Development and Responsible Living in Home Economics," in Ulf Schrader et al., eds., *Enabling Responsible Living* (Berlin: Springer, 2013).

6. Suzanne Piscopo and Karen Mugliett, "Redefining and Repackaging Home Economics: Case of a Mediterranean Island," *Victorian Journal of Home Economics* 53, no. 1 (2014): 2; Japan Association of Home Economics Education, *Home Economics Education in Japan 2012* (Tokyo: 2012); Finnish National Board of Education, "Part IV: Chapters 7.10–7.21," in *National Core Curriculum for Basic Education 2004* (Vammala, Finland: 2004); Skolverket (Swedish National Agency for Education), *Sweden: Curriculum for the Compulsory School, Preschool Class and the Recreation Centre 2011* (Stockholm: 2011).

7. Consumer Classroom, "Resources," www.consumerclassroom.eu/online-teaching-resources.

8. University of Kentucky, College of Agriculture, Food and Environment, School of Human Environmental Sciences, Family & Consumer Sciences Extension, *Building Strong Families for Kentucky: 2014* (Lexington, KY: 2014). Box 14–1 from the following sources: Isadore Reaud, personal communication with author, September 19, 2016; Pornpimol Kanchanalak, "It's a Bird, It's a Plane, It's the Bamboo School," *The Nation*, October 9, 2014; Mechai Viravaidya Foundation, "Mechai Bamboo School," www.mechaifoundation.org/index2.php, viewed October 15, 2016; Mechai Viravaidya, "Mechai Pattana Bamboo School – Buriram, Thailand," video, September 14, 2012, www.youtube.com/watch?v=kpuPr54kJBU.

9. Presentation Secondary Mitchelstown, "Green Schools," http://presmitchelstown.ie/?page_id=1805.

10. Glenamaddy Community School, "Home Economics," www.glenamaddycs.ie/index.php/subject-depart ments/home-economics.

11. Gaelic Athletic Association (GAA), "Healthy Club Focus: GAA Recipes for Success!" March 2, 2016, www

.gaa.ie/gaa-tv/healthy-club-focus-gaa-recipes-for-success/.

12. Utah Education Network, "Family and Consumer Sciences: Classroom & Laboratory Management," www .uen.org/cte/family/class.

13. Adam Vaughan, "Failure to Teach Cooking at School 'Contributing to £12bn a Year Food Waste,'" *The Guardian* (U.K.), July 13, 2016; Hedmark University of Applied Sciences, "The Partnership for Education and Research About Responsible Living (PERL/UNITWIN)," http://eng.hihm.no/project-sites/living-respon sibly; Little Flower Girls' School, "Fair Trade Does Great Trade," www.littleflowerschool.co.uk/about/latest -news/241-fair-trade-does-great-trade.

14. St. Aidan's Comprehensive School, "First Year Textile Projects," www.staidans.ie/first-year-textile-projects .html.

15. Heathcote High School, "Home Economics," www.heathcote-h.schools.nsw.edu.au/curriculum-activities /faculties/home-economics.

16. Donna Pendergast, "Sustaining the Home Economics Profession in New Times: A Convergent Moment," in Anna-Liisa Rauma, Sinikka Pöllänen, and Pirita Seitamaa-Hakkarainen, eds., *Human Perspectives on Sustainable Future,* Proceedings of the 5th International Household and Family Research Conference (Joensuu, Finland: University of Joensuu, 2006), 3–32; Donna Pendergast, "The Intention of Home Economics Education: A Power-ful Enabler for Future Proofing the Profession," in Donna Pendergast, Sue L. T. McGregor, and Kaija Turkki, *Creating Home Economics Futures: The Next 100 Years* (Samford Valley, Queensland, Australia: Australian Academic Press, 2012), 12–24; Terttu Tuomi-Grohn, "Everyday Life as a Challenging Sphere of Research, An Introduction," in Terttu Tuomi-Grohn, ed., *Reinventing Art of Everyday Making* (Frankfurt: Peter Lang, 2008), 7.

17. Tuomi-Grohn, "Everyday Life as a Challenging Sphere of Research," 9.

18. Box 14–2 from the following sources: World Health Organization (WHO), "Global Database of Age-Friendly Practices," https://extranet.who.int/datacol/custom_view_report.asp?survey_id=3536&view_id=6301&display _filter=1; Tine Buffel et al., "Promoting Sustainable Communities Through Intergenerational Practice," *Procedia – Social and Behavioral Sciences* 116 (February 21, 2014): 1,785–91; Ann Kristin Boström, *Lifelong Learning, Intergenerational Learning, and Social Capital* (Stockholm: Institute of International Education, Stockholm University, 2003); Alan Hatton-Yeo and Clare Batty, "Evaluating the Contribution of Intergenerational Practice," in Peter Ratcliffe and Ines Newman, *Promoting Social Cohesion: Implications for Policy and Evaluation* (Bristol, U.K.: Policy Press, 2011); Mariano Sanchez et al., "Intergenerational Programmes: Towards a Society for All Ages," *Journal of Intergenerational Relationships* 6, no. 4 (2008): 485–87; Judi Aubel, "Elders: A Cultural Resource for Promoting Sustainable Development," in Worldwatch Institute, *State of the World 2010: Transforming Cultures* (Washington, DC: Island Press, 2010); Zohl de Ishtar, "Elders Passing Cultural Knowledge to Their Young Women," Kapululangu Aboriginal Women Law and Culture Centre, December 9, 2012; Wendy Stueck, "Seabird Island Band's Walks in Woods Aim to Pass Down Aboriginal Heritage," *Globe and Mail* (Toronto), April 13, 2016; Jayalaxshmi Mistry and Andrea Berardi, "Bridging Indigenous and Scientific Knowledge," *Science* 352, no. 6291 (June 10, 2016): 1,274–75; Ben Goldfarb, "Researchers Around the World Are Learning from Indigenous Communities. Here's Why That's a Good Thing," *Ensia*, May 31, 2016; Nathalie Fernbach and Harriet Tatham, "Indigenous Knowledge and Western Science Unite to Save the Reef," *ABC News*, June 2, 2016; Donald Huis-ingh, "New Challenges in Education for Sustainable Development," *Clean Technology and Environmental Policy* 8, no. 15 (February 3–8, 2006); D'Vera Cohn and Jeffrey S. Passel, "A Record 60.6 Million Americans Live in Multigenerational Households," Pew Research Center, August 11, 2016; International Longevity Centre Global Alliance, *Global Perspectives on Multigenerational Households and Intergenerational Relations* (London: International Longevity Centre–UK, March 2012); Sally Newman and Alan Hatton-Yeo, "Intergenerational Learning and the Contributions of Older People," *Ageing Horizons* 8 (2008): 31–39; WHO, "WHO Global Network for Age-friendly Cities and Communities," www.who.int/ageing/projects/age_friendly_cities_network/en/; Tiffany R. Jansen, "The Nursing Home That's Also a Dorm," CityLab.com, October 2, 2015; Lacy Cooke, "New Dutch

Housing Model Lets Students Stay at a Senior Living Home for Free," Inhabit, September 23, 2016; European Map of Intergenerational Learning website, www.emil-network.eu; Kyle Wiens, "Why Seniors Are the Heroes of the Fixer Movement," iFixit.org, June 14, 2014; Martin Charter and Scott Keiller, *Grassroots Innovation and the Circular Economy: A Global Survey of Repair Cafés and Hackerspaces* (Surrey, U.K.: Centre for Sustainable Design, University for the Creative Arts, 2014); Repair Café, "About Repair Café," https://repaircafe.org/en/about/; WHO, *World Health Report: Research for Universal Health Coverage* (Geneva: 2013); Donald Ropes, "Intergenerational Learning in Organizations: An Effective Way to Stimulate Older Employee Learning and Development," *Development and Learning in Organizations* 28, no. 2 (2014): 7–9; Lisa Quast, "Reverse Mentoring: What It Is and Why It Is Beneficial," *Forbes*, January 3, 2011; Jane Wakefield, "Technology in Schools: Future Changes in Classrooms," *BBC News*, February 2, 2015.

Chapter 15. Our Bodies, Our Future: Expanding Comprehensive Sexuality Education

1. Saskia de Melker, "The Case for Starting Sex Education in Kindergarten," PBS Newshour, May 27, 2015, www .pbs.org/newshour/updates/spring-fever/.

2. Future of Sex Education, "Definition of Comprehensive Sex Education," www.futureofsexed.org/definition .html; United Nations Educational, Scientific and Cultural Organization (UNESCO), *International Technical Guidance on Sexuality Education*, Vol. 1 (New York: 2009), 2. Box 15–1 from United Nations Population Fund (UNFPA), *UNFPA Operational Guidance for Comprehensive Sexuality Education: A Focus on Human Rights and Gender* (New York: 2014), 6.

3. Anastasia Moloney, "Incest, Lack of Sex Education Drive Teen Pregnancies in El Salvador," *Reuters*, May 2, 2016; International Planned Parenthood Federation, Western Hemisphere Region, "Five Sex Education Successes in Latin America," April 1, 2014, www.ippfwhr.org/en/blog/five-sex-education-successes-in-latin-america.

4. World Health Organization, "Sexual and Reproductive Health: Defining Sexual Health," www.who.int /reproductivehealth/topics/sexual_health/sh_definitions/en/.

5. Sonia Perilla and José Alberto Mojica, "'Decir Ideología de Género Es Tergiversación,' Dice Naciones Unidas," *El Tiempo*, August 24, 2016; "La Encrucijada de la Ministra Gina Parody," *Semana*, August 13, 2016; Wolfgang Lutz, Director, World Population Program, International Institute for Applied Systems Analysis, "Population, Education and the Sustainable Development Goals," presented at 2016 meeting of the American Association for the Advancement of Science, Washington, DC, February 11–15, 2016; Mary A. Ott and John S. Santelli, "Abstinence and Abstinence-only Education," *Current Opinion in Obstetrics & Gynecology* 19, no. 5 (October 2007): 446–52.

6. United Nations Sustainable Development Knowledge Platform, "Goal 4, Target 4.7," https://sustainabledevel opment.un.org/sdg4; Restless Development, "Have You Seen My Rights?" December 21, 2015, restlessdevelop ment.org/news/2015/12/21/spotlight-on-have-you-seen-my-rights.

7. Population data calculated from United Nations, Department of Economic and Social Affairs, Population Division, *World Population Prospects: The 2015 Revision*, DVD Edition, Excel workbooks "WPP2015_POP _F01_1_TOTAL_POPULATION_BOTH_SEXES" and "WPP2015_POP_F07_1_POPULATION_BY_AGE _BOTH_SEXES," available at https://esa.un.org/unpd/wpp.

8. Gilda Sedgh et al., "Intended and Unintended Pregnancies Worldwide in 2012 and Recent Trends," *Studies in Family Planning* 45, no. 3 (September 2014): 301–14; Paul A. Murtaugh and Michael G. Schlax, "Reproduction and the Carbon Legacies of Individuals," *Global Environmental Change* 19, no. 1 (February 2009): 14–20.

9. Richard Kollodge, ed., *The Power of 1.8 Billion: Adolescents, Youth and the Transformation of the Future* (New York: United Nations, 2014).

10. Rutgers WPF, "Sexuality Education for Young Children," 2015, www.rutgerswpfindo.org/en/module/sexual

ity-education-for-young-children; UNFPA, *The Evaluation of Comprehensive Sexuality Education Programmes: A Focus on the Gender and Empowerment Outcomes* (New York: 2015), 34–35.

11. UNFPA, *UNFPA Operational Guidance for Comprehensive Sexuality Education*, 6; Lesley R. Craft, Heather M. Brandt, and Mary Prince, "Sustaining Teen Pregnancy Prevention Programs in Schools: Needs and Barriers Identified by School Leaders," *Journal of School Health* 86, no. 4 (April 2016): 258–65.

12. UNESCO, *Education 2030: Incheon Declaration and Framework for Action: Towards Inclusive and Equitable Quality Education and Lifelong Learning for All* (Paris: 2015).

13. Federal Centre for Health Education (BZgA), *Standards for Sexuality Education in Europe: A Framework for Policy Makers, Educational and Health Authorities and Specialists* (Cologne, Germany: 2010).

14. UNFPA, *UNFPA Operational Guidance for Comprehensive Sexuality Education*; Nicole Haberland, "The Case for Addressing Gender and Power in Sexuality and HIV Education: A Comprehensive Review of Evaluation Studies," *International Perspectives on Sexual and Reproductive Health* 41, no. 1 (2015): 31–42; UNFPA, unpublished report of a conference on CSE advocacy, 2016.

15. Individual studies are cited in UNESCO, *International Technical Guidance on Sexuality Education*, 15, 17; Deborah Rogow and Nicole Haberland, "Sexuality and Relationships Education: Toward a Social Studies Approach," *Sex Education* 5, no. 4 (November 2005): 333–44.

16. UNESCO, *Emerging Evidence, Lessons and Practice in Comprehensive Sexual Education: A Global Review* (Paris: 2015), 38.

17. *Ministerial Declaration: Preventing Through Education*, first meeting of Ministers of Health and Education to Stop HIV and STIs in Latin America and the Caribbean (Mexico City: 1 August 2008); Maria Antonieta Alcalde, presentation on comparative progress in the implementation of *Preventing Through Education*, 2008–2015, in UNFPA, unpublished report of a conference on CSE advocacy.

18. United Nations, *Montevideo Consensus on Population and Development* (Montevideo, Uruguay: August 12–15, 2013).

19. UNESCO, *Young People Today. Time to Act Now.* (Paris: 2013).

20. UNFPA, "Y-PEER: Empowering Young People to Empower Each Other," May 26, 2006, www.unfpa.org /news/y-peer-empowering-young-people-empower-each-other.

21. Authors' calculation based on United Nations, *World Population Prospects: The 2015 Revision*, Excel workbook "WPP2015_POP_F07_1_POPULATION_BY_AGE_BOTH_SEXES"; UNESCO, *Sexuality Education in Asia and the Pacific: Review of Policies and Strategies to Scale Up* (Bangkok: 2012); Jo Sauvarin, UNFPA Bangkok office, personal communication with Mona Kaidbey; Abby Young-Powell, "Six of the Best Sex Education Programmes Around the World, *The Guardian* (U.K.), May 20, 2016.

22. UNESCO, *School-based Sexuality Education Programmes: A Cost and Cost-effectiveness Study in Six Countries* (Paris: 2011); Evert Ketting, on behalf of the European Expert Group on Sexuality Education, "Impact Assessment of Holistic Sexuality Education Programme in Estonia," in UNFPA, *The Evaluation of Comprehensive Sexuality Education Programs* (New York: 2015); Kai Haldre, Kai Part, and Evert Ketting, "Youth Sexual Health Improvement in Estonia, 1990–2009: The Role of Sexuality Education and Youth-friendly Services," *European Journal of Contraception & Reproductive Health Care* 17, no. 5 (2012): 351–62.

23. Evert Ketting, Minou Friele, and Kristien Micheilsen, "Evaluation of Holistic Sexuality Education: A European Expert Group Consensus Agreement," *European Journal of Contraception & Reproductive Health Care* 21, no. 1 (2016): 68–80.

24. Advocates for Youth, *Adolescent Sexual Health in Europe and the United States* (undated), www.advocates

foryouth.org/storage/advfy/documents/adolescent_sexual_health_in_europe_and_the_united_states.pdf; World Bank, "Adolescent Fertility Rate (Births per 1,000 Women 15–29)," data.worldbank.org/indicator/SP.ADO .TFRT.

Chapter 16. Suddenly More Than Academic: Higher Education for a Post-Growth World

1. Anthony Cortese, "Afterword," in Mitchell Thomashow, *The Nine Elements of a Sustainable Campus* (Cambridge, MA: The MIT Press, 2014), 212.

2. Will Steffen et al., "Planetary Boundaries: Guiding Human Development on a Changing Planet," *Science* 347, no. 6223 (January 15, 2015): 736–46.

3. Juliet Schor, "What Is Sustainable Consumption," presentation at No Growth: Slower by Design, Not Disaster Roundtable, York University, Toronto, Canada, May 2010; Juliet Schor, *True Wealth: How and Why Millions of Americans Are Creating a Time-Rich, Ecologically Light, Small-Scale, High-Satisfaction Economy* (New York: Penguin, 2011); Richard Norgaard, "The Church of Economism and Its Discontents," The Great Transition Network, December 2015; Gus Speth, "Manifesto for a Post-Growth Economy," *Yes! Magazine*, September 19, 2012.

4. Speth, "Manifesto for a Post-Growth Economy."

5. Neil Irwin, "A Low Growth World: One Key to Persistent Economic Anxiety," *New York Times*, August 7, 2016, A1. Box 16–1 from the following sources: Richard Dobbs et al., *Poorer Than Their Parents? A New Perspective on Income Inequality* (McKinsey Global Institute, July 2016); Robinson Meyer, "Donald Trump Is the First Demagogue of the Anthropocene," *The Atlantic*, October 19, 2016; Christopher Hayes, *Twilight of the Elites: America After Meritocracy* (New York: Broadway Books, 2013); Michael Maniates, "Teaching for Turbulence," in Worldwatch Institute, *State of the World 2013: Is Sustainability Still Possible?* (Washington, DC: Island Press, 2013), 255–68; Karen Litfin, "Framing Science: Precautionary Discourse and the Ozone Treaties," *Millennium: Journal of International Studies* 24, no. 2 (1995): 251–77.

6. Robert Costanza et al., "Time to Leave GDP Behind," *Nature* 505 (January 16, 2014): 283–85.

7. John Gill, "Universities and Economic Growth Go Together," *Times Higher Education*, March 31, 2016; Elisa Stephens, "Higher Education and America's Economic Growth," *Huffington Post*, May 25, 2011; Johan Rockström and Mattias Klum, *Big World, Small Planet: Abundance Within Planetary Boundaries* (New Haven, CT: Yale University Press, 2015).

8. Kevin Krizek et al., "Higher Education's Sustainability Imperative: How to Practically Respond?" *International Journal of Sustainability in Higher Education* 13, no. 1 (2012): 19–33; Nancy Kurland, "Evolution of a Campus Sustainability Network: A Case Study in Organizational Change," *International Journal of Sustainability in Higher Education* 12, no. 4 (2011): 395–429.

9. Association for the Advancement of Sustainability in Higher Education (AASHE), *Sustainable Campus Index: 2016 Top Performers and Highlights* (Philadelphia, PA: October 2016); UI Green Metric website, http://greenmet ric.ui.ac.id; AASHE website, www.aashe.org.

10. Box 16–2 based on Eve Bratman et al., "Justice Is the Goal: Divestment as Climate Change Resistance," *Journal of Environmental Studies and Sciences* 6, no. 4 (2016): 1–14. See also Fossil Free, "Divestment Commitments," www.GoFossilFree.org/commitments.

11. Robert Frank, *Are Arms Races in Higher Education a Problem?* (Boulder, CO: EDUCAUSE Center for Analysis and Research (ECAR), 2004).

12. Ronald Yanosky, *Shelter from the Storm: IT and Business Continuity in Higher Education* (Boulder, CO: ECAR, 2007); Elizabeth and Chris Smith, "Integrating Resilience Planning into University Campus Planning: Measuring Risks and Leveraging Opportunities," *Planning for Higher Education* 44, no. 1 (2015): 10–19; Ann Waple,

"How Higher Ed Works with Communities to Build Resilience," Greenbiz.org, August 11, 2014; Second Nature, "Second Nature Announces College Leadership in Resilience; Alliance for Resilient Campuses, Partnership with Cities," press release (Boston, MA: May 5, 2014).

13. Gar Alperovitz, "The Political-Economic Foundations of a Sustainable System," in Worldwatch Institute, *State of the World 2014: Governing for Sustainability* (Washington, DC: Island Press, 2013), 191–202; Marjorie Kelly, *Owning Our Future: The Emergent Ownership Revolution* (San Francisco, CA: Berrett-Koehler Publishers, 2012).

14. Joe Guinan, Sarah McKinley, and Benzamin Yi, *Raising Student Voices: Student Action for University Community Investment* (Brooklyn, NY and Takoma Park, MD: Responsible Endowments Coalition and The Democracy Collaborative, 2013).

15. Peter Victor, *Managing Without Growth: Slower by Design, Not Disaster* (Cheltenham, U.K.: Edward Elgar Publishing, 2008); Tim Jackson, *Prosperity Without Growth: Economics for a Finite Planet* (Abingdon, Oxon, U.K.: Earthscan/Routledge, 2011); Schor, *True Wealth*; Shana Lebowitz, "Here's How the 40-hour Workweek Became the Standard in America," *Business Insider*, October 24, 2015.

16. Rebecca Greenfield, "The Six-Hour Workday Works in Europe. What About America?" *Bloomberg*, May 10, 2016; Isabel Sawhill, "Is It Time for a Shorter Workweek?" *Washington Post*, May 13, 2016.

17. Michael Maniates, "Editing Out Unsustainable Behavior," in Worldwatch Institute, *State of the World 2010: Transforming Cultures* (New York: W. W. Norton & Company, 2010), 119–26; Lindsey Ramsey, "Sustainability in Practice: Bottled Water Bans Gain Traction," FoodService Director, May 9, 2012; Yingchen Kwok et al., *Yale-NUS College Dining Hall Report 2015* (Singapore: Yale-NUS College, 2016); Tom Van Heeke, Elise Sullivan, and Phineas Baxandall, *A New Course: How Innovative University Programs Are Reducing Driving on Campus and Creating New Models for Transportation* (Washington, DC: U.S. PIRG Education Fund, February 2014).

18. Jennifer Washburn, *University, Inc.: The Corporate Corruption of Higher Education* (New York: Basic Books, 2008).

19. Box 16–3 from the following sources: The Land Institute website, https://landinstitute.org; Wes Jackson, "The Serious Challenge of Our Time," *Resilience*, May 31, 2013; "Paradigm Shift U: Talk of an Education and Worldview Called Ecospheric, in Results If Not in Name," *Land Report* (The Land Institute) 112 (Summer 2015).

20. Robert Reich, *Supercapitalism: The Transformation of Business, Democracy, and Everyday Life* (New York: Vintage Books, 2008).

Chapter 17. Bringing the Classroom Back to Life

1. Swaraj University, "Year 1," www.swarajuniversity.org/year-wise-flow.html.

2. Deep Time Walk website, http://deeptimewalk.org.

3. "Free Home University," www.facebook.com/freehomeuni/.

4. Stephen Sterling, "Sustainable Education: Towards a Deep Learning Response to Unsustainability," *Policy and Practice*, no. 6 (Spring 2008): 64.

5. Judi Marshall, Gill Coleman, and Peter Reason, eds., *Leadership for Sustainability: An Action Research Approach* (Sheffield, U.K.: Greenleaf Publishing, 2011), 14; Donna Trueit, ed., *Pragmatism, Post-Modernism and Complexity Theory* (London: Routledge, 2012), 197–98; Richard Buchanan, "Wicked Problems in Design Thinking," *Design Issues* 8, no. 2 (Spring 1992): 5–21.

6. Peter Reason and Hilary Bradbury, eds., *Handbook of Action Research: Participative Inquiry and Practice* (London: Sage, 2015).

7. Allan Kaplan and Sue Davidoff, *A Delicate Activism: A Radical Approach to Change* (Johannesburg: Proteus

Initiative, 2014), 26.

8. Trueit, ed., *Pragmatism, Post-Modernism and Complexity Theory*, 239. Box 18–1 based on the following sources: Richard J. Light, *Making the Most of College: Students Speak their Mind* (Boston: Harvard University Press, 2001); Eric Mazur, *Peer Instruction: A User's Manual Series in Educational Innovation* (Upper Saddle River, NJ: Prentice Hall, 1997); Jane Eberle and Marcus Childress, "Heutagogy: It Isn't Your Mother's Pedagogy Any More," *National Social Science Journal* 28, no. 1 (2007): 28–32; Sabine O'Hara, *The UNISA Signature Curriculum Project: Implementing UNISA's Academic Identity as The African University in the Service of Humanity* (Washington, DC: Global Ecology LLC, 2012); Sabine O'Hara, "Local Commitments, Global Reach: Advancing an Ethic of Sustainability," in Divya Singh, ed., *Globethics: Responsible Leadership in Higher Education* (forthcoming 2017); Lisa Marie Blaschke, Program Director, Carl von Ossietzky University of Oldenburg, "Sustaining Lifelong Learning: A Review of Heutagogical Practice," EDEN presentation in Dublin, Ireland, 2011. See also Sabine O'Hara, "Sustainable Urban Agriculture," UDC course numbers ENVS 452 and ENVS 453, and E. Harrison, "Exploration and Inquiry: Capstone," UDC course numbers IGED 391 and IGED 392.

9. "An Invitation: Eco-versities Network: A Gathering of Kindred Folk Reimagining Higher Education," April 10, 2015, www.swaraj.org/shikshantar/Newecoversitiesinvite.pdf.

10. Gaia University website, www.gaiauniversity.org.

11. Red Crow Community College website, www.redcrowcollege.com.

12. Universidades de La Tierra website, http://universidadesdelatierra.org; Mike Emiliani, "In Rural Mexico, Student-Led Education Heals Old Wounds," *Yes! Magazine*, January 11, 2013.

13. Parker J. Palmer, *The Courage to Teach: Exploring the Inner Landscape of a Teacher's Life* (San Francisco, CA: Wiley and Sons, 2007), 25.

14. John Dewey, "The Quest for Certainty," quoted in Buchanan, "Wicked Problems in Design Thinking," 6.

15. Francisco Varela et al., *The Embodied Mind: Cognitive Science and Human Experience* (Cambridge, MA: The MIT Press, 1993); George Lakoff and Mark Johnson, *The Metaphors We Live By* (Chicago, IL: University of Chicago Press, 2003).

16. Daniella Tilbury, *Education for Sustainable Development: An Expert Review of Processes and Leaning* (Paris: United Nations Educational, Scientific and Cultural Organization (UNESCO), 2011).

17. Efrat Eilam and Tamar Trop, "ESD Pedagogy: A Guide for the Perplexed," *Journal of Environmental Education* 42, no. 1 (2011): 43–64.

18. George Monbiot, "Why We Fight for the Living World: It's About Love, and It's Time We Said So," *The Guardian* (U.K.), June 16, 2015.

19. Jonathan Dawson, "Bringing the Classroom to Life," TEDxFindhorn, May 29, 2015, www.youtube.com/watch?v=q3jxrT9VQ_w.

20. Kenneth J. Gergen, *Social Construction and Pedagogical Practice*, unpublished paper, 10, www.swarthmore.edu/Documents/faculty/gergen/Social_Construction_and_Pedagogical_Practice.pdf.

21. Stephen Allen, "Reflexivity for Sustainability: Appreciating Entanglement and Becoming Relationally Reflective," *International Journal of Work Innovation* 1, no. 2 (January 2015): 240–50.

22. Innovation Hub website, www.theinnovationhub.com.

23. Gaia Education website, www.gaiaeducation.org/index.php/en.

24. CEAL-Network website, http://ceal.eu.

25. Box 18–2 from the following sources: Morten Asfeldt and Glen Hvenegaard, "Perceived Learning, Critical Elements and Lasting Impacts on University-based Wilderness Educational Expeditions," *Journal of Adventure Education and Outdoor Learning* 14, vol. 2 (2014); Lisbeth Grundy and Bud Simpkin, "Working with the Youth Service," in John Huckle and Stephen Sterling, eds., *Education for Sustainability* (London: Earthscan, 2001); Archibald Sia, Harold Hungerford, and Audrey Tomera, "Selected Predictors of Responsible Environmental Behavior: An Analysis," *Journal of Environmental Education* 17, no. 2 (1985/86): 31–41; Takako Takano, "A 20-year Retrospective Study of the Impact of Expeditions on Japanese Participants," *Journal of Adventure Education and Outdoor Learning* 10, no. 2 (2010): 77–94; Takako Takano, *Nohsanson-no-hito-to-kurashi-ga-sasaeru-chiiki-no-kyoikuryoku (Educational Power of Community Based on People and Life in Rural Villages)* (Tokyo: Ecoplus, 2011), 1–26.

26. Gregory Bateson, *Steps to an Ecology of Mind* (Aylesbury, U.K.: Intertext, 1972), 17.

27. Jim Garrison, "The Holy Grail in Education," *Huffington Post*, March 4, 2014. A review of a survey of 42,257 students aged 18–25 (the millennial generation) from one hundred countries identified substantial disaffection with conventional formal university education. It found that 53 percent of interviewees see a disconnection between what they are learning today versus what they will need tomorrow. The review concluded that, "Universities will need to transform themselves into a place where young people can not only study and take exams, but learn from doing. To provide them with real-world experiences that are relevant." See YouthSpeak and AISEC (in partnership with PriceWaterhouseCoopers), *Improving the Journey from Education to Employment: YouthSpeak Survey Millennial Insight Report* (Rotterdam: 2015), 32; Stefan Collini, *What Are Universities For?* (London: Penguin Books, 2012); Cristina Escrigas, *A Higher Calling for Higher Education* (Great Transition Initiative, June 2016).

28. Box 18–3 from the following sources: Edgar Morin, *Homeland Earth* (New York: Hampton Press, 2002); Moacir Gadotti, *Pedagogia da Terra*, second edition (São Paulo, Brazil: Peirópolis, 2010). The Paulo Freire Institute in São Paulo launched the *Ecopedagogy Letter* at an international meeting in 1999; Howard Gardner, *Multiple Intelligences* (New York: Basic Books, 2006); Daniel Goleman, *Ecological Intelligence* (New York: Broadway Books, 2009); Daniel Goleman, Lisa Bennett, and Zenobia Barlow, *Ecoliterate: How Educators Are Cultivating Emotional, Social and Ecological Intelligence* (Berkeley, CA: Center for Ecoliteracy, 2012); UNESCO, *Learning: The Treasure Within* (Paris: 1996); Rafael Díaz-Salazar, *Educación y cambio ecosocial* (Madrid: PPC Editorial, 2016).

Chapter 18. Preparing Vocational Training for the Eco-Technical Transition

1. Kevin Carey, *The End of College: Creating the Future of Learning and the University of Everywhere* (New York: Riverhead Books, 2015); David N. F. Bell and David G. Blanchflower, *Young People and the Great Recession* (Bonn, Germany: Institute for the Study of Labor, April 2011); International Labour Organization (ILO), *Global Employment Trends for Youth 2015: Scaling Up Investments in Decent Jobs for Youth* (Geneva: 2015), 1–4.

2. Organisation for Economic Co-operation and Development (OECD), *Skills Beyond School Synthesis Report* (Paris: 2014), 14.

3. Nancy Lee Wood, "Community Colleges: A Vital Resource for Education in the Post-Carbon Era," in Richard Heinberg and Daniel Lerch, eds., *The Post-Carbon Reader: Managing the 21st Century's Sustainability Crises* (Healdsburg, CA: Watershed Media, 2010).

4. United Nations Educational, Scientific and Cultural Organization (UNESCO), "Technical Vocational Education and Training," www.unesco.org/new/en/education/themes/education-building-blocks/technical-vocational-education-and-training-tvet/; UNESCO, "Education for Sustainable Development," http://en.unesco.org/themes/education-sustainable-development; United Nations Sustainable Development Goals, "Goal 4: Ensure Inclusive and Quality Education for All and Promote Lifelong Learning," www.un.org/sustainabledevelopment/education.

5. Tamar Jacoby, "Why Germany Is So Much Better at Training Its Workers," *The Atlantic*, October 16, 2014.

6. Box 18–1 from Barefoot College, "The Barefoot Story," www.barefootcollege.org/about, and from Bunker Roy, "Learning from a Barefoot Movement," TED Talk, July 2011, www.ted.com/talks/bunker_roy.

7. Green Worker Cooperatives website, www.greenworker.coop.

8. International Renewable Energy Agency, *Renewable Energy Jobs – Annual Review 2016* (Abu Dhabi, United Arab Emirates: 2016).

9. Renewable Energy Jobs, "Complete Guide to Renewable Energy Training and Education," www.renewableen-ergyjobs.com/content/complete-guide-to-renewable-energy-training-and-education; Solar Living Institute website, www.solarliving.org; University of Central Lancashire, "Energy and Environmental Management," www.uclan.ac.uk/courses/msc__pgdip_pgcert_energy_and_environmental_management.php.

10. Bristol Community College (BCC), "Programs of Study," http://bristolcc.smartcatalogiq.com/en/2016-2017/Catalog/Programs-of-Study.

11. BCC, "EGR 102 Introduction to Sustainable and Green Energy Technologies," http://bristolcc.smartcatalogiq.com/2016-2017/catalog//courses/egr-engineering/100/egr-102; BCC, "EGR 123 Green Building Practices," http://bristolcc.smartcatalogiq.com/2016-2017/catalog//courses/egr-engineering/100/egr-123.

12. American Society of Civil Engineers, "Guided Online Courses," www.asce.org/continuing-education/guided-online-courses; Green Training, "Advanced Building Assessment & Retrofit Strategies," www.greentrainingusa.com/advanced-building-assessment-and-retrofit-strategies.html.

13. Massachusetts Institute of Technology, Transportation@MIT, "About," http://transportation.mit.edu/about.

14. University of California–Davis Graduate Studies, "Transportation Technology and Policy," https://gradstudies.ucdavis.edu/programs/gttp; Judith Cruz et al., *Sustainable Transportation Curricula* (Davis, CA: National Center for Sustainable Transportation, 2015).

15. Biodiversity for a Livable Climate website, www.bio4climate.org; Healthy Soils Australia website, www.healthysoilsaustralia.org.

16. Box 18–2 based on Mark Ritchie, "Sustainability Education, Experiential Learning, and Social Justice: Designing Community Based Courses in the Global South," *Journal of Sustainability Education* 5 (May 11, 2013), and on communications with community leaders Padti Saju in Huay Hee Village, Mae Hong Son, Northern Thailand, and Loong Prapat, Mae Taa Community, Lampang, Northern Thailand.

17. The Savory Institute website, http://savory.global/institute.

18. The Savory Institute, *Annual Report 2015: Growth Acceleration* (Boulder, CO: 2015); Africa Centre for Holistic Management, "What We Do," www.africacentreforholisticmanagement.org/what-we-do.html.

19. Growing Power website, www.growingpower.org.

20. Trees for the Future website, www.treesforthefuture.org.

21. Box 18–3 from the following sources: Roosevelt quote from *The New York Times*, May 7, 1934, as cited in Allida Black, *Casting Her Own Shadow: Eleanor Roosevelt and the Shaping of Postwar Liberalism* (New York: Columbia University Press, 1996), 29–33; Eleanor Roosevelt Papers Project, George Washington University, "Teaching Eleanor Roosevelt Glossary: National Youth Administration," www.gwu.edu/~erpapers/teachinger/glossary/nya.cfm, viewed September 19, 2016; Civilian Conservation Corps Legacy, "CCC Brief History," www.ccclegacy.org/CCC_Brief_History.html, viewed September 19, 2016; U.S. Bureau of Labor Statistics, "Labor Force Statistics from the Current Population Survey," www.bls.gov/web/empsit/cpseea03.htm, updated September 2, 2016; Federal Reserve Bank of New York Research and Statistics Group, *Microeconomic Studies, Quarterly Report on*

Household Debt and Credit (New York: May 2016); Ipsos Public Affairs, "Rock the Vote/USA Today Millennial Survey," conducted January 4–7, 2016; The Corps Network website, www.corpsnetwork.org; Green Corps website, www.greencorps.org; Corporation for National & Community Service, "Resilience AmeriCorps," www .nationalservice.gov/programs/americorps/americorps-initiatives/resilience-americorps.

22. Peace and Justice Studies Association, "Academic Programs," www.peacejusticestudies.org/academic-pro grams; GradSchools.com, "Conflict and Peace Studies," www.gradschools.com/programs/conflict-peace-studies; Dana Micucci, "Peace Studies Take Off," *New York Times*, October 14, 2008; UWC, "UWC-USA," www.uwc.org /school/uwc-usa-0.

23. Nonviolence 365 website, http://choosenonviolence.org/nonviolence365.

Chapter 19. Sustainability Education in Prisons: Transforming Lives, Transforming the World

1. Officer Jeffrey Swan and Technician Nick Hacheney, Monroe Correctional Complex, Monroe, WA, personal communications with Joslyn Rose Trivett, September 9, 2014 and October 12, 2016.

2. Ibid.

3. Ibid.

4. Unpublished material, Sustainability in Prisons Project (SPP), 2016.

5. Christopher Ingraham, "The U.S. Has More Jails Than Colleges," *Washington Post*, January 6, 2015.

6. Daniella Tilbury, "Environmental Education for Sustainability: Defining the New Focus of Environmental Education in the 1990s," *Environmental Education Research* 1, no. 2 (1995): 195–212; multinational plea from Meg Keen, Valerie A. Brown, and Rob Dyball, *Social Learning in Environmental Management: Towards a Sustainable Future* (New York: Earthscan, 2005), 4; Daniella Tilbury, *Education and Sustainability: Responding to the Global Challenge* (Gland, Switzerland: Commission on Education and Communication, International Union for Conservation of Nature, 2002); Noel Gough, "Thinking/acting Locally/globally: Western Science and Environmental Education in a Global Economy," *International Journal of Science Education* (Special Edition: Environmental Education and Science Education) 24, no. 11 (2002): 1,217–37; National Institute of Corrections, *The Greening of Corrections: Creating a Sustainable System* (Washington, DC: U.S. Department of Justice (DOJ), March 2011).

7. Danielle Kaeble et al., *Correctional Populations in the United States, 2014* (Washington, DC: DOJ, Bureau of Justice Statistics, December 2015), Table 2; Michael D. Sinclair, *Survey of State Criminal History Information, 2008* (Washington, DC: DOJ, Bureau of Justice Statistics, October 2009), 3–4; Pew Charitable Trusts, *Collateral Costs: Incarceration's Effect on Economic Mobility* (Washington, DC: 2010); Caroline Wolf Harlow, *Education and Correctional Populations* (Washington, DC: DOJ, Bureau of Justice Statistics, 2003); Lois M. Davis et al., *Evaluating the Effectiveness of Correctional Education: A Meta-analysis of Programs That Provide Education to Incarcerated Adults* (Santa Monica, CA: RAND Corporation, 2013). Box 19–1 from the following sources: Jeff Romm, "The Coincidental Order of Environmental Justice," in Kathryn M. Mutz, Gary C. Bryner, and Douglas S. Kenney, *Justice and Natural Resources: Concepts, Strategies, and Applications* (Washington, DC: Island Press, 2002); Keen, Brown, and Dyball, *Social Learning in Environmental Management*; United Nations Educational, Cultural and Scientific Organization (UNESCO), *UNESCO World Report: Investing in Cultural Diversity and Intercultural Dialogue* (Paris: 2009); Julian Agyeman and Tom Evans, "Toward Just Sustainability in Urban Communities: Building Equity Rights with Sustainable Solutions," *Annals of the American Academy of Political and Social Science* 590, no. 1 (2003): 35–53.

8. Davis et al., Evaluating the Effectiveness of Correctional Education; Erin L. Castro et al., "Higher Education in an Era of Mass Incarceration: Possibility Under Constraint," *Journal of Critical Scholarship on Higher Education and Student Affairs* 1, no. 1 (2015); Kaia Stern and Bruce Western, "National Directory of Prison Education

Programs," Prisons Studies Project, 2016, http://prisonstudiesproject.org/directory.

9. Davis et al., *Evaluating the Effectiveness of Correctional Education.*

10. Sustainable Practices Lab from Rob Branscum, Washington State Penitentiary, Walla Walla, WA, personal communication with Joslyn Rose Trivett, July 28, 2016; Casa del la Paz from John Stubbs, student at Harvard College, Cambridge, MA, personal communication with Joslyn Rose Trivett, July 3, 2015.

11. Roots of Success website, rootsofsuccess.org.

12. Liliana Caughman, *Diversity in Sustainability Education: Investigating Programs for Underserved Populations,* paper presented at the Just Sustainability Conference, Seattle, WA, August 7–9, 2016.

13. Heidi B. Carlone and Angela Johnson, "Understanding the Science Experiences of Successful Women of Color: Science Identity as an Analytic Lens," *Journal of Research in Science Teaching* 44, no. 8 (2007): 1,187–1,218; David S. Yeager and Gregory M. Walton, "Social-psychological Interventions in Education: They're Not Magic," *Review of Educational Research* 81, no. 2 (June 2011): 267–301; George Marshall, "Losing Alaska," *Hidden Brain,* National Public Radio, April 18, 2016.

14. National Institute of Corrections, *The Greening of Corrections.*

15. Roots of Success website, rootsofsuccess.org.

16. Unpublished material, SPP, 2016; classroom dynamics from Eugene Youngblood, "Each One, Teach One," sustainabilityinprisons.org, February 12, 2015.

17. Cyril Delanto Walrond, "Reaching the Unreachable," sustainabilityinprisons.org, September 9, 2016.

18. Office of the United Nations High Commissioner for Human Rights, "Your Human Rights," www.ohchr .org/EN/Issues/Pages/WhatareHumanRights.aspx; Hanna Graham and Rob White, *Innovative Justice* (London: Routledge, 2015); inmate quote from Jose Morales, "Santa Clara County Sheriff's Office Sustainability Program— Lawn Conversion Project," video, October 28, 2015, https://vimeo.com/143950115.

19. United Nations, "The International Bill of Human Rights" (New York: December 10, 1948); Timothy Buchanan, *Leadership for Sustainability: Manifesting Systems Change in Corrections* (Boston, MA: American Correctional Association Conference, August 2016); Cheryl Young et al., *Keeping Prisons Safe: Transforming the Corrections Workplace* (Olympia, WA: Gorham Publishing, 2014); Martin Seligman, *Flourish: A Visionary New Understanding of Happiness and Well-being* (New York: Simon and Schuster, 2012).

20. Kristofer Bret Bucklen and Gary Zajac, "But Some of Them Don't Come Back (to Prison)! Resource Deprivation and Thinking Errors as Determinants of Parole Success and Failure," *The Prison Journal* 89, no. 3 (2009): 239–64; Kathryn Waitkus, *The Impact of a Garden Program on the Physical, Environmental, and Social Climate of a Prison Yard at San Quentin State Prison,* master of science thesis (Malibu, CA: Pepperdine University, 2004).

21. The Horticultural Society of New York, "Horticultural Therapy Partnership," http://thehort.org/horttherapy_htp.html; Carolyn M. Tennessen and Bernadine Cimprich, "Views to Nature: Effects on Attention," *Journal of Environmental Psychology* 15 (March 1995): 77–85; Charles A. Lewis, *Green Nature/Human Nature: The Meaning of Plants in Our Lives* (Chicago, IL: University of Illinois Press, 1996); Seiji Shibata and Naoto Suzuki, "Effects of Indoor Foliage Plants on Subjects' Recovery from Mental Fatigue," *North American Journal of Psychology* 3, no. 2 (2001): 385–96; Sander van der Linden, "Green Prison Programmes, Recidivism and Mental Health: A Primer," *Criminal Behaviour and Mental Health* 25, no. 5 (December 2015): 338–42; Jody M. Hines, Harold R. Hungerford, and Audrey N. Tomera, "Analysis and Synthesis of Research on Responsible Environmental Behavior: A Meta-analysis," *Journal of Environmental Education* 18, no. 2 (1997): 1–8; Sebastian Bamberg and Guido Möser, "Twenty Years After Hines, Hungerford, and Tomera: A New Meta-analysis of Psycho-Social Determinants of Pro-environmental Behaviour," *Journal of Environmental Psychology* 27, no. 1 (2007): 14–25; Lance Schnacker, *Nature Imagery in Prisons Project at the Oregon Department of Corrections* (Salem, OR: Oregon

Youth Authority Research Brief, 2016).

22. Ibid.; Timothy Hughes and Doris James Wilson, *Reentry Trends in the United* States (Washington, DC: DOJ, Bureau of Justice Statistics, 2003).

23. Barb Toews, Assistant Professor in Criminal Justice, University of Washington-Tacoma, personal communication with Joslyn Rose Trivett and Kelli Bush, September 24, 2015.

24. Carri J. LeRoy et al., *The Sustainability in Prisons Project Handbook: Protocols for the SPP Network,* first edition (Olympia, WA: Gorham Publishing, 2013).

25. Shannon Swim, Sagebrush Coordinator, Lovelock Corrections Center, Lovelock, NV, personal communication with Joslyn Rose Trivett, September 19, 2016.

26. Unpublished materials, SPP, 2014 and 2015; Robert Mayo, "Roots of Success Graduation Speech," sustainabilityinprisons.org, September 30, 2015.

Chapter 20. Bringing the Earth Back into Economics

1. International Student Initiative for Pluralism in Economics, "An International Student Call for Pluralism in Economics," May 5, 2014, www.isipe.net/open-letter.

2. Paul Krugman, "The Profession and the Crisis," *Eastern Economic Journal* 37, no. 3 (2011): 307–12; Edward Fullbrook, ed., *Real World Economics: A Post-Autistic Economics Reader* (London: Anthem Press, 2007).

3. Fullbrook, ed., *Real World Economics*, 475; Joseph Stiglitz, "There Is No Invisible Hand," *The Guardian* (U.K.), December 20, 2002; Paul Krugman, "How Did Economists Get It So Wrong?" *New York Times*, September 2, 2009; U.S. House of Representatives, U.S. Congress, "Energy Reorganization Act of 1973: Hearings, Ninety-third Congress, first session" (Washington, DC: 1973), 248; George A. Akerlof and Robert J. Schiller, *Phishing for Phools: The Economics of Manipulation and Deception* (Princeton, NJ: Princeton University Press, 2015); Dan Ariely, *Predictably Irrational: The Hidden Forces That Shape Our Decisions* (New York: Harper Collins, 2008); Daniel Kahneman, *Thinking, Fast and Slow* (New York: Farrar, Straus and Giroux, 2011); Richard H. Thaler, *Misbehaving: The Making of Behavioral Economics* (New York: W. W. Norton & Company, 2015).

4. David Colander, "The Aging of an Economist," *Journal of the History of Economic Thought* 25, no. 2 (2003): 157–76; Marion Fourcade, Etienne Ollion, and Yann Algan, "The Superiority of Economists," *Journal of Economic Perspectives* 29, no. 1 (2015): 89–114.

5. Will Steffen et al., "The Anthropocene: Conceptual and Historical Perspectives," *Philosophical Transactions of the Royal Society A: Mathematical, Physical and Engineering Sciences* 369, no. 1938 (2011): 842–67; J. Bradford DeLong, *Macroeconomics* (Burr Ridge, IL: McGraw-Hill Higher Education, 2002); Johan Rockström et al., "A Safe Operating Space for Humanity," *Nature* 461 (September 24, 2009): 472–75.

6. Paul J. Crutzen, "Geology of Mankind," *Nature* 415 (January 3, 2002): 23; Joshua Farley and Alexey Voinov, "Economics, Socio-ecological Resilience and Ecosystem Services," *Journal of Environmental Management* 183, no. 2 (2016): 389–98.

7. Harold J. Barnett and Chandler Morse, *Scarcity and Growth: The Economics of Natural Resource Availability* (Baltimore, MD: Johns Hopkins University Press, 1963); The President's Materials Policy Commission, *Resources for Freedom: A Report to the President. Volume I: Foundations for Growth and Security* (Washington, DC: U.S. Government Printing Office, 1952); R. David Simpson, Michael A. Toman, and Robert U. Ayres, eds., *Scarcity and Growth Revisited: Natural Resources and the Environment in the New Millennium* (Washington, DC: Resources for the Future, 2005); Jeroen C. J. M. van den Bergh, "Externality or Sustainability Economics?" *Ecological Economics* 69, no. 11 (2010): 2,047–52.

8. Economics for the Anthropocene website, e4A-net.org.

9. Herman E. Daly, *Steady-State Economics: The Political Economy of Bio-physical Equilibrium and Moral Growth* (San Francisco, CA: W. H. Freeman and Co., 1977); Nicholas Georgescu-Roegen, *The Entropy Law and the Economic Process* (Cambridge, MA: Harvard University Press, 1971); British Petroleum, *Statistical Review of World Energy, Full Report 2015* (London: 2015); Rajendra K. Pachauri and Leo A. Meyer, eds., *Climate Change 2014: Synthesis Report. Contribution of Working Groups I, II and III to the Fifth Assessment Report of the Intergovernmental Panel on Climate Change* (Geneva: 2014); Joan Martinez-Alier, "The Environmentalism of the Poor," *Geoforum* 54 (July 2014): 239–41.

10. Howard T. Odum, *Environment, Power, and Society* (New York: Wiley-Interscience, 1971); Herman E. Daly and Joshua C. Farley, *Ecological Economics: Principles and Applications*, second edition (Washington, DC: Island Press, 2010); Jianguo Liu et al., "Complexity of Coupled Human and Natural Systems," *Science* 317, no. 5844 (2007): 1,513–16.

11. Roldan Muradian, "Ecological Thresholds: A Survey," *Ecological Economics* 38, no. 1 (2001): 7–24; R. A. Kerr, "Climate Tipping Points Come in from the Cold," *Science* 319 (January 11, 2008): 153; Timothy M. Lenton and Hywel T. P. Williams, "On the Origin of Planetary-scale Tipping Points," *Trends in Ecology & Evolution* 28, no. 7 (2013): 380–82.

12. Herman Daly, "A Further Critique of Growth Economics," *Ecological Economics* 88 (2013): 20–24.

13. Ibid.

14. Joshua Farley and Skyler Perkins, "Economics of Information in a Green Economy," in Robert Robertson, ed., *Building a Green Economy* (East Lansing, MI: Michigan State University Press, 2013), 83–100.

15. Robert M. Axelrod, *The Evolution of Cooperation* (New York: Basic Books, 1984); Anatol Rapoport and Albert M. Chammah, *Prisoner's Dilemma* (Ann Arbor, MI: University of Michigan Press, 1965).

16. Yoram Bauman and Elaina Rose, "Selection or Indoctrination: Why Do Economics Students Donate Less Than the Rest?" *Journal of Economic Behavior & Organization* 79, no. 3 (2011): 318–27; Robert H. Frank, Thomas Gilovich, and Dennis T. Regan, "Does Studying Economics Inhibit Cooperation?" *Journal of Economic Perspectives* 7, no. 2 (1993): 159–71; Andrew L. Molinsky, Adam M. Grant, and Joshua D. Margolis, "The Bedside Manner of Homo economicus: How and Why Priming an Economic Schema Reduces Compassion," *Organizational Behavior and Human Decision Processes* 119, no. 1 (2012): 27–37; Björn Frank and Günther G. Schulze, "Does Economics Make Citizens Corrupt?" *Journal of Economic Behavior & Organization* 43, no. 1 (2000): 101–13; Axelrod, *The Evolution of Cooperation*; Joseph Henrich and Natalie Henrich, *Why Humans Cooperate: A Cultural and Evolutionary Explanation* (New York: Oxford University Press, 2007); Martin Nowak and Roger Highfield, *SuperCooperators: Altruism, Evolution, and Why We Need Each Other to Succeed* (New York: Free Press (Simon Schuster), 2011).

17. Joshua Farley et al., "Extending Market Allocation to Ecosystem Services: Moral and Practical Implications on a Full and Unequal Planet," *Ecological Economics* 117 (2015): 244–52.

18. Joshua Farley, Jon Erickson, and Herman E. Daly, *Ecological Economics: A Workbook for Problem-Based Learning* (Washington, DC: Island Press, 2005); Joshua Farley, Lyudmyla Zahvoyska, and Lyudmyla Maksymiv, "Transdisciplinary Paths Towards Sustainability: New Approaches for Integrating Research, Education and Policy," in Ihor P. Soloviy and William S. Keeton, eds., *Ecological Economics and Sustainable Forest Management: Transdisciplinary Approach to the Carpathian Mountains* (Lviv, Ukraine: Ukrainian National Forestry University Press/Liga-Press, 2009), 40–54.

19. Leslie Picker, Danny Hakim, and Michael J. de la Merced, "Bayer Deal for Monsanto Follows Agribusiness Trend, Raising Worries for Farmers," *New York Times*, September 14, 2016; Eric Lipton and Rachel Abrams, "EpiPen Maker Lobbies to Shift High Costs to Others," *New York Times*, September 16, 2016; Andrew Pollack, "Drug Goes from $13.50 a Tablet to $750, Overnight," *New York Times*, September 20, 2015; Akerlof and Schiller,

Phishing for Phools; Raquel Meyer Alexander, Stephen W. Mazza, and Susan Scholz, "Measuring Rates of Return for Lobbying Expenditures: An Empirical Case Study of Tax Breaks for Multinational Corporations," *Journal of Law and Politics* 25, no. 401 (2007); Gina Chon, "Rising Drug Prices Put Big Pharma's Lobbying to the Test," *New York Times*, September 1, 2016.

20. Maura Borrego and Lynita K. Newswander, "Definitions of Interdisciplinary Research: Toward Graduate-Level Interdisciplinary Learning Outcomes," *The Review of Higher Education* 34, no. 1 (2010): 61–84; Farley, Erickson, and Daly, *Ecological Economics*; National Science Foundation, *National Science Foundation Investing in America's Future Strategic Plan FY 2006–2011* (Arlington, VA: 2006).

21. John Gowdy and Jon D. Erickson, "The Approach of Ecological Economics," *Cambridge Journal of Economics* 29 (2005): 207–22.

Chapter 21. New Times, New Tools: Agricultural Education for the Twenty-First Century

1. University of the District of Columbia (UDC), College of Agriculture, Urban Sustainability and Environmental Sciences (CAUSES), "East Capitol Urban Farm," www.udc.edu/college_of_urban_agriculture_and _environmental_studies/east_capitol_urban_farm.

2. Dwane Jones, Director, Center for Sustainable Development, UDC CAUSES, personal communication with author, September 28, 2016.

3. Johan Rockström et al., "Planetary Boundaries: Exploring the Safe Operating Space for Humanity," *Ecology and Society* 14, no. 2 (2009): 32; Pete Smith et al., "Agriculture, Forestry and Other Land Use (AFOLU)," in Ottmar Edenhofer et al., eds., *Climate Change 2014: Mitigation of Climate Change. Contribution of Working Group III to the Fifth Assessment Report of the Intergovernmental Panel on Climate Change* (Cambridge, U.K. and New York, NY: Cambridge University Press, 2014); Polly Erickson, Hans-Georg Bohle, and Beth Stewart, "Vulnerability and Resilience in Food Systems," in John Ingram, Polly Erickson, and Diana Liverman, eds., *Food Security and Global Environmental Change* (Washington, DC: Earthscan, 2010), 203–11; Beverly D. McIntyre et al., eds., *Agriculture at a Crossroads: International Assessment of Agricultural Knowledge, Science and Technology for Development, Synthesis Report* (Washington, DC: Island Press, 2009). See, for example: Millennium Ecosystem Assessment, *Ecosystems and Human Well-being: Synthesis* (Washington, DC: Island Press, 2005); National Research Council, *Transforming Agricultural Education for a Changing World* (Washington, DC: National Academies Press, 2009). Note: Although no simple typology or set of categories can capture the complexity of farming systems, we use the term *sustainable* to refer to farming systems that emphasize the use of natural processes within the farming system, often called "ecological" or "ecosystem" strategies, which build efficiency (and ideally resilience) through complementarities and synergies within fields, on the entire farm, and at larger scales across the landscape and community. Such farming systems represent a major departure from the key features which characterize *industrial agriculture*: large size combined with a high degree of specialization, reliance on off-farm and synthetic inputs, and the production of commodities under contract to food processors and handlers. A *food system* is the complex set of actors, activities, and institutions that link food production to food consumption. Food systems differ from farming systems in that the primary focus is beyond the farm gate. See the following: National Research Council, "A Pivotal Time in US Agriculture," in *Toward Sustainable Agricultural Systems in the 21st Century* (Washington, DC: The National Academies Press, 2010); Daniele Giovannucci et al., *Food and Agriculture: The Future of Sustainability. A Strategic Input to the Sustainable Development in the 21st Century (SD21) Project* (New York: United Nations Department of Economic and Social Affairs, Division for Sustainable Development, 2012); International Panel of Experts on Sustainable Food Systems (IPES-Food), *From Uniformity to Diversity: A Paradigm Shift from Industrial Agriculture to Diversified Agroecological Systems* (Louvain-la-Neuve, Belgium: 2016); Johan Rockström et al., "Sustainable Intensification of Agriculture for Human Prosperity and Global Sustainability," *Ambio* (2016): 1–14.

4. Ray V. Herren and M. Craig Edwards, "Whence We Came: The Land-Grant Tradition—Origin, Evolution,

and Implications for the 21st Century," *Journal of Agriculture Education* 43, no. 4 (2002): 88–98; McIntyre et al., eds., *Agriculture at a Crossroads*.

5. National Research Council (NRC), *Agricultural Education and the Undergraduate* (Washington, DC: National Academies Press, 2003); NRC, *Colleges of Agriculture at the Land Grant Universities* (Washington, DC: National Academies Press, 1996); NRC, *Transforming Agricultural Education for a Changing World*. For examples around the world, see the following: Arjen Wals and Richard Bawden, *Integrating Sustainability into Agricultural Education: Dealing with Complexity, Uncertainty, and Diverging World Views* (Ghent, Belgium: Interuniversity Conference for Agricultural and Related Sciences in Europe, AFANet, 2000); Francisco Carlos T. Leite and Rama B. Radhakrishna, "Profile of Agricultural Education and Extension: Challenges from a Changing Brazilian Rural Milieu," *Journal of International Agricultural & Extension Education* 11, no. 3 (2004): 13–21; Liu Yonggong and Zhang Jingzun, *A Reform of Higher Agricultural Education Institutions in China* (Rome: United Nations Food and Agriculture Organization (FAO) and International Institute for Educational Planning, 2004); David Atchoarena and Keith Holmes, "The Role of Agricultural Colleges and Universities in Rural Development and Life-long Learning in Asia," *Asian Journal of Agricultural Development* 2, nos. 1–2 (2004): 15–24); Tiffany J. Freer, *Modernizing the Agricultural Education and Training Curriculum* (Blacksburg, VA: U.S. Agency for International Development's InnovATE Program, 2015).

6. NRC, *Agricultural Education and the Undergraduate*; NRC, *Colleges of Agriculture at the Land Grant Universities*; Damian M. Parr et al., "Designing Sustainable Agriculture Education: Academics' Suggestions for an Undergraduate Curriculum at a Land Grant University," *Agriculture and Human Values* 24, no. 4 (2007): 523–33; IPES-Food, *From Uniformity to Diversity*; Mark Balschweid, Neil A. Knobloch, and Bryan J. Hains, "Teaching Introductory Life Science Courses in Colleges of Agriculture: Faculty Experiences," *Journal of Agricultural Education* 55, no. 4 (2014): 162–75.

7. Laura Sayre, "The Student Farm Movement in Context," in Laura Sayre and Sean Clark, eds., *Fields of Learning: The Student Farm Movement in North America* (Lexington, KY: University of Kentucky Press, 2011). For examples of nonprofits, see the following: Practical Farmers of Iowa website, www.practicalfarmers.org; Land Stewardship Association's Farm Beginnings Collaborative, http://landstewardshipproject.org/morefarmers /fbotherregions/thecollaborative; Ecology Action's Grow Biointensive Program, www.growbiointensive.org /grow_main.html; The Savory Institute, http://savory.global/institute. Box 21–1 from the following sources: FAO, "Pistachios, FAOSTAT Production/Crops," http://faostat3.fao.org/browse/Q/QC/E, viewed September 7, 2016; I. Açar, ed., *Proceedings of the Panel Pistachio Production and Consumption: From Problems to Solutions* (Istanbul: TEMA, 2014); Nestlé, "Sustainable Pistachio Production. Country: Turkey," September 18, 2012, www.nestle .com/csv/case-studies/allcasestudies/sustainable-pistachio-production-turkey.

8. See, for example: Miguel A. Altieri and Charles A. Francis, "Incorporating Agroecology into the Conventional Agricultural Curriculum," *American Journal of Alternative Agriculture* 7, nos. 1–2 (1992): 89–93; Charles Francis et al., "Impact of Sustainable Agriculture Programs on U.S. Landgrant Universities," *Journal of Sustainable Agriculture* 5, no. 4 (1995): 19–33; Geir Lieblein et al., "Future Education in Ecological Agriculture and Food Systems: A Student-Faculty Evaluation and Planning Process," *Journal of Sustainable Agriculture* 16, no. 4 (2000): 49–69; "Facilitating Sustainable Agriculture: A Participatory National Conference on Postsecondary Education," Asilomar Conference Grounds, Pacific Grove, California, January 24–25, 2006, Executive Summary; Center for Agroecology & Sustainable Food Systems, University of California–Santa Cruz, http://casfs.ucsc.edu; Cary J. Trexler, Damian M. Parr, and Navina Khanna, "A Delphi Study of Agricultural Practitioners' Opinions: Necessary Experiences for Inclusion in an Undergraduate Sustainable Agriculture Major," *Journal of Agricultural Education* 47, no. 4 (2006): 15–25; Parr et al., "Designing Sustainable Agriculture Education"; John M. Gerber, *Communiversities: Beyond the Land Grant* (Amherst, MA: University of Amherst, 2011); N. Jordan et al., "New Curricula for Undergraduate Food-Systems Education: A Sustainable Agriculture Education Perspective," *NACTA Journal* 58, no. 4 (2014): 302–10; Kathleen Hilimire, "Theory and Practice of an Interdisciplinary Food Systems Curriculum," *NACTA Journal* 60, no. 2 (2016): 227–33; Charles Francis et al., "Agroecologist Education

for Sustainable Farming and Food Systems," *Agronomy Journal*, in press 2016.

9. Thomas F. Patterson, Jr., "The Rise and Fall of Innovative Education: An Australian University Case Study," *Innovations in Higher Education* 32, no. 2 (2007): 71–84; Richard Bawden, "Systemic Development at Hawkesbury: Some Personal Lessons from Experience," *Systems Research and Behavioral Science* 22, no. 2 (2005): 151–64.

10. Julie Cotton, Chair, Sustainable Agriculture Education Association Steering Council, personal communication with author, September 29, 2016.

11. Charles Francis et al., "Phenomenon-Based Learning in Agroecology: A Prerequisite for Transdisciplinarity and Responsible Action," *Agroecology and Sustainable Food Systems* 37, no. 1 (2013): 60–75; Charles Francis et al., "Innovative Education in Agroecology: Experiential Learning for a Sustainable Agriculture," *Critical Reviews in Plant Sciences* 30, no. 1–2 (2011): 226–37; Charles Francis, personal communication with author, September 27, 2016.

12. Francis et al., "Agroecologist Education for Sustainable Farming and Food Systems." Box 21–2 from the following sources: Laura Lengnick, *Resilient Agriculture: Cultivating Food Systems for a Changing Climate* (Gabriola Island, BC: New Society Press, 2015). For examples of resilience curricula in sustainable food systems programs, see: Montana State University, "Sustainable Food and Bioenergy Systems," http://sfbs.montana.edu; Kansas Wesleyan University, "Ecospheric Studies and Community Resilience B.S.," www.kwu.edu/academics/academic-departments/department-of-biology/ecospheric-studies-and-community-resilience; University of Wisconsin Steven's Point, "Sustainable and Resilient Food Systems–Master of Science Degree," www.uwsp.edu/HPHD/Pages/graduateDegrees/srfsDegree/default.aspx. For examples of resilience curricula in environmental studies programs, see the following: Marianne E. Krasny, Keith G. Tidball, and Nadarajah Sriskandarajah, "Education and Resilience: Social and Situated Learning Among University and Secondary Students," *Ecology and Society* 14, no. 2 (2009): 38; Ioan Fazey, "Resilience and Higher Order Thinking," *Ecology and Society* 15, no. 3 (2010): 9; Monique R. Myers, "A Student and Teacher Watershed and Wetland Education Program: Extension to Promote Social-ecological Resilience," *Journal of Extension* 50, no. 4 (2012). For examples of resilience curricula in community-based education programs, see the following: Rob Hopkins, *Transition Handbook: From Oil Dependency to Local Resilience* (Cambridge, U.K.: UIT Cambridge Ltd., 2008); Bryce Dubois and Marianne E. Krasny, "Educating with Resilience in Mind: Addressing Climate Change in Post-Sandy New York City," *Journal of Environmental Education* 47, no. 4 (2013): 255–70; Laura Lengnick et al., *An Energy Descent Plan for Warren Wilson College* (Swannanoa, NC: 2009), Executive Summary.

13. Francis, personal communication; NRC, *Transforming Agricultural Education for a Changing World*.

14. See, for example: Martha W. Gilliland and Amelia A. Tynan, "Transforming Higher Education: Overcoming the Barriers to Better Education," *Solutions* 1, no. 6 (2010): 56–61; Arjen E. J. Wals et al., "Education for Integrated Rural Development: Transformative Learning in a Complex and Uncertain World," *Journal of Agricultural Education & Extension* 10, no. 2 (2004): 89–100; Cotton, personal communication.

15. Jones, personal communication.

Chapter 22. Educating Engineers for the Anthropocene

1. The Famous People, "John Smeaton," www.thefamouspeople.com/profiles/john-smeaton-5400.php.

2. Paul Crutzen, "Geology of Mankind," *Nature* 415 (January 3, 2002): 23.

3. Royal Academy of Engineering, *Engineers for Africa: Identifying Engineering Capacity Needs in Sub-Saharan Africa* (London: 2012)

4. Daniel Hoornweg, *Cities and Sustainability: A New Approach* (London: Routledge, 2016); Daniel Hoornweg and Kevin Pope, "Population Predictions for the World's Largest Cities in the 21st Century," *Environment and Urbanization* (September 23, 2016); Engineers Canada, "National Membership Report," https://engineerscanada

.ca/reports/national-membership-report.

5. Hoornweg, *Cities and Sustainability*.

6. David E. Goldberg and Mark Somerville, *A Whole New Engineer* (Douglas, MI: ThreeJoy Associates, Inc., 2014).

7. Brad Allenby, "Educating Engineers in the Anthropocene," in *2008 IEEE International Symposium on Electronics and the Environment May 19–21, 2008, San Francisco, CA* (2008).

8. Daniel Hoornweg et al., "Meeting the Infrastructure Challenges of African Cities," in John Crittenden, Chris Hendrickson, and Bill Wallace, *ICSI 2014: Creating Infrastructure for a Sustainable World* (2014): 471–81; Deb Niemeier, Harry Gombachika, and Rebecca Richards-Kortum, "How to Transform the Practice of Engineering to Meet Global Health Needs," *Science* 345, no. 6202 (September 12, 2014): 1,287–90.

9. Cesar A. Poveda and Michael G. Lipsett, "An Integrated Approach for Sustainability Assessment: The Wa-Pa-Su Project Sustainability Rating System," *International Journal of Sustainable Development & World Ecology* 21, no. 1 (2014): 85–98.

10. University of Prince Edward Island, "Sustainable Design Engineering," www.upei.ca/programsandcourses /engineering; "EESD Barcelona Declaration," settled at the 2nd International Conference on Engineering Education for Sustainable Development, October 2004, http://eesd15.engineering.ubc.ca/declaration-of-barcelona/.

11. James Gover and Paul Huray, *Educating 21st Century Engineers* (Washington, DC: IEEE–USA, 2007); George Bugliarello, "The Engineering Challenges of Urban Sustainability," *Journal of Urban Technology* 15, no. 1 (2008): 53–83; Elinor Ostrom, "A General Framework for Analyzing Sustainability of Social-Ecological Systems," *Science* 325, no. 5939 (July 24, 2009): 419–22.

12. "Canada's Best Schools," *Maclean's*, November 7, 2016.

13. Peter Murray, "Eight Out of China's Top Nine Government Officials Are Scientists," SingularityHub, May 17, 2011.

14. Hoornweg, *Cities and Sustainability*.

15. Nicholas A. Robinson, "Beyond Sustainability: Environmental Management for the Anthropocene Epoch," *Journal of Public Affairs* 12, no. 3 (2012): 181–94.

16. Wikipedia, "Engineers Without Borders," https://en.wikipedia.org/wiki/Engineers_Without_Borders, viewed November 11, 2016.

17. Calestous Juma, "Redesigning African Economies: The Role of Engineering in Development" (London: Royal Academy of Engineering, October 3, 2006); United Nations Educational, Scientific and Cultural Organization (UNESCO), *Engineering: Issues, Challenges, and Opportunities for Development* (Geneva: 2010); Hoornweg et al., "Meeting the Infrastructure Challenges of African Cities."

18. Nadine Ibrahim et al., "Engineering Education for Sustainable Cities in Africa," presented at 8th Conference on Engineering Education for Sustainable Development, Bruges, Belgium, September 4–7, 2016; A. Kumar, Aoyi Ochieng, and Maurice S. Onyango, "Engineering Education in African Universities: A Case for Internationalization," *Journal of Studies in International Education* 8, no. 4 (2004): 377–89.

19. Royal Academy of Engineering, *Engineers for Africa*; UNESCO/IGU Workshop on Women in Engineering in Africa and the Arab States, Paris, France, December 10, 2013; World Bank, *A Decade of Development in Sub-Saharan African Science, Technology, Engineering & Mathematics Research* (Washington, DC: 2014).

20. Engineers Canada, "Women in Engineering," www.engineerscanada.ca/diversity/women-in-engineering; Engineers Canada, *Canadian Engineers for Tomorrow* (Ottawa, ON: 2016).

21. Ibid.

22. Women in Engineering website, www.womeng.org; David E. Winickoff, Jane A. Flegal, and Asfawossen Asrat, "Engaging the Global South on Climate Engineering Research," *Nature Climate Change* 5 (2015): 627–634; Irina Bokova, Director General of UNESCO, presentation at UNESCO/IGU Workshop on Women in Engineering in Africa and the Arab States.

23. United Nations Population Division, World Population Prospects: The 2015 Revision (New York: 2015).

Chapter 23. The Evolving Focus of Business Sustainability Education

1. World Commission on Environment and Development, *Our Common Future* (Oxford, U.K.: Oxford University Press, 1987).

2. John Ehrenfeld, *Sustainability by Design* (New Haven, CT: Yale University Press, 2008).

3. John Ehrenfeld and Andrew Hoffman, *Flourishing: A Frank Conversation About Sustainability* (Palo Alto, CA: Stanford University Press, 2013).

4. United Nations Millennium Ecosystem Assessment, *Ecosystems and Human Well-Being: Synthesis Report* (Washington DC: Island Press, 2005); Barbara Crossette, "Kofi Annan's Astonishing Facts," *New York Times*, September 27, 1998.

5. Greenbiz, *State of the Profession 2016* (San Francisco, CA: 2006); Peter Lacy et al., *A New Era of Sustainability: UN Global Compact–Accenture CEO Study* (New York: Accenture Institute for High Performance, 2010).

6. Net Impact, *Business as Unusual: The Social and Environmental Impact Guide to Graduate Programs—For Students by Students* (San Francisco, CA: 2014); Yale University and World Business Council for Sustainable Development (WBCSD), *Rising Leaders on Environmental Sustainability and Climate Change: A Global Survey of Business Students* (New Haven, CT: Yale University Center for Business and the Environment, 2015).

7. Figure 23–1 from Trucost, *The Greening of Higher Education's Academic Agenda: Teaching and Research on Corporate Sustainability and Natural Capital* (London: 2013), 8; Aspen Institute, *Beyond Grey Pinstripes 2011–2012, Top 100 MBA Programs* (New York: 2012); Timothy Hart et al., "Do, But Don't Tell. The Search for Social Responsibility and Sustainability in the Websites of the Top-100 US MBA Programs," *International Journal of Sustainability in Higher Education* 16, no. 5 (2015): 706–28.

8. Andrew Hoffman, *Competitive Environmental Strategy: A Guide to the Changing Business Landscape* (Washington DC: Island Press, 2000).

9. Joseph Schumpeter, *Capitalism, Socialism, and Democracy* (New York: Harper, 1975). Figure 23–2 from Hoffman, *Competitive Environmental Strategy*.

10. Amanda Albright, "Sustainable-finance MBAs Struggle to Find Jobs," *Bloomberg*, July 24, 2015.

11. Paul Crutzen, "Geology of Mankind," *Nature* 415, no. 23 (January 2002): 23; Richard Monastersky, "Anthropocene: The Human Age," *Nature* 519, no. 7542 (March 2015): 144–47.

12. Ehrenfeld, *Sustainability by Design*; Andrew J. Hoffman and John R. Ehrenfeld, "The Fourth Wave: Management Science and Practice in the Age of the Anthropocene," in Susan Albers Mohrman, James O'Toole, and Edward E. Lawler, eds., *Corporate Stewardship: Achieving Sustainable Effectiveness* (Sheffield, U.K.: Greenleaf Publishing, 2015), 228–46.

13. Net Impact and Aspen Institute, *New Leaders, New Perspectives. A Survey of MBA Student Opinions on the Relationship Between Business and Social/Environmental Issues* (San Francisco, CA: Net Impact, 2009); Lynn Stout, "The Problem of Corporate Purpose," *Issues in Governance Studies* (Brookings) 48 (June 2012); Harvard Business School Course #1524, "Reimagining Capitalism: Business and Big Problems," www.hbs.edu/coursecat

alog/1524.html.

14. University of Cambridge, *Rewiring the Economy: Ten Tasks, Ten Years* (Cambridge, U.K.: University of Cambridge Institute for Sustainability Leadership, 2015).

15. MIT Sloan School of Management Sustainability Initiative, http://mitsloan.mit.edu/sustainability/; World Environment Center and Net Impact, *Business Skills for a Changing World: An Assessment of What Global Companies Need from Business Schools* (Washington, DC: World Environment Center, 2011).

16. Yale School of Management, Center for Business and the Environment, http://cbey.yale.edu.

17. Weatherhead School of Management, Case Western Reserve University, "Appreciative Inquiry Certificate in Positive Business and Society Change," https://weatherhead.case.edu/executive-education/certificates/appreciative-inquiry.

18. Stanford Graduate School of Business, "Certificate in Corporate Innovation," www.gsb.stanford.edu/exed/lead/LEAD-CourseDescriptions.pdf.

19. Net Impact, "About Us," https://www.netimpact.org/about; Frederick A. and Barbara M. Erb Institute on Global Sustainable Enterprise, University of Michigan, http://erb.umich.cdu.

20. University of Michigan Ross School of Business, Center for Positive Organizations, http://positiveorgs.bus.umich.edu; University of Michigan Ross School of Business, "Positive Change," https://michiganross.umich.edu/about/positive.

21. Sumantra Ghoshal, "Bad Management Theories Are Destroying Good Management Practices," *Academy of Management Learning and Education* 4, no. 1 (2005): 75–91; Eric Beinhocker and Nick Hanauer, "Redefining Capitalism," *McKinsey Quarterly*, September 2014; Stout, "The Problem of Corporate Purpose"; Nicholas Stern, *The Global Deal: Climate Change and the Creation of a New Era of Progress and Prosperity* (New York: Public Affairs, 2009); Joseph Stiglitz, Amartya Sen, and Jean-Paul Fitoussi, *Mismeasuring Our Lives: Why GDP Doesn't Add Up* (New York: The New Press, 2015); Ivey School of Business, Building Sustainable Value Research Centre, Western University, www.ivey.uwo.ca/sustainability.

22. Net Impact and Aspen Institute, *New Leaders, New Perspectives*; Yale University and WBCSD, *Rising Leaders on Environmental Sustainability and Climate Change*.

23. Andrew Hoffman, *Finding Purpose: Environmental Stewardship as a Personal Calling* (Leeds, U.K.: Greenleaf Publishing, 2016).

Chapter 24. Teaching Doctors to Care for Patient and Planet

1. Allison Crimmins et al., *The Impacts of Climate Change on Human Health in the United States: A Scientific Assessment* (Washington, DC: U.S. Global Change Research Program, 2016), Executive Summary, 2; Suzanne Goldenberg, "Climate Change Threat to Public Health Worse Than Polio, White House Warns," *The Guardian* (U.K.), April 4, 2016.

2. United Nations Sustainable Development Goals, "Goal 13: Take Urgent Action to Combat Climate Change and Its Impacts," www.un.org/sustainabledevelopment/climate-change-2/.

3. K. S. Gehle, J. L. Crawford, and M. T. Hatcher, "Integrating Environmental Health into Medical Education," *American Journal of Preventive Medicine* 41, no. 4, supplement 3 (October 2011): S296–301.

4. Ibid., S297.

5. Caroline Tomes, "Teaching Sustainable Healthcare to Tomorrow's Doctors: A Mixed Method Analysis of Medical School Innovations in England," poster (Cambridge, U.K.: University of Cambridge, 2011); Jenny Hellsing and Marion Carey, "Are Australia's Future Doctors Being Educated About Climate Change and Health?"

poster (Sydney, Australia: University of Sydney, undated).

6. Sarah C. Walpole et al., "Exploring Emerging Learning Needs: A UK-wide Consultation on Environmental Sustainability Learning Objectives for Medical Education," *International Journal of Medical Education* 6 (2015): 200.

7. Ibid., 195.

8. Sarah C. Walpole et al., "What Do Tomorrow's Doctors Need to Learn About Ecosystems? A BEME Systematic Review: BEME Guide No. 36," *Medical Teaching* 38, vol. 4 (April 2016): 345.

9. Ibid.

10. Ibid.

11. See, for example: Practice Greenhealth website, https://practicegreenhealth.org; Health Care Without Harm website, https://noharm.org.

12. William A. Haseltine, *Affordable Excellence: The Singapore Healthcare Story* (Singapore and Washington, DC: Ridge Books and Brookings Institution Press, 2016); Greg Connolly, Global Health Council, *Costa Rican Health Care: A Maturing Comprehensive System Right to Health Care: Moving from Idea to Reality*, background paper on "Right to Health Care: Moving from Idea to Reality" presented at the Asian Social Forum, Hyderabad, India, January 3–4, 2003; Koji Nabae, "The Health Care System in Kerala: Its Past Accomplishments and New Challenges," *Journal of the National Institute of Public Health* (Japan) 52, no. 2 (2003): 140–45. Box 24–1 from the following sources: Consejo Mayor de Médicos Mayao'b' por Nacimiento, *Raxnaq'il Nuk'aslemal: Medicina Maya' en Guatemala* (Guatemala City: Cholsamaj, 2016), 22–47; Monica Berger-González et al., "Transdisciplinary Research on Cancer Healing Systems Between Biomedicine and the Maya of Guatemala: A Tool for Reciprocal Reflexivity in a Multi-epistemological Setting," *Qualitative Health Research* 26, vol. 1 (2016): 77–91; Monica Berger-González, Eduardo Gharzouzi, and Christoph Renner, "Maya Healers' Conception of Cancer as Revealed by Comparison to Western Medicine," *Journal of Global Oncology* 2, no. 2 (April 2016): 56–67; Monica Berger-González et al., "Relationships That Heal: Going Beyond the Patient-Healer Dyad in Mayan Therapy," *Journal of Medical Anthropology: Cross-cultural Studies in Health and Illness* 35, no. 4 (2016): 356–67.

13. Jessica Pierce and Andrew Jameton, *The Ethics of Environmentally Responsible Health Care* (New York: Oxford University Press, 200), 106–07.

14. World Bank, "Health Expenditures Per Capita (Current US$)," http://data.worldbank.org/indicator/SH.XPD .PCAP, viewed August 24, 2016; World Bank, "Mortality Rate, Infant (per 1,000 Live Births)," http://data .worldbank.org/indicator/SP.DYN.IMRT.IN, viewed August 24, 2016.

15. Box 24–2 from Pierce and Jameton, *The Ethics of Environmentally Responsible Health Care*.

16. Douglas Klahr, "Sustainability for Everyone: Trespassing Disciplinary Boundaries," in Kirsten Allen Bartels and Kelly A. Parker, eds., *Teaching Sustainability, Teaching Sustainably* (Sterling, VA: Stylus Publishing, 2012), 19–30.

Chapter 25. The Future of Education: A Glimpse from 2030

1. United Nations, Department of Economic and Social Affairs, Population Division, *World Population Prospects: The 2015 Revision, Key Findings and Advance Tables* (New York: 2015); "Summary for Policymakers," in T. F. Stocker et al., eds., *Climate Change 2013: The Physical Science Basis. Contribution of Working Group I to the Fifth Assessment Report of the Intergovernmental Panel on Climate Change* (Cambridge, U.K. and New York, NY: Cambridge University Press, 2013), 20; Brady Dennis and Chris Mooney, "Scientists Nearly Double Sea Level Rise Projections for 2100, Because of Antarctica," *Washington Post*, March 30, 2016; Suzanne Goldenburg, "Arctic Sea Ice Extent Breaks Record Low for Winter," *The Guardian* (U.K.), March 28, 2016; Michael D. Lemonick, "The

Future of Mountain Glaciers Is Bleak," Climate Central, April 6, 2015; Tim Radford, "Speed of Glacier Retreat Worldwide 'Historically Unprecedented,' Says Report," *The Guardian* (U.K.), August 4, 2015.

2.   Gerard Wynn, "Long-term Sea Level Rise Will Be Much Higher, But Barely Studied – IPCC," Climate Home, March 26, 2014; Joanna Peasland, "Major Cities Threatened by Rapid Sea Level Rise, New Reports Find," Climate Home, February 22, 2016; John Vidal, "Global Warming Could Create 150 Million 'Climate Refugees' by 2050," *The Guardian* (U.K.), November 2, 2009; Stephane Hallegatte et al., "Future Flood Losses in Major Coastal Cities," *Nature Climate Change* 3 (August 18, 2013): 802–06.

3.   David Orr, "What Is Education For? Six Myths About the Foundations of Modern Education, and Six New Principles to Replace Them," *In Context: A Quarterly of Humane Sustainable Culture* 27 (Winter 1991); Michael Maniates, "Teaching for Turbulence," in Worldwatch Institute, *State of the World 2013: Is Sustainability Still Possible?* (Washington, DC: Island Press, 2013), 255–68.

4.   Chayan MacIntyre, "Bamboo Schools and Community Resilience," *International Journal of Earth Education* 8 (Winter 2027): 67–78.

5.   Arwa Mahdawi, "Can Cities Kick Ads? Inside the Global Movement to Ban Urban Billboards," *The Guardian* (U.K.), August 12, 2015.

6.   David Sobel, "Outdoor School for All: Reconnecting Children to Nature," in Worldwatch Institute, *State of the World 2017: EarthEd: Rethinking Education on a Changing Planet* (Washington, DC: Island Press, 2017); Cynthia Okoroafor, "Does Makoko Floating School's Collapse Threaten the Whole Slum's Future?" *The Guardian* (U.K.), June 10, 2016; Semester at Sea, "Our Organization," www.semesteratsea.org/info-hub/our-organization, viewed October 23, 2016; Kei Franklin, "Providing Environmental Consciousness Through Life Skills Training," in Worldwatch Institute, *State of the World 2017*; Bunker Roy, "The Barefoot Model," in idem.

7.   Erik Assadourian, "The Rise and Fall of Consumer Cultures," in Worldwatch Institute, *State of the World 2010: Transforming Cultures* (New York: W. W. Norton & Company, 2010).

Index